The India Handbook

Regional Handbooks of Economic Development
Prospects onto the 21st Century

The China Handbook
The India Handbook

Forthcoming

The Japan Handbook
The Southeast Asia Handbook

The India Handbook

Edited by

C. Steven LaRue

Lloyd I. Rudolph
University of Chicago

Susanne Hoeber Rudolph
University of Chicago

Philip Oldenburg
Southern Asian Institute
Columbia University

FITZROY DEARBORN PUBLISHERS
CHICAGO · LONDON

 For information write to:

FITZROY DEARBORN PUBLISHERS
70 East Walton Street
Chicago, Illinois 60611
USA

or

11 Rathbone Place
London W1P 1DE
England

British Library and Library of Congress Cataloging in Publication Data is available

ISBN 1884964-89-3

First published in the USA and UK 1997
Typeset by Print Means Inc., New York, New York
Printed by Braun-Brumfield, Inc., Ann Arbor, Michigan

Contents

Editor's Note

The India Handbook, the second volume in the series entitled *Regional Handbooks of Economic Development: Prospects onto the 21st Century,* provides an overview of India's development since independence (1947). The individual chapters focus on a wide range of economic, political, social, and trade topics. In common with the other volumes in the series, *The India Handbook* is designed to address complex development issues in a manner accessible to academic and nonacademic audiences alike. It is our special aim to make India's economic development understandable to college-level students.

The contents of this book were chosen with the help of three advisers: Lloyd I. Rudolph and Susanne Hoeber Rudolph of the University of Chicago, and Philip Oldenburg of Columbia University. Topics range from economic planning and policy to the diverse influences that affect economic development in India, including political organization, population growth, human development, technology transfer, trade, and multiculturalism. In addressing these topics, contributors were asked to consider past policies and performance as well as future prospects. The first part, History and Context, provides two chapters that place India's development since independence in a historical context. The book is then divided into the following parts: Economic Policy, Social and Cultural Aspects of Economic Growth, International Political Economy, and Future.

Each chapter has been prepared by a recognized expert in the field and consists of an essay on the topic (which in many cases includes tables) and a further reading list with annotations. In-text references to sources listed in "Further Reading" are given as author's last name followed by date and page number; sources not listed in "Further Reading" are given more complete details within the text.

The volume closes with a series of appendices, including a detailed chronology of events, a glossary of terms, biographical entries on key personalities, and an annotated bibliography of secondary sources.

Throughout this volume, abbreviations and acronyms have been spelled out the first time they appear in each chapter. The city names Bombay and Madras have been used throughout rather than their recent designations Mumbai and Chennai.

The editor would like to thank the advisers and the contributors for their cooperation and efforts. Special thanks belong to Paul Schellinger and Christopher Hudson for their invaluable help at every stage of this project, and Dobby Gibson for his research assistance.

Note on the Advisers

Philip Oldenburg is adjunct assistant professor of political science and director of the Southern Asian Institute at Columbia University. His research has focused on grassroots government in India and electoral politics and foreign policy in South Asian countries. His publications include *Big City Government in India: Councilor, Administrator, and Citizen in Delhi* (1976) and numerous articles in scholarly journals. Since 1988, he has been the editor or coeditor of *India Briefing,* an annual volume of scholarly articles published for the Asia Society.

Lloyd I. Rudolph is professor of political science and former chair of the Committee on International Relations at the University of Chicago. His most recent book, *In Pursuit of Lakshmi: The Political Economy of the Indian State* (1987), was coauthored with Susanne Hoeber Rudolph. He has also published extensively in numerous scholarly journals.

Susanne Hoeber Rudolph is professor of political and social sciences at the University of Chicago. Currently she is director of the South Asia Language and Area Center. Her most recent book, *In Pursuit of Lakshmi: The Political Economy of the Indian State* (1987), was coauthored with Lloyd I. Rudolph. Since 1971, she had been working on editing the Amar Singh diary, the 90-volume work of a North Indian nobleman written between 1898 and 1942.

History and Context

Chapter One

Independence to the Mid-1970s

John Adams

More than any other emergent new nation of the early postwar era, India engaged the hopes and fears of a watching world. Mohandas Gandhi, Jawaharlal Nehru, and senior leaders of the Indian National Congress had used irresistible moral authority, brilliant nonviolent tactics, and skilled mass mobilization to bring the globe's second most populous nation, and its largest democracy, to independence from once-powerful Britain. In the hearts of an aspiring Indian populace and in the thoughts of fascinated observers, the question of the hour was whether promises of civil liberties, social progress, and economic development could be redeemed in one of the world's poorest countries, amid a people riven by caste, religion, and regional identities, and wholly unschooled in popular self-governance. As the dividing lines of the Cold War hardened, the United States and its European allies were concerned about the stability of India's polity and the advancement of its economy. For millions yet to be freed from the lingering toils of colonialism, India served as a beacon and a model. A quick collapse or a wayward step would dash the prospects of those attentive Asian and African peoples who were themselves poised on the doorstep of freedom and self-determination. Doubts lingered about the ability of non-European peoples to rule themselves.

There are two ways to assess how well India's leadership and peoples succeeded in meeting the vows made in 1947. The first is to employ the wisdom of a full half-century of hindsight and draw on recent advances in economic policy analysis and political theory; this path has the merit of permitting the analyst to appear sagacious and superior to earlier mortals. It has the drawback that it does little to help fathom how choices were made, and outcomes realized, in an environment of social ideology, international relations, and economic principles much different than those of the mid-1970s, much less of the year 2000. A second avenue is to voyage back in time to the salad days of India's independence and seek to recapture the enthusiasms, to plumb the canons of then current economic and social knowledge, and to feel the constraining realities of this formative phase of modern India. Of course, one cannot fully jettison the freshest contents of one's mind, replete with superior comprehension of the processes of economic growth and social development and with awareness of ultimate results. Nonetheless, the second tack is at once the more challenging, the fairest in premising its judgments, and the most likely to provide insights into the motives and aims of the actors of that now distant epoch. This said, it is as obvious which road we should traverse as it was to India's new guides when, after long struggle, they at last took command of their nation's destiny.

The Socialist Pattern of Society

Late on the evening of 14 August 1947, Nehru addressed the Constituent Assembly at the stroke of midnight, the moment when India became a free nation, but his thoughts were probably more preoccupied by the reports of vicious communal bloodletting in Lahore than by the address to the nation that he had to make. Notwithstanding, his address was steadfast and eloquent: "Long years ago we made a tryst with destiny, and now the time comes when we shall redeem our pledge not wholly or in full measure, but very substantially."

The pledge of which Nehru spoke was threefold. Having enlisted the efforts of millions of Indians in the movement for independence, he and India's new leadership had to repay the confidence and trust reposed in them by the expectant masses. The vehicle for this mobilization, the Congress Party, was committed by its principles to the highest ideals of disinterested public service, to leveling India's intolerable caste hierarchy, and to resisting the blandishments of powerful landlords and capitalists. Most widely, the pledge incorporated a social compact among India's peoples themselves that transcended any leader or party. Their mutual covenant incorporated the political dimension of securing a secular, rights-based democracy, but it also encompassed a material tenet, to launch an urgent, multifront attack on poverty, ignorance, and economic stagnation.

The political structures of modern India did not spring forth from Nehru's forehead, or anyone else's, in complete and elegant form, and, without settled governance, collective economic initiatives could not be pursued. India's constituent assembly met for two and a half years, sitting from 1947 until 1950, when the Constitution of India was at last approved. The preamble confirms that India will be a "sovereign, socialist, secular, democratic republic" that will provide justice and social, economic, and political security to all its citizens. The Fundamental Rights of part 3 of the Constitution and the less-binding Directive Principles of State Policy strongly dedicate the Indian government to pursuing economic ends for the nation's citizens–employment, education, and well-being–while concurrently reducing concentrations of landholdings, industrial wealth, and incomes, always with an eye toward the common good. This vision was explicitly endorsed in the preamble to India's Second Five-Year Plan (1955–56 to 1960–61) under the heading "the Socialist Pattern of Society," the "basic criterion" of which was that "lines of advance" must not be shaped by "private profit, but social gain." Further, the "pattern of development . . . should be so planned that [it results] not only in appreciable increases in national income and employment but also in greater equality in incomes and wealth. Major decisions regarding production, distribution, consumption, and investment . . . must be made by agencies informed by social purpose."

From 1947 to 1975, the consensus of India's leadership and people that the government should provide the impetus for economic development and social reform was robust and centrist, although never free from challenge. Socialist and communist parties on the left of the Congress Party favored even stronger action in support of the poor or unpropertied groups, such as the landless laborers and urban workers. On the right wing, there was sentiment for a more favorable treatment of large farmers, traders, and industrialists and for reliance on a predominantly free-enterprise paradigm. The Gandhian strain of Indian political economy remained strong even after the Mahatma's assassination on 30 January 1948. His plea to adopt simplicity in dress and austerity in diet was village based and arose from moral doubts about wholesale conversion to industrial and urban ways of life. Despite the intrinsic power of these competing ideologies, the socialist industrial economic programs of the Congress Party were overwhelmingly endorsed in the five national elections in the years 1952, 1957, 1962, 1967, and 1971, in which the party always polled about four times as many votes as its nearest competitor and only once failed to control more than two-thirds of the seats in the Lok Sabha, India's parliament. The socialist pattern of society was neither alien to India's masses nor imposed from the top by a narrow oligarchy. Through the mid-1970s, no alternative was seriously considered.

At the end of World War II, India's national economy was crippled and stagnant. At one time threatened from the east by a Japanese invasion, an India primed to be freed of colonial overlordship had reluctantly followed Britain into war. India's railways and mills suffered from overintensive use. Never abundant, British investment in Indian industry, even for replacements for depreciated machines and plants, dried up during the conflict. Since about 1920, India's foodgrain output had been shrinking, reducing the nation's per capita supplies of food. Partition separated the subcontinent's raw jute tracts, which went to East Pakistan, from the jute manufacturing mills in Calcutta; likewise, a sizable acreage planted in cotton lay in West Pakistan, depriving the spinning and weaving plants of the Gujarat and Maharashtra states in western India of fiber input. In 1950–51, India produced only 52 million tons of foodgrains and applied a minuscule 55,000 tons of nitrogenous fertilizer. Installed electricity capacity was 2.3 million kilowatts and of India's more than 550,000 villages, merely 3,700 were electrified. The Indian nation of 359 million people produced essentially no machine tools, no tractors, no tires, no pharmaceuticals, no motorcycles, no petroleum products, and no industrial chemicals. There were 168,000 telephones, a ratio of one for every 2,137 persons. Only 43% of children aged six to 11 were in school; 5.3% of potential students over the age of 14 were in secondary institutions. Not surprisingly, the national literacy rate was 18%. The average Indian's life expectancy at birth was 32 years. Out of every 1,000 live births, 146 infants died in their first year.

When India's policy makers juxtaposed this profile of economic stagnation and social destitution with the high hopes incorporated in the social compact embodied in the Constitution and the Congress Party's platform, the need for urgency was impelling. As they saw matters, there were three obstacles that had to be surmounted if the nation was to generate self-sustaining momentum. The inert and inept colonial approach to economic and social problems had to be supplanted by a much more activist stance. India's villagers, caught in age-old patterns of tradition, casteism, and fatalism, had to be awakened and two-way links established between them and their government. Lastly, those landlords and other propertied interests that might oppose rapid progress in agriculture, commerce, and industry had to be confronted and neutralized, if not vanquished.

National Planning

The means chosen to shock and energize India's economy was national planning, which was thought to be the best means of establishing sequential paths of accomplishment that would lead cumulatively toward the realization of national goals. A coordinated planning drive that struck with unified force at India's interrelated economic and social problems promised rapid progress. In choosing an Indian path to socialism, the nation believed that it was not only making the right ethical choice but was premising its future on the scientifically best, most rational principles. Although in retrospect it is easy to aver that India could have taken a more capitalistic route, this must be balanced by recognizing that the communist alternative was at least as plausible. Actually, neither was attractive or likely. The right option appeared too close to colonialistic laissez-faire, and the left path seemed antithetical to India's participatory democracy. India's middle way was perceived to take the best elements of both extreme modes of operating an economy and adapt them to Indian conditions. Among other things, this meant state leadership of modern industry but retention of private ownership for farming and most types of commercial activity and banking. There was no intent to nationalize India's existing private manufacturing firms.

Current world history and postwar economic theorizing were compellingly supportive of India's mixed socialist experiment, with planning as its matrix. India's ongoing experiment had powerful feedback effects on the formulation of ideas in the emerging field of development economics. Nehru had visited Moscow as a young man, in 1927, and like many, he was powerfully affected by the Soviet model as it evolved. Fabianism was attractive to many Indian students educated in England. The long tradition of anti-imperialistic thought stretched from John A. Hobson, V. I. Lenin, and Karl Marx to Indian intellectuals who associated capitalism with colonialism.

Not by any measure was suspicion of the market economy and its prime agents predicated entirely on foreign notions. India's Brahmin elite habitually scorned the trading castes. The farming communities found it easy to distrust the moneylenders and wholesale merchants who appeared to make endless profit out of the toil of the peasant. Dadabhai Naoroji and M. G. Ranade were nineteenth-century Indian writers who opposed England's policies toward India, and the latter, who was among the leaders of the early Congress Party, articulated an infant economy analysis of India's state that drew heavily on the work of Friedrich List, a German critic of Britain's free trade persuasion. The wartime experience of the Allies demonstrated how quickly nations could mobilize resources under the pressures of a compelling exigency. The US-sponsored Marshall Plan and Point Four initiatives extended the wartime reliance on state leadership in economic initiatives into the peacetime years. All arrows pointed in the same direction: a commanding governmental presence could propel the rapid deployment of economic resources, secure high rates of economic growth, and enkindle substantial changes in India's ancient traditions and institutions.

The new field of development economics had roots in earlier thought and found ready audiences in the developing world and in Western policy circles. From various perspectives, the common theme in this literature was antipathy to unbridled market forces and fear of pervasive market failures that could be rectified only by public policy. John Maynard Keynes's analysis of the Great Depression strongly supported the intervention of an activist state in economic policy. His promotion of the International Monetary Fund and the World Bank established two quasi-governmental international institutions to deal with global monetary stability and imperfections in worldwide capital markets. The Keynesian theory was pushed forward by Roy Harrod and Evsey Domar in their parallel formulations of the practicalities of estimating the capital needed to achieve self-sustaining rapid growth. The management of national savings and investment flows in order to achieve rapid capital formation and growth seemed readily feasible, even simplistically mechanistic.

In 1955, W. Arthur Lewis brought many ideas together in a book, *The Theory of Economic Growth,* that confidently laid out what was known. His classical formulation in 1954 of the process of industrialization in a labor-abundant agricultural economy was the most important single article in the new field. During the war, Paul Rosenstein-Rodan had argued that rebuilding and developing national economies would require a high level of state intervention because crucial components of the economy, such as roads, power stations, and even large-scale consumer enterprises, would be underprovided by the private sector. The reason was that only government could assess the net social benefits of investments since the existence of externalities caused a divergence of social returns and private profit. All of these notions were ideally suited to trial in India, a newly independent nation that possessed a huge effectuating asset: the steel frame of the Indian civil service, long accustomed to managing governmental activities at all levels of administration, stretching from cabinet offices in Delhi to India's far-flung villages. The task was to shift the nature of rule from the colonial injunctions to collect taxes and preserve law and order to distributing resources and taking leadership in rural reform and industrial management. The perks and power associated with these new tasks elicited an enthusiastic response from most members of the newly indigenized bureaucracy.

The concrete realization of planning in 1950 was based on years of preparatory debate within the Congress Party and in the retreating colonial government. In 1938, the Congress Party had created a National Planning Committee following its election to power in eight provinces in 1937, when its representatives initially faced the practical choices and tasks of governance. For the first time, the apex members of the Congress Party had to think their way through problems of agrarian development, of providing sufficient power, and of relations with private business. Consistent policies had to be devised on the basis of compromises within the party. Tension between Gandhian exponents of small-scale cottage industries and Nehruvian advocates of large-scale manufacturing enterprises was conspicuous. Sardar Patel was sometimes successful in moderating the most extreme rhetoric of confrontation with the propertied groups.

Right after the war, the colonial government itself issued the *Second Report on Reconstruction Planning,* which made far-reaching proposals. Twenty major industries were to be taken over by the central administration. Rural development and social programs were to be pushed by the state. A rival Bombay Plan was devised by some of India's leading industrialists; this in turn was countered by a People's Plan promoted by the Indian Federation of Labor. The Industrial Policy Resolution of 1948, issued by the interim Indian government, rejected nationalization of industries but allotted a large role to the state in new initiatives, especially in key industries such as coal, petroleum, iron and steel, shipbuilding, and telecommunications. Gandhi's notions were not swept aside. Village outreach and nurturing of small-scale industry and India's historical crafts, especially hand spinning and weaving, found places in the Congress Party's development program.

Despite Nehru's commanding presence and the large majorities the Congress Party enjoyed in parliament, achieving harmony in the fundamentals of India's planned, socialist economy was never automatic, even after planning began formally in 1950. The British-style parliamentary system yielded a cabinet of powerful ministries, such as those of finance, defense, and industries, all with varying claims to and elements of control over the national purse. India's central bank, the Reserve Bank of India, was the cockpit of monetary control and banking oversight. The state governments had their own economic responsibilities in such important fields as agriculture and education. On 15 March 1950, the linchpin of India's planning and economic policy apparatus was put in place with the creation of the Planning Commission by means of a cabinet resolution. This clandestine birthing meant that neither the Constitution nor parliamentary action sanctioned what immediately became India's most formidable economic policy organization. With the prime minister as its chairman, cabinet members rotating in and out, and prestigious economists on board, the Planning Commission immediately became the hub of India's plan designs. Its mission was not only to engage in the technical art of forecasting and plan construction but also to reconcile the many competing claims for projects within the framework of national goals.

In order to bring the important political players into the discussion of priorities and the regional dispersion of projects, the National Development Council was constituted in August 1952. Participation in the council involved the prime minister, the chief ministers of the states, members of the Planning Commission, and could include center and state cabinet ministers, economic and technical experts, and representatives of the Reserve Bank. Its open-ended functions and participation meant that it served simultaneously as a sounding board for the proposals of the Planning Commission and as a crucible of discussion, reconciliation, and recalcitrance. The parallel existence of the Planning Commission and the Development Council bore witness to the obvious fact that the plans were at once technical economic programs and intensely political documents.

The First Four Five-Year Plans 1951–74

The degree of planning authority sought and obtained by the new Congress government was unprecedented in a peacetime democracy, much less one that was just finding its feet. The rush to independence and the transition to power left little leeway for preparatory thought and construction of a consistent plan based on a set of economic principles. In order to initiate the planning process and at the same time permit ample room for discussions among experts, in parliament, and by the general public, events moved on parallel tracks. The First Five-Year Plan was set to cover the period 1951–52 to 1955–56, but the first two years were to see only a cobbling together of various projects and initiatives, while the final three years were to operate with a revised, detailed, and expanded agenda, once this could be devised. There was sufficient initial success, with good monsoons and abatement of price inflation, that the size of the plan in its final three years was much enlarged. This euphoria carried over into the much more complex Second Five-Year Plan (1956–57 to 1960–61), which drew heavily on the contributions of P. C. Mahalanobis, a socialist, physicist, and systems analyst on whom Nehru elected to rely heavily in devising an analytical core to serve as the frame for the plan.

As the design of the second plan reached its final stages, the government adopted the Second Industrial Policy Resolution in April of 1956. This policy statement superseded the 1948 manifesto on industrial policy. The second resolution divided the major industrial sectors into three categories. The state would have exclusive responsibility for such key industries as defense, atomic energy, iron and steel, coal and lignite, most minerals, heavy electricals, the railways, telephone communications, and electricity. In most of these, existing private companies could continue to operate or even expand if licensed to do so. The second category covered industries that would have a mixed public and private character: aluminum and other nonferrous metals, machine tools, antibiotics and other essential drugs, fertilizers, and road and sea transport. In all remaining areas, the private sector would take the lead, subject to potential state entry and to comprehensive regulation and oversight.

With evolutionary changes, the planning and policy regime put in place at the beginning of the second plan continued in operation through 1975, and indeed, it has retained its broad contours into the 1990s, even though the plans have had a progressively smaller impact on the directions of movement in the economy. The Third Five-Year Plan (1961–62 to 1965–66) again redoubled expenditures. Rainfall and foreign exchange crises derailed long-range planning before the Fourth Five-Year Plan (1969–70 to 1973–74) finally came on line. For three years, 1966–67 to 1968–69, India relied on stopgap annual plans. The evolution of the planning system, the emphases and details of each plan, and the achievements of each cycle relative to initial targets are extremely complicated subjects about which there remains dispute.

By the mid-1970s, rates of economic change and growth in India were disappointing to everyone, which is not to say that planning had failed or had been a bad idea from the beginning. A broad view of the character of each of the plans and the accomplishments of each quinquennium can only highlight the coarser patterns. Nonetheless, even a defense of India's dirigiste policy regime through this quarter century must recognize a distinct decline in the effectiveness of planning in the sense that outcomes increasingly diverged from intentions. The plans showed themselves to be too rigid to allow midcourse reactions to such shocks as crop failures, wars, or foreign exchange shortages. The nature of Indian politics changed so that more groups sought special favors; also, state governments were less commonly willing to fall in line with Delhi's dictates as the Congress Party's dominance waned. Line ministries acquired influence relative to the Planning Commission in setting economic priorities.

These tendencies were in play even before Nehru's death in May of 1964 but quickly accelerated afterward.

The summary contours of the first four plans are provided in Table 1.1, which shows the planned and actual shares of total spending allocated to six sectors. Total expenditures in rupee terms are given in the last two rows. These indicate that each plan was, roughly speaking, about twice as large as the one before, so that the percentage share of a sector must be cut in half for there to be a decline in the absolute level of resources being committed. It is traditional to use these shares to identify the priorities built into each plan, and they provide other information as well. The first plan is generally thought to have put emphasis on agriculture (14.9%) and irrigation (19.7%) compared to industry (7.9%) and power (7.5%). Refurbishing the depleted railway system absorbed the great bulk of spending on transportation and communication (24.0%), and concern with education and health (26.0%) was strong. The second and later plans switched the focus to industry, and the shares of spending devoted to agriculture, irrigation, and social services fell correspondingly. Chronic power shortages led to rising investments in this sector through the 1960s and into the 1970s. This profile of spending accords well with the Nehru-Mahalanobis emphasis on developing core industries supported by ample power, transportation, and communications services.

The six categories can be rearranged to yield an interpretation different than that conveyed by the conventional wisdom. Agriculture and industry may be clubbed together as the directly productive sectors, while irrigation, power, and transport are combined into a social overhead capital category. When this is done, the contrast between the first and later plans is muted. Comparing the first to the second plan, the share of social capital spending falls only from 51.2% to 46.9%, and the directly productive sectors' combined fraction rises from 22.8% to 32.4%. Social service disbursement can be regarded as a residual. By the third plan, the agriculture-industry and social capital shares stabilize. On this reading, India's first four plans were consistently tilted toward the provision of conventional public goods such as power, roads, railways, communications, and irrigation, which is hardly an exceptional practice for any government.

Another picture emerges by combining spending on rural development (agriculture, irrigation, and cottage industries) and urban-industrial development (industry and power) and setting transportation apart because of its function as linking overhead capital. In plans two through four, the rural development share remains in the 24% to 26% range. The urban-industrial sphere claims a proportion rising from 23% to 36%, equally at the expense of transportation and social services.

Putting the conventional pattern together with these two alternatives yields a complex but more accurate appraisal of what India was trying to accomplish. The first priority was to provide an adequate infrastructure, and this capital-intensive activity absorbed the largest fraction of the four plans' allocations. After the ad hoc first plan, the industry-first strategy acquired weight, but weak agricultural performance made it desirable to increase the share of the farm sector in plan spending after 1961. Social services were never given pride of place, and in the first three plan periods, actual spending was noticeably below plan targets, indicating poor follow-on implementation.

In addition to setting expenditure benchmarks, each plan established a number of physical targets. These pertained to rice, wheat, commercial crops, minerals, textiles, paper, chemicals, fertilizers, metals, engineering goods, and others. As an illustration, the fourth plan took the 1968–69 level of rice production of 39.8 million tons and set a target of 52.0 million tons in 1973–74, implying an increment of 12.2 million tons, or 30.7%. In the event, actual output in the final year was only 44.0 million tons, which was 84.6% of the total target, but this meant that only 4.2 million tons, or 34.4% of the anticipated increment was achieved (4.2/12.2). The inability to hit incremental targets, even with a generous plus-or-minus margin of error of 20%, declined appreciably with each plan. Applying this method of scoring, over the first, second, and third plans, respectively, the percentage of targets hit was 38%, 30%, and 24%.

The intricacy of the fourth plan makes calculation of a comparable measure hard, but some representative attempts to hit incremental targets are wheat (211%), sugarcane (36%), iron ore (38%), cotton yarn (21%), man-made fibers (100%), and sulfuric acid (25%). By 1975, the effort to plan and manage the economy through a combination of financial allocations and the setting of physical targets was floundering. The steel frame verged on becoming an elastic bandage.

Economic Performance, 1950–75

The influence of planning was only one force determining achievements in sectors of the Indian economy. The condition of the world economy and the impact of the monsoons were parameters over which India had no control. Plan spending was a shrinking part of the central budget, as defense spending, salaries, and debt service took greater shares. Despite the expansion of the state-owned industrial base, which eventually led to public management of over two-thirds of India's large-scale manufacturing assets, agriculture, commerce, and much industry remained in private hands. In 1970–71, only 14.9% of India's economic activity sprang from the public sector, far below the proportion in developed market economies and not out of line with the share in other less-developed countries. Indeed, the central and state governments controlled the commanding heights of the economy, as they had sought, but the minions of private enterprise had ample scope to determine aggregate outcomes in many fields. At the beginning of the fourth plan, public investment constituted only about 35% of total national capital formation. It is true, though, that public spending was complementary to private spending, so that government decisions mattered disproportionately. To some extent as well, the public-private dichotomy masks the state's true role, which was magnified by extensive regulation of private activities in India's bureaucratic maze that implemented what became known as the "permit raj."

Because Indians in the high and low reaches of society judged how well the economy was performing within the framework of the five-year plans, it makes sense to adopt this point of view. In a developed economy, the primary markers of economic success are the growth rates of national product and industrial production, the rate of inflation, and the level of unemployment. Under Indian conditions, this last variable is essentially irrelevant because degrees of underemployment are pervasive. The most important single indicator of economic attainment is the output of foodgrains; closely related are the prices of foodstuffs, cooking oils, petroleum fuels, and sugar, the everyday staples of the ordinary family. Assessing the economic record in India means examining a mixture of conventional and particularly Indian auguries. Table 1.2 shows indexes of change in each of the four plan periods for selected items. An approximation of the annual growth rates in each quinquennium may be garnered from dividing the difference between the first year of a plan (=100) and the reported final-year indexes by 5. Outcomes for the annual plan years, 1966–67 to 1968–69, are omitted.

India's gross domestic product (GDP) shows modest expansion in each of the plan periods, but at no time was the rate of economic growth eye opening. Population multiplied at about 2.1% per year on average, rising from 359 million in 1950–51 to 607 million in 1975–76. The average Indian experienced a gain in income of about 1.0% per year and in 1975 had about one-third again as much income as in 1950. Industrial output showed sharp gains in the postwar recovery period of the first plan and in response to plan initiatives in the second and third plans. The range of industrial products expanded as well. From about 1965 through 1975, India's industrial growth slowed, and this is captured in the deceleration of this sector's increase in the fourth plan. Since industrialization was a prime intent of the plans, a slowdown at this juncture was disheartening. The planning operation and public-sector management of state enterprises came under increasing fire.

Prices had actually fallen as India benefited from good rainfall in the first plan, and inflation was moderate in the second- and third-plan periods. During the fourth plan, price rises were higher than many poor Indian families could bear without erosion in their meager standards of living. A feature of the 1950–75 era was resolute control of the supply of money and credit by India's central bank, the Reserve Bank of India. Comparative price stability and monetary discipline distinguished India from many less-developed nations, most notably those of Latin America.

Agricultural performance was solid but not outstanding between 1950 and 1975. Growth in food and commercial crop turnout moved up at a rate of about 3.0% or a little more, providing a margin of +1.0% over population increase. The results for the first two plans were satisfactory, but there were troublesome years in the mid-1950s when output stagnated. A good year at the end of the second plan helped farmers reach the target of about 76 million tons of foodgrains. Then, in the third plan, some of the most severe monsoon failures ever recorded cumulated into a frightening overall slump. Because planned control was supposed to deliver what was expected, the vicissitudes of the monsoon were not a sufficient excuse. An intensive district development strategy was adopted, in effect backing away from the multifaceted community development approach that emphasized providing a menu of social services in the villages. American and other foreign experts worked closely with Indian advocates of a focused, input-intensive strategy that relied on new high-yielding seed strains that responded well to fertilization and timed irrigation. Recognition that India's larger farmers were, after all, going to have to be encouraged with adequate incentives to improve India's agricultural record meant that vigorous land reform was defenestrated. Important commercial crops, such as cotton, jute, and oilseeds, turned in a mixed performance after 1950. The continuous investment in major and minor irrigation works shows up in the expansion of irrigated acreage in each plan, but as the food crop and commercial crop instability shows, it did not ensure secure outcomes.

In most industries, the four plan periods show strong results, consistent with the typical 6.0+% per year expansion of this sector. Electrical generation capacity rose by a larger fraction in each quinquennium. As the big steel mills came on line in the second and third plans, output rose commensurately. Cement and fertilizer output grew strongly, from low bases. Personal transportation, the bicycle in India's early days, became more available to middle-income and working-class people. A jarring weakness was the failure of cotton textile output from the large mills to expand. This was damaging to poorer families since traditional cotton saris and dhotis or modern shirts and pants are common garments. Policies that discouraged imports of and investment in modern machinery handicapped the modern sector, and Gandhian favoritism to the handloom industry stifled policy reform in India's huge textile sector. Sugar, often mixed in potent tea and coffee brews, was more commonly consumed in low-income families.

Table 1.2 closes with two social measures. The share of eligible children in primary schools rose, but even at the end of the fourth plan, at least one-quarter of potential students were not in classrooms, and probably not much more than half were regular attendees. Discrimination against untouchables, who were often excluded from village schools, remained strong. The number of hospital beds rose compared to the needful population, but no one could pretend that most Indians had ready access to adequate medical and hospital services.

By the 1970s, so many dimensions had been woven into India's plans and related policies that it is not always easy to identify continuing priorities. In a sense, everything was a priority and everything was important, at least in principle. Nonetheless, there were three consistent imperatives that followed from India's commitments to self-reliance and to the socialist pattern of society. Each was vital to India's internal political stability and to the aim of assuring the nation's national integrity and regional influence in international relations. The first was to grow enough food. The second was to create an industrial base sufficient to provide rising volumes of capital equipment and consumer goods and to create for India a strategically critical defense sector to supply an increasingly advanced air force, navy, and land army. The third was to boost the nation's social agenda.

From 1950 to 1975, India's food policy must be judged a success. The expansion of irrigation and the impact of the Green Revolution after 1967–68 were the underlying factors. In industry, India likewise reached its major goals by 1975. Indians now produced virtually a full array of producer and consumer goods, albeit of sometime inferior quality and higher cost than prevailed externally. It is true that public and private sector efficiencies were low, management was weak, and workers were not productive by world standards. The maze of licenses, controls, and regulations exacted a high price. Corrosive corruption increasingly permeated the licensing and permit system as the noble sentiments of the early administrative cadres yielded to the irresistible incentives created by regulatory authority. If India's socialism failed, it failed, paradoxically, exactly where its ideals should have yielded the most salubrious outcomes. In 1975, India was still very far from attaining universal primary education. Literacy rates for women and in rural areas were low, health services were thinly spread, and clean water supplies and sanitary facilities in the villages were gravely deficient.

Economic Policies and Growth Processes

Growth is a dynamic process characterized by the emergence of leading sectors and the interactive behavior of constituent components. Policy decisions can have substantial effects on the rate and sources of growth. Outcomes in the Indian economy between 1950 and 1975 and beyond were affected by such exogenous factors as the monsoon, wars with Pakistan in 1965 and 1971, and China in 1962, and international conditions, exemplified by the oil shocks of the 1970s. As flawed and as imperfect as it was in many aspects of its operation, the planning process was fundamental to establishing momentum. Shocks and plans do not embrace all the elements that determined India's economic development progress after 1950. This contention can be illustrated by examining three policy interventions that set off chain reactions in crucial areas of the economy at exigent junctures. These interventions had the common feature that they transcended the planning mechanism itself, although they interacted with it and complemented it in helping India pursue its national priorities. In the late 1960s, India's changing leadership pressed to the fore three policy measures that had immediate and substantial effects on the supply of wheat, the national savings rate, and the level of exports.

The bad monsoons of the 1960s added urgency to mounting concerns inside and outside India that not enough was being done to raise food output. Given India's ample water supplies and fertile soils, low yields could be doubled or trebled by the application of scientific agriculture and still not become comparable to the world's best. In 1959, a Ford Foundation report on India's food crisis recommended much keener focus on research, extension, intensifying modern input use, and incentives. Land reforms had not been effective in changing India's agrarian relations, although ceilings on the amount of land a farmer could hold had some effects in limiting the size of holdings. The cooperative movement had never caught fire because the hierarchies of caste and power in India's villages made it difficult to get all cultivators to work together to channel credit and inputs collectively or to market their crops jointly.

Especially in Punjab and Haryana states, the new extension strategy found sympathetic response among independent farmers. A key innovation was the creation of a number of regional agricultural universities, modeled on the public land-grant schools of the United States. The use of fertilizers, power tillers, pumps, and credit soared as high-yielding varieties of wheat were adopted, first by larger farmers but rather soon by almost all farmers. The Green Revolution's input- and incentive-oriented strategy was at odds with the plans' stress on institutional change and community development; it also favored what was already India's richest region and, at least at first, the bigger farmers rather than the marginal cultivators and landless laborers. Pragmatism overwhelmed ideological bias.

From the beginnings of the plans, it was intended that a large portion of the gains of growth be soaked up as taxes by the government or as deposits by the banking system from whence government could recycle these forms of savings into new capital formation. This was intrinsic to the Mahalanobis stratagem and wholly consistent with Lewis's development model. Yet from an average rate of savings of 10.3% of national income in the first plan, India's propensity to save climbed only to 13.2% during the third plan (Table 1.3). Arguably, a rate over 16%, or even 20%, was needed. The failure to elevate the nation's savings effort was substantive in itself, but it was also symbolic of the weakened planning system, the food crisis, and sluggish overall growth. The policy consensus that Nehru had been able to sustain while alive was fragmenting; India was about to experience swings in policy orientation as numerous ideas, interests, and individuals jousted for influence.

After Nehru's death in 1964, Lal Bahadur Shastri held power until he, too, died in office in January 1966. After a leadership tussle in the Congress Party in which the party's chief barons reached a standoff, Shastri was succeeded by Nehru's daughter, Indira Gandhi, who was thought to be a pliable compromise choice. Through the economic crises of the mid-1960s, policy swung from a mild relaxation in controls on prices and investments toward more radical politics as Indira Gandhi took control of a wing of the Congress Party and sought to marshal popular support before and after the election of 1971. At her initiative, parliamentary steps were taken to make it easier to acquire large landholdings or industrial assets without full compensation, threatening private property rights. An antimonopolies act was passed, the biggest industrial houses were to have less access to licenses for new manufacturing capacity, and the insurance and coal industries were nationalized. In 1969, the largest private banks were taken over and put in the public sector with the intent of directing them to provide more credit to rural and small-business clients.

Although Indira Gandhi's impatient and radical economic policies can be faulted on a number of grounds, the nationalization of banks was followed quickly by a huge increase in the number of rural and suburban branches. India's nationalized banks had 8,262 branches in June 1969 and 62,100 branches in 1995, with much of the expansion coming in the 1970s. The earmarking of certain proportions of credit to farmers and other groups traditionally neglected by private bankers shunted huge flows of loans to entirely new clients. The chief consequential effect of the opening of branch banks, including savings windows offering a mix of account options, was to draw a large inflow of deposits from India's households. The numbers of accounts rose sharply, as did the amount of savings. As Table 1.3 demonstrates, after 1968–69, India's savings rate climbed sharply to the levels thought necessary to sustain investment deployments that would in turn put India's economy on a higher growth track.

The Mahalanobis plan frame was predicated on the assumption that exports did not matter or were not necessary. Such export neglect or export pessimism was consistent with most thinking about economic development, particularly as applied to a large country (such as India) that had diverse resources and a potentially large domestic market. The economic problems of the mid-1960s, and considerable rethinking about policies, led to modest reforms in trade procedures that were favorable to exporters. The centerpiece of the policy changes was the devaluation of the rupee in 1966, which brought India's overvalued currency into line with its likely market value and made India's exports more attractive.

Although it is not widely appreciated, India experienced significant export growth in the early 1970s, setting in motion a process that has continued into the 1990s. Table 1.3 shows the flatness of India's exports in the first and second plans, and modest growth in the third plan and annual plans. In the fourth and fifth plans, India's exports take off rapidly. The impact on the national economy as a whole was not great because exports comprised only 3.6% of GDP in 1970–71, but this breakthrough does mark a turning point. Interestingly, growth comes almost entirely from new exports, such as iron and steel, chemicals, engineering goods, seafood products, transport equipment, and gems and jewelry (mostly polished diamonds). India's 20-year import-substitution policy had begun to yield export advantages.

The lesson of the stories of wheat, savings, and exports is that policy matters and problem-solving reform has been a significant constituent of India's development strategy. This pragmatism is wholly consistent with a democratic mode of governance, which often involves trial and error. What is involved is not simply a choice between state and market mechanisms but a consideration of the relations and coordination between them within the broader legal, political, and cultural environment. The gains in wheat came about through shrewd exploitation of available technologies and effective extension bolstered by government pricing policies for inputs and selected crops. The intrinsic character and skills of the farmers of northwestern India were cardinal. The visible hand of the state is as evident as the invisible hand of the market, and neither mattered as much as the hand on the plow. Bank nationalization stemmed from a wave of populist excess and was based on no sound economic premise, but Indira Gandhi's gut feeling was correct: conservative bankers needed to be prodded in the most commanding way to get them to make their banks truly national institutions, serving all clienteles. The devaluation of 1966 and modest simplification of the trade regime had positive effects on exports, but this too reflects a symbiosis of public and private action. Tinkering with commercial and exchange-rate policy was to continue afterward as a predictable component of the minister of finance's annual budget message.

Reprise

In 1975, Indians could look back on 25 years of economic gains. The colonial shackles had been broken and stagnation turned into momentum. The nation's industrial sector had been broadened and deepened beyond recognition, with average rates of growth of 6.0% or so. Military production was sufficient to support India's claim to be the primary regional power in South Asia and a midrange world power. The Green Revolution had transformed India's agricultural position into one of increasing but still modest abundance. Exports had begun to expand as segments of Indian industry and agriculture emerged from the cocoon of protectionism and established competitive world positions. Continued population growth eroded moderate income expansion, limiting per capita gains to about 1.0% per year. The weakest, almost shameful, showing appeared on the education, health, and rural welfare gauges. Urban slums were much in evidence and underemployment remained acute. The plans had never actually incorporated a consistent employment strategy, despite repeated expressions of concern.

As of 1975, much remained to be done. Certainly the following 25 years, through the end of the century, did and will see more political turmoil, more social change, more policy experiments, and more economic development, but a foundation was certainly in place in 1975.

Further Reading

Arndt, H. W., *Economic Development: The History of an Idea,* Chicago: University of Chicago Press, 1987

A sympathetic and knowledgeable review of the foundational ideas and chief contributors to the field of economic development. It can be consulted for references to the writings of economists mentioned in this chapter.

Bhagwati, Jagdish H., and Padma Desai, *India: Planning for Industrialization, Industrialization and Trade Policies since 1951,* London: Oxford University Press, Development Center of the Organization for Economic Co-operation and Development, Paris, 1970

The first comprehensive attempt to bring economic analysis together with the best available data on the early years of India's planning system. A balanced landmark treatment of shortcomings with laments for opportunities missed.

Chandhok, H. L., and the Policy Group, *India Database: The Economy, Annual Time Series Data,* volumes 1 and 2, New Delhi: Living Media, 1990

An invaluable and comprehensive systematization of India's principal economic series.

Collins, Larry, and Dominique Lapierre, *Freedom at Midnight,* London: Book Club Associates, and New York: Simon and Schuster, 1975

A novelistic rendering of the transfer of power from Britain to India and Pakistan at partition in 1947. An extremely readable and credible history.

Ford Foundation, Agricultural Production Team, *India's Food Crisis and Steps to Meet It,* Delhi: Government of India, Ministry of Food and Agriculture, 1959

One of the key documents in the maneuvering to shift India's rural development strategy in the direction of focus on incentives and inputs.

Frankel, Francine R., *India's Political Economy, 1947–1977: The Gradual Revolution,* Princeton, N.J.: Princeton University Press, 1978

A detailed treatment of the politics of India's plans and economic policies.

Government of India, *Second Five-Year Plan,* New Delhi: Government of India, Planning Commission, 1956

The flavor and techniques of Indian planning can only be appreciated by reading the sequence of plan documents, which combine reviews of the past record with the layout of programs for the next five years and beyond.

Hanson, A. H., *The Process of Planning: A Study of India's Five-Year Plans, 1950–1964,* London: Oxford University Press, 1966

A carefully reasoned and judicious account of the making of India's first three five-year plans that centers on administrative processes rather than political infighting or economic results.

Lewis, John Prior, *India's Political Economy: Governance and Reform,* Delhi: Oxford University Press, 1995

This book and Lewis's *Quiet Crisis in India* (Washington, D.C.: The Brookings Institution, 1965) put bookends around the experiences and perspectives of one of the United States' most empathetic and informed Indian spectators.

Lewis, W. Arthur, *The Theory of Economic Growth,* London: Ruskin House, and Homewood, Ill.: R. D. Irwin, 1955

Magisterial synthesis of the incentive, institutional, and input factors in economic growth. Coupled with Lewis's "Economic Development with Unlimited Supplies of Labor," in *The Manchester School* 22 (May 1954): 139–91, this volume set the terms of reference for the analysis and planning of economic growth in the less-developed world.

Mellor, John, Thomas F. Weaver, Uma J. Lele, and Sheldon R. Simon, *Developing Rural India: Plan and Practice,* Ithaca, N.Y.: Cornell University Press, 1968

Assesses India's rural development program through the fourth plan. Includes field studies of farmers, traders, and the village of Senapur.

Nair, Kusum, *Blossoms in the Dust: The Human Factor in Indian Development,* New York: Frederick A. Praeger, 1963

Based on a year of travel throughout the country, Nair captures in evocative prose the aspirations and fears of India's diverse rural peoples.

Rosen, George, *Democracy and Economic Change in*

India, Bombay: Vora, and Berkeley, Calif.: University of California Press, 1966

Excellent review of India's economic development through the early 1960s. Adept evaluation of rural and urban winners and losers, and the political implications.

Rudolph, Lloyd I., and Susanne Hoeber Rudolph, *In Pursuit of Lakshmi: The Political Economy of the Indian State,* Chicago: University of Chicago Press, 1987

India's political and economic development through 1985. Examines the weakening of the Congress Party, the growth of interest-group politics, and the impact of these on economic policies.

John Adams is professor of economics and chairman of the economics department at Northeastern University, Boston, Massachusetts.

Table 1.1: India's First Four Five-Year Plans
Sectoral planned and actual expenditures (% of total) 1951/52–1973/74

Sector	*First plan* *1951/52–1955/56*	*Second plan* *1956/57–1960/61*	*Third plan* *1961/62–1965/66*	*Annual plans* *1966/67–1968/69*	*Fourth plan* *1969/70–1973/74*
Agriculture and allied sectors					
planned	14.9	11.3	14.2	n/a	17.2
actual	14.8	11.8	12.7	16.7	14.7
Irrigation and flood control					
planned	19.7	9.7	8.7	n/a	6.8
actual	22.0	9.3	7.8	7.1	8.6
Power					
planned	7.5	7.1	13.5	n/a	15.4
actual	7.7	9.5	14.6	18.3	18.6
Industry and minerals					
planned	7.9	21.1	23.8	n/a	22.8
actual	4.9	24.1	22.9	24.7	19.7
Transport and communications					
planned	24.0	29.8	19.8	n/a	20.4
actual	26.4	27.0	24.6	18.4	19.5
Social and community services					
planned	26.0	21.0	20.0	n/a	17.4
actual	24.1	18.3	17.4	14.7	18.9
Total (million Rs.)					
planned	23,780	45,000	75,000	n/a	159,020
actual	19,600	46,720	85,770	66,250	157,790

Source: Calculations by the author based on plan documents and H. L. Chandhok and the Policy Group, *India Database,* New Delhi: Living Media, 1990. Every effort has been made to use the estimates available at the time, not as restated by later adjustments.

Table 1.2: Growth Indexes for Plan Periods

Series	*First plan*	*Second plan*	*Third plan*	*Fourth plan*
Gross domestic product	119	122	114	118
Population	110	108	112	112
Industrial output	136	137	154	126
Agricultural output	117	115	96	114
Exports (US$)	100	106	126	179
Wholesale prices	86	135	132	153
Foodgrains	120	123	90	121
Rice	130	121	87	111
Wheat	130	124	95	125
Raw cotton	135	133	92	127
Raw jute	127	98	110	193
Oilseeds	111	125	90	138
Irrigated area	127	112	116	120
Electricity capacity	147	168	179	137
Steel	130	177	196	116
Cement	174	170	135	131
Cotton cloth (mills)	137	107	95	99
Bicycles	508	209	147	132
Ammonium fertilizer	826	402	230	215
Sugar	173	158	117	121
Children 6–11 in school	126	118	123	114
Hospital beds/1,000	111	149	129	112

Source: Calculations by the author based on plan documents and H. L. Chandhok and the Policy Group, *India Database,* New Delhi: Living Media, 1990. Every effort has been made to use the estimates available at the time, not as restated by later adjustments.

Table 1.3: Performance for Plan Periods—wheat, savings, exports

	First plan *1951/52–1955/56*	*Second plan* *1956/57–1960/61*	*Third plan* *1961/62–1965/66*	*Annual plans* *1966/67–1968/69*	*Fourth plan* *1969/70–1973/74*	*Fifth plan* *1974/75–1978/79*
Wheat (1969/70=100)	44.4	54.0	61.3	83.9	129.5	165.4
Savings rate (% of GDP)	10.3	11.7	13.2	13.7	16.1	20.4
Exports (US$) (1966/67–1968/69)	75.9	76.1	94.4	100.0	141.7	334.6

Source: H. L. Chandhok, and the Policy Group, *India Database: The Economy, Annual Times Series Data,* volumes 1 and 2, New Delhi: Living Media, 1990

Chapter Two

Mid-1970s to the Present

Christopher Candland

When independent India turns 50, in August 1997, its population will be somewhere around 940 million; estimates vary. At projected growth rates, India's population may exceed that of China in two generations. In addition to being one of the world's most populous countries, India is also one of the world's poorest. Average annual per capita income in 1997 is Rs. 11,200, or US$330 by prevailing exchange rates (US$1,500 by purchasing power parity). Some Indian development achievements are considerable, including self-sufficiency in foodgrain production and an excellent system of higher education and training in sciences and technology. The most glaring need at independence, however, the lifting of a widespread and crushing poverty, remains unfulfilled. A significantly smaller proportion of the population lives in poverty today than 20 years ago, but the number of individuals falling below the poverty line has increased. There are important disagreements about how to define and measure poverty, but hundreds of millions of Indians subsist under extreme forms of deprivation of nutrition, shelter, employment, and health.

Since independence in 1947, the central government has been committed to an economic development strategy that would promote social welfare (Maddison, 1971: 86–91). Precise methods for the uplift of the masses have been opaque, however. In general, economic policy has been aimed at promoting economic and social welfare while not abridging civil and political rights. The priority placed on economic self-reliance did help to foster a diversified domestic private and public sector, and maintain low import dependence and foreign indebtedness (Burki, 1984). However, the emphasis on self-reliance also contributed to low productivity and technological obsolescence in many industries. Since the late 1970s, and with greater purpose since 1991, the government has attempted to reorient the inward economy. While it is too early to say whether India's economic reforms since 1991 will place India on a sustained high growth trajectory, it is clear that efforts to alleviate poverty and to raise the low levels of human development that perpetuate poverty–from low literacy and primary school enrollment to access to clean drinking water and basic medical care–must be redoubled.

When Indian development objectives were defined, especially in the decade before independence (Thakurdas et al., 1944–45) and the decade following (Industrial Policy Resolution, 1956), it was generally believed, in India and throughout the world, that economic development was conferred by industrialization. In a country scarce of capital and industry, surpluses from agriculture, supplemented where possible by foreign assistance, appeared to be the only sources for the necessary investments in industry. Thus, in India, agriculture was used to finance industrialization, though not to the extremes to which the strategy was pursued in the Soviet Union, China, and North Korea.

India's legacy of democratic institutions is unusual among large developing countries. The regular holding of general franchise elections among competitive political parties is an intractable feature of Indian development. However strained, democratic governance is remarkably resilient, which is suggested even by the brevity of India's experience with authoritarianism in the late 1970s. Party competition, substantial federalism, and the opportunities afforded to relatively advantaged segments of society to influence government programs, have particular consequences for economic development. Well-entrenched democracies may be anathema to the developmental policies that promoted rapid growth and surprisingly egalitarian distribution in some East Asian economies after World War II. Land reforms, for example, are more readily opposed or subverted in a democratic polity. In addition to the challenge of raising one-third of the population out of poverty, therefore, there are the strictures of a well-entrenched democracy. Indian development in all likelihood will retain its popular, democratic character and continue to rely on incentives rather than coercion to achieve economic and social policy objectives, such as lowering birth rates. Indian efforts made and lessons learned in adopting development policies and implementing development programs therefore will continue to make substantial contributions to the field of economic development.

Politics of Economic Adjustment since 1975

India was faced with serious economic crises in the early 1970s. With average annual gross national product (GNP) growth between 1970 and 1975 of less than 1.3% and with population growth at nearly 2.5%, per capita income was declining. The United States suspended food aid and other economic assistance in response to India's intervention in the brutal civil war in East Pakistan, now Bangladesh. The war and nearly 10 million refugees further burdened the Indian economy. The 1973 Organization of Petroleum Exporting Countries (OPEC) price hike added to inflationary pressures and government expenditure. The first wave of green technologies, implemented in the north (Punjab, Haryana, and western Uttar Pradesh) in the late 1960s led to greater agricultural productivity. But similar results were not as easily attained in the second wave of the Green Revolution in the 1970s, when new inputs were introduced in the southern and eastern states, and income inequalities and land dispossession increased as relatively better-off farmers capitalized on rural inputs.

Indira Gandhi's Congress government also faced serious political challenges. Student agitation helped to topple the Congress government in Gujarat, and the subsequent elections brought to power one of the prime minister's chief rivals, Morarji Desai. Popular political movements threatened the Congress Party. In May 1974, the socialist trade union leader George Fernandes led a paralyzing railway strike, which the government crushed (30,000 workers were arrested). In July 1974, the government impounded state employee's wage and cost-of-living increases.

In 1975, Indira Gandhi declared a state of emergency. She claimed that the emergency would protect against external imperialist threats and internal capitalist ones. Careful studies of the political economy of the emergency could not confirm a link between the turn to authoritarianism and the economic crises of the 1970s. Barnet Rubin, for example, found no significant increase in public capital expenditure under the emergency ("Private Power and Public Investment in India," Ph.D. dissertation, University of Chicago, 1982). And just as André Gunder Frank declared that a permanent emergency had been ushered in by the exhaustion of import substitution and the economic cost of populism (Frank, 1977), the emergency was lifted. Whatever the precise relationship between populism, stagnation, and authoritarianism, by 1975, Indian economic development efforts had clearly failed to meet their objectives. Average annual per capita income in 1975 at current prices was Rs. 1,032, fewer than three rupees (US$1.75 in purchasing price parity) per day. Fewer than 3% of the land designated for redistribution in 1972 land-reform regulations had been distributed to the landless (Toye, 1977: 305). Far from having achieved national literacy, as designed a generation earlier, two-thirds of the adult population were illiterate. Infant mortality to age one was 130 for every 1,000 deliveries, fewer than one-third of which were attended by any health care personnel.

Directly precipitating the declaration of emergency in the early hours of 26 June 1975 was the Allahabad High Court's order that Indira Gandhi resign and be barred from holding office for six years (Hart, 1976: 3). Indira Gandhi used the Maintenance of Internal Security Act to have thousands of politicians and other detractors arrested and jailed, the media censored, and industrial strikes and political gatherings banned. She had the president dismiss legislative assemblies in the two non-Congress-controlled states, Gujarat and Tamil Nadu, and with a stronger than two-thirds majority in both houses of parliament, she amended the Constitution to remain in power.

Indira Gandhi touted a 20-point program with such populist features as lowering retail prices, seizing the luxury goods of tax evaders, and enforcing the ceiling on land holdings. The emergency, however, marked not a deepening of the economic populism that helped her to retain control of the Congress organization in the 1960s but a discernible initiative to dismantle a heavily regulated economy. The first efforts at economic adjustment had come in the wake of the failure of the nationalization of the wholesale wheat trade in January 1974. In 1975, the government permitted industries to expand capacity and produce in related areas (a practice called *broadbanding*) and lifted the ceilings of the Monopolies and Restrictive Trade Practices (MRTP) Act.

Persuaded by her advisers that she would win, the prime minister scheduled national elections for March 1977. After an ignominious 20-month emergency rule, Indira Gandhi was soundly defeated by a hastily assembled coalition of political parties and protest movements. The Janata (People) Party professed Gandhian development sentiments and generally regarded the public sector with the same suspicions as it did big business. Under Prime Minister Morarji Desai, an economic conservative (but a liberal in the language of his day), and under Desai's successor and rival in the United Front, Charan Singh, the United Front continued to deregulate the economy while encouraging rural development. The Janata government, however, suffered from "debilitating . . . personal aggrandizement," and the coalition fell in August 1979 (Rudolph and Rudolph, 1987: 177).

The view had emerged by the late 1970s among businesspeople, journalists, government officials, and academic economists that the Indian economy could perform better domestically and internationally. Indian economists attributed the economic decline, which began in the mid-1960s, to (1) low growth in agriculture (Chakravarty, 1979), (2) worsening income inequality and consequent weak consumer demand (Nayyar, 1978), (3) unequal terms of trade between agricultural and manufacturing goods (Mitra, 1979), and (4) a decline in capital investment (Bardhan, 1984). While the need for adjustment has long been recognized, adjustment itself has been gradual. Policy makers began to reconsider the emphasis on minimizing imports as a means to increase self-reliance more than two decades ago.

On returning to power in 1980, the Congress adopted a new Industrial Policy Resolution designed to bring coherence to delicensing, broadbanding, capacity expansion, and the other ad hoc adjustment measures already announced. The Industrial Policy Resolution of 1980 marked a break with the 1956 Industrial Policy. It dropped reference to building a strong public sector that would occupy the "commanding heights" of the Indian economy in favor of reference to a public sector that would serve as the "pillars of infrastructure."

To relieve its balance-of-payments problem (which was exacerbated by the rising cost of oil imports), India negotiated a US$5.8 billion loan from the International Monetary Fund (IMF) in 1981, which gave new impetus to economic reforms. The loan, the IMF's largest at the time, consumed one-sixth of its hard currency reserves and came with few conditions. It was nevertheless vigorously opposed in the Indian parliament as an affront to Indian sovereignty, and the Minister of Finance, Rawaswami Venkataraman, was at pains to assure members of parliament that the conditions agreed to were in India's own interests; the scene would be repeated in response to the IMF agreement in 1991. The government did not agree to a currency devaluation in 1981; but it did initiate substantial changes in its financial and import policies. Later, the government took anti-inflation measures, dropping the bank lending rates and raising minimum reserve requirements, and, during a year of record wheat harvests, purchased more than 1 million tons of US wheat. The government could procure only 6.3 million tons of the more than 33 million tons domestically produced, indicating the inadequate level of procurement prices relative to market prices.

The government ban on strikes in 1981 foretold the heightened conflict between the government and organized labor in the early 1980s. The ordinance, promulgated in July 1981 while parliament was out of session, applied to government designated "essential industries," including the railway, electrical, telephone, and postal services, ports and airlines, banks, hospitals, and defense-related industries, including petrochemicals industries. In response, eight national trade unions called a nationwide general strike to protest the National Security and Essential Services Act. In police raids starting on the eve of the strike, over 25,000 activists and striking workers were jailed.

On 31 October 1984, ten weeks before the completion of her third term as prime minister, Indira Gandhi was assassinated in response to the Indian Army's storming of the Golden Temple, the sacred site occupied at that time by Sikh militants. Indira Gandhi's younger son, Rajiv Gandhi, a member of parliament groomed for succession since his brother Sanjay's death in 1980, was quickly chosen to be prime minister by the parliament. In 1985, Prime Minister Rajiv Gandhi began to promote market-oriented reforms. New items were made available for import, and import tariffs were lowered on the grounds that higher technology would be India's vehicle to rapid growth into the twenty-first century. While Rajiv Gandhi's efforts at economic adjustment were not the first nor the most extensive, he is regarded by some as the chief proponent of economic reform among India's prime ministers because he articulated the position that the development strategy of Jawaharlal Nehru, his grandfather, had outlived its usefulness and that government-business alliances should be formed.

The Indian National Congress (Indira)(Congress [I]) government lost the November 1989 general election. A coalition government, headed by the Janata Dal leader and former Congress finance minister, V. P. Singh, held office until the professedly Hindu Bharatiya Janata Party (BJP) withdrew its support in October 1990, causing the Janata Dal to lose a vote of confidence in parliament in November 1990. The Janata Dal's National Front coalition government was succeeded by a new coalition government, headed by Chandra Shekhar, the leader of a faction of the Janata Dal, the Janata Dal (Socialist). His government attempted to shift attention to agriculture, as had Charan Singh's government in the late 1970s. The brevity of both governments helped to explain why the shift did not have sustained effects. Chandra Shekhar's government lost the outside support of the Congress (I) in May 1991, making way for the return of the Congress in the May-June 1991 elections. In the midst of these elections, Rajiv Gandhi was assassinated in retaliation for the Indian intervention in northern Sri Lanka. P. V. Narasimha Rao took control of the Congress Party and the new Congress government.

Adjustment since 1991

The new Congress government found itself in an unsustainable fiscal situation when it took office in June 1991. The international credit-rating agencies Standard and Poors and Moody's had lowered their rating of short-term debt to Government of India financial institutions, including the State Bank of India, the Industrial Credit and Investment Corporation of India, and the

Oil and Natural Gas Commission. Six factors were involved in the assessment, including (1) the increase in public debt, (2) the increase in external commercial borrowing and subsequent higher interest payments, (3) the increase in the external debt-to-export ratio, (4) the adverse impact on export earnings and foreign remittances of the Iraqi invasion of Kuwait, (5) the increase in budget deficits and the subsequent increase in interest payments and inflation, and (6) the recession in the Organization for Economic Cooperation and Development (OECD) countries and the subsequent decrease in export potential (Sen, 1994: 808). The fiscal deficit for 1990–91 was estimated at 8% of gross domestic product (GDP), having climbed from roughly 4% in the mid-1970s. Interest payments on internal debt alone constituted nearly 20% of total central government expenditure. Inflation had reached double digits, a historically high and politically dangerous level for India. The consumer price index for fiscal year 1990–91 increased by 13.6%, with the sharpest rises in foods and the other essential commodities that compose the bulk of the poor people's expenditure. The balance-of-payments situation was serious. Foreign exchange reserves had dwindled to an amount sufficient for only two weeks of imports, threatening India's first foreign loan default.

Minister of Finance Manmohan Singh announced in July 1991 in the Lok Sabha, the lower house of parliament, the removal of most industrial licensing requirements and the lifting of location and capacity restrictions on industry. The new industrial policy reduced the number of industrial sectors reserved for public sector investment from 17 to eight and abolished requirements for government approval of domestic investment in all but 18 sensitive areas. Foreign investment of up to 51% of equity was granted in 34 sectors, as was automatic approval of foreign technology agreements. The government declared two major devaluations in early July 1991, amounting to more than 20%, followed by a further devaluation. In October 1991, the Government of India signed an agreement with the IMF for a standby loan of approximately US$2.1 billion.

In response to criticism inside parliament and out, the government described the adjustment measures and the IMF loan decision as the unavoidable response to a serious fiscal crisis. The crisis can be traced to a combination of domestic and international factors. Although India had been financially prudent with foreign and domestic borrowing since independence, it began running fiscal deficits in excess of 6% of GDP in the late 1970s. The deficits were based on sharp rises in interest payments on domestic and international borrowing, defense purchases, government salaries, and domestic subsidies in the 1980s. Large current account deficits, of 25% of exports in the early 1980s and 40% of exports by the late 1980s, were met through the 1981 IMF loan, dispersed between 1982 and 1984, and large commercial borrowings. To manage its fiscal deficits, the government reduced the growth of real spending on capital investment and increased its short-term external commercial borrowing. The conflict in the Persian Gulf made this precarious situation worse. The return of Indian workers from Kuwait and the loss of their remittances added to the foreign exchange crisis. The price of petroleum and petroleum products, India's single most costly import item, increased sharply.

Until the implementation of structural adjustment programs in South Asia, its governments were relatively conservative in foreign-exchange borrowing and were generally regarded as protected from debt crises of Latin American proportions. Between 1990–91 and 1994–95, India's external debt as a percentage of GNP rose from 30.6% to 34.2% and then dropped in 1993–94 to 35.3%. While India's debt to GNP is not high in comparison with other major developing countries and short-term debt was lower in 1993–94 than in 1990–91, India's debt service ratio (external debt as a percentage of export earnings) of nearly 290% in 1995–96 (Government of India, 1997: S87, S108) is among the highest in the world.

Since the 1991 adjustment measures,

inflation too has risen. Historically, inflation has been kept under tight control in South Asian economies. The high rates of inflation that Indians have faced over the past few years seriously threaten the reform process and the Congress government. The Ministry of Finance regards inflation control as a priority and a requirement if further economic reforms are to be politically feasible. The government, however, is seriously constrained as the adjustment program is, in part, based on adjustments in administered prices, including of food, fertilizers, and energy. Large government deficits in the 1980s helped to provoke the decline in capital expenditure and the external debt crisis, and they fueled inflation. After reducing the fiscal deficit as a percentage of GDP from 8.3% in 1990–91 to 5.9% in 1991–92 and 5.7% in 1992–93, the government allowed the fiscal deficit to rise to 7.5% in 1993–94 but reduced it again, to 5.0%, by 1996–97.

The Indian economy has not undergone structural adjustment in the conventional sense. Indeed, Indian policy makers, and Indian economic and political realities, have modified what is now understood to constitute structural adjustment. The IMF has expressed its appreciation for what is actually an unusual form of adjustment, without labor law reform, economic austerity, or privatization. For example, although the IMF often makes as a loan condition that revenues from the sale of public industries and government enterprises not be used against fiscal deficit targets, privatization has been central to most countries' adjustment programs. In India, privatization has not been a significant source for government revenue. Some state governments in India, which operate the bulk of the country's public sector enterprises, have pursued privatization better than the central government, but most of the shares sold in central government units have gone to government financial institutions, effectively transferring public debt from public-sector industry to public-sector financial institutions. While in numerous cases, a relatively small change in equity has led to significant industrial reorganization, few public sector enterprises have been subject to any disinvestment, and these at low rates of equity.

Reducing government expenditure on social welfare and capital investment will only endanger long-term growth. A better candidate for deep reduction is government expenditure on defense. Government defense expenditure is seven times greater than expenditure by both the central and state governments on education. The challenges to India's external security are considerable, but grossly inadequate commitments to human development are a greater threat to India's security than are its neighbors. Bangladesh, Indonesia, and Thailand, by contrast to India, spend more on education than on defense in absolute terms (World Bank, 1995: 126; World Bank, 1996: 214).

State expenditure on social development, of course, depends not only on public commitments but also on government revenue. Greater domestic tax revenues have been proposed (Narasimham, 1992). Tax revenues rose from 7% of national income in 1950–51 to 17% in 1990–91: "In terms of both the level of taxes and their growth, the performance of India's tax system has been quite satisfactory" (Mundle and Rao, 1992: 235–36). The average tax to GNP ratio for countries at comparable levels of per capita income is 12%, but the Indian tax system is unduly complicated, too narrowly based, and increasingly dependent on inegalitarian indirect taxes (Mundle and Rao, 1992). From 1950–51 to 1980–81, indirect tax revenues as a portion of total tax revenue increased from 58% to 80% (Ahmed and Stern, 1989: 1017). Through the 1980s, the proportion of central government tax revenue collected through direct taxation (including corporate taxes, personal income taxes, interest taxes, estate duties, wealth taxes, and gift taxes) dwindled further. A taxation system that facilitates poverty reduction, rather than one with biases against the poorer segments of society, is clearly required.

It is methodologically difficult to distinguish between the economic effects of a structural adjustment program and the economic conditions that make it necessary.

The recessions, oil price shocks, monetarist shifts, declining developing economy export commodity prices, falling international development assistance, and rising interest rates lead to a reversal of net financial transfers between developed and developing economies. The Indian structural adjustment program stands in contrast to the structural adjustment programs initiated in Latin America and Africa in the 1980s, in which negative economic growth and high inflation had devastating consequences for the poor and for productive investment. There is little evidence yet of widespread immiseration of the populace through economic austerity. The government has recognized the need to revamp the public food and essential commodities distribution system so as to extend its use by the poor and thus limit its use by the nonpoor (Shankar, 1997). Nevertheless, there are no adequate social safety nets for those who are negatively affected by economic adjustments such as labor force restructuring.

Assessing Indian Performance

As others have noted, assessments of development accomplishments are necessarily comparative. Paul Brass provides five ways of assessing Indian economic development performance: one can (1) compare performance after independence to performance prior, (2) evaluate GNP growth rates in light of population growth rates, (3) compare achievements to goals, (4) compare India to other large developing countries, or (5) evaluate India's changed "status in the world economy" (Brass, 1994: 281).

Economic Development

Economic performance after independence improved considerably over performance under the last five decades of British colonial rule, when British trade and fiscal policy undermined economic development. While it has been contested, per capita income probably declined in the first half of the century, the last 50 years of colonial rule (Maddison, 1971: 43–70).

Growth since 1975 was more impressive than in the two decades directly after independence. In its per capita income group, India outperforms other economies in GDP and gross domestic investment (GDI) growth as well as in basic human development indicators. Comparison with other lower per capita income countries may, however, not be altogether appropriate, for average performance in the group was lowered considerably by a number of economies in prolonged recession. The middle-income economies provide a better comparative reference. Against middle-income countries, India performs better on key economic indicators (except domestic investment) and nearly as well on key social indicators (see Table 2.1).

Of course, not all countries have similar development opportunities or objectives. Specific goals are shaped by country resources, including human resources and technologies, as well as political and geostrategic imperatives and other country-specific factors. It may be useful, therefore, to compare India with the other populous, developing economies with largely rural populations, namely, Argentina, Brazil, China, Indonesia, Mexico, and Pakistan.

India, China, and Pakistan, unlike the other populous and land-scarce countries of Argentina, Brazil, Mexico, and Indonesia, do not possess large natural endowments, such as minerals or fuels, on a per capita basis. India and China, unlike Pakistan and Indonesia, have received little foreign development assistance on a per capita basis. And China and India, the two most populous countries, are each more than three times as populous as the third most populous country, the United States. For these and other reasons, among the other populous rural economies, China makes for a useful comparison (see Table 2.2).

China's growth in all sectors of the economy has been stronger than India's. China's agricultural growth in the 1980s and its industrial growth in the 1990s has been particularly strong. In comparison with China, India's improvements in primary and secondary school education and in literacy are weak. The lesson typically drawn from the

Chinese experience is that deregulation, foreign investment, and an export orientation (begun in China in the mid-1980s) promotes growth. China's economic growth since the early 1980s, it should be noted, began with the reform of collective agriculture and the liberalization of rural and agricultural market activity in the late 1970s. Indian agriculture, of course, cannot boost agricultural production through decollectivization. The Chinese experience, however, does suggest that in largely agrarian economies, higher food availability and higher rural incomes may be the most effective means of stimulating broad economic growth through the creation of new demand from rural consumers.

There are obvious limits to the lessons to be drawn from Chinese experiences. China comprises several significant minority communities, but India is considerably more diverse. Its more than two dozen states, 14 of which have populations over 20 million, possess substantive decentralization of political and economic authority. Distinct regional economies and configurations of political power operate, and major economic actors are entrenched in a vibrant and competitive political party democracy. For these reasons, development lessons may be best drawn from intracountry comparison.

Despite the country's highest level of unemployment and declining levels of foodgrain production, the southern state of Kerala has performed well in quality of life measures, many of which are attributable to public investment in human development (see Table 3.3). Much of that performance, it should be noted, is also attributable to unusually effective and egalitarian land reforms and high levels of labor remittances from overseas workers.

Kerala has outperformed India as a whole (and China) in the provision of social services and enhancement of human capabilities. Kerala's higher social and economic standards of living follow directly from higher commitment to and expenditure on human development. Commitments to education, especially the education of girls, and to public health, including women's participation in public health programs, did not solve the problem of tightly limited employment opportunities; unemployment is a persistent problem in the state. However, social investment did lead to lower infant mortality, lower fertility, and higher life expectancy. Kerala was able to lower fertility better than China without resorting to coercion (Sen, 1996).

Despite a population density (1 March 1991) of 749 per square kilometer–even higher than India as a whole, which at 274 persons per square kilometer is one of the world's most densely populated countries–Kerala enjoys the advantages that progress in education and literacy earlier in the century have brought, a committed communist party that mobilized the rural poor since the days of independence, and considerable labor remittances and labor out-migration today. A number of factors, including the political mobilization of the poor, helped to raise literacy and school attendance, for females as well as males, to levels elsewhere achieved only in high income economies. It was also important that the government of Kerala raised incomes by implementing rigorous land reforms and fostering agricultural production through public investment in agriculture. Kerala's example to India is that high quality of life is possible without heavy industrialization or progressive dismantling of government regulation. While sluggish growth in agriculture productivity and in employment are continuing problems (Mencher, 1980), Kerala had the highest agricultural productivity of all Indian states until the early 1980s, when it was surpassed by Punjab.

Considering the challenges, Indian economic development since 1975 appears impressive. However, if one compares achievements to what might have been possible or to stated goals, especially the goals of poverty alleviation and the development of basic human capabilities and skills, Indian performance is disappointing. Poverty remains widespread and is virtually guaranteed by the severe neglect of basic human development. Poverty is still overwhelmingly the burden of Scheduled Castes (former "Untouchables") and Scheduled

Tribes. Half of each community falls under the poverty line. There are important debates over how poverty is best measured, but the development of elementary human capacities and skills, such as the ability to read and write, is clearly the major deficiency in the Indian economy.

Rural Development

Structural adjustment has drawn attention away from rural development and needed changes in the relations of production in agriculture. From 1990 to 1996, agricultural production grew by an average of 2.43% per year, only slightly higher than the population growth rate (Ghosh, 1997). Public investment in agriculture has consistently declined since 1980 (Dhawan and Yadav, 1997). From 1990 to 1996, agricultural growth was on average 3.4% annually while industrial growth was 4.7% annually, deepening the trend since independence of growth in industry outstripping agriculture by more than 2.5%. Government policies have tilted the value of production against agriculture. One estimate, for the 1970s, is that government policies deprived the agricultural sector of 10% of its income annually (Swami and Gulati, 1986).

Some have argued that intensifying cropping and high yield inputs, increasing rural credit and private investment in agriculture, and other efforts will not lead to high sustained growth without changes in patterns of land ownership. Earlier land reform did undercut *zamindari nizam* (the absentee landlord system), but a second set of reforms only instituted partial land ceilings and distributed excess (often marginal and nonviable) land to tenant farmers. As agriculture is constitutionally both a state and a central government concern, and as state governments pursued the reforms with limited enthusiasm, land reform has not substantially changed ownership patterns in agricultural. The net result has been the economic and political strengthening of high-status landowning castes and middle-status cultivating castes, varying from state to state, and relatively slow growth of agricultural productivity.

It is a well-known strategy of opposition to argue that if some consequences of a proposed solution are anathema to the ultimate goal, then the proposed solution should be abandoned (Hirschman, 1991: 43–80). Similar arguments have been made concerning Green Revolution solutions to low agricultural productivity and incomes, which have been described as heightening income inequality. Knowledge that a rural investment program may disproportionately benefit the nonpoor, however, or even benefit the economically and politically empowered, is not grounds for dismissing it outright. Institutions involving public access to information and accountable decision making can limit corruption and other rent-seeking behavior. Moreover, higher agricultural productivity is essential to poverty alleviation (Kakwani and Subbarao, 1993: 453). Higher agricultural productivity has a strong positive influence on poverty reduction through greater wage employment to the landless. Increasing rural wages through increased agricultural productivity is argued to be the best means of alleviating poverty (Martin Ravallion and Gaurav Datt, "Growth and Poverty in Rural India," August 1994, mimeographed).

In practice, increased agricultural production has focused on improving the inputs to existing landholders rather than redistributing land, which would thereby threaten patterns of economic and political power in the countryside. In the low productivity areas of the country, particularly the east, where employment elasticity is high, estimates are that "with concerted efforts in small irrigation, input supplies and marketing, the overall growth of agricultural employment is likely to be over 2 per cent per annum as against the one per cent in recent years." Combined with "diversification of agriculture into high value items like fruits and vegetables, particularly in the agriculturally developed regions," the overall employment growth rate in agriculture is likely to rise to 2.5% (Papola, 1992: 312).

Recent attention has focused on ensuring the economic success and social returns of rural development programs. In a study of new agricultural technologies and social and economic consequences in the second wave of the Green Revolution, Peter Hazell and C. Ramasamy found that under Green Revolution technologies in North Arcot district (Tamil Nadu), "the small-scale farmer did lag behind their larger brethren during the initial phase of the green revolution, but they subsequently caught up, thereby preventing the undesirable inequity effects that were initially feared." Hazell and Ramasamy's Cambridge-Madras universities study also confirms the necessity of four other institutional features: (1) the involvement of "small-scale, owner-occupied farms" and mechanisms to lower "the likelihood that small farmers would be bought out or evicted from their land" as land values increase, (2) the availability of nonagricultural outlets for surplus rural labor, (3) local research institutions capable of adapting high yielding variety (HYV) seeds to local conditions and tastes, and (4) government infrastructural development and direct assistance (Hazell and Ramasamy, 1991: 251).

Whether ownership patterns in agriculture are to be restructured or whether high yield inputs will lead to increased production and labor absorption, it is clear that the majority of India's poor are the landless laborers. Kirit Parikh and T. N. Srinivasan (1993) use a model constructed by them and N. S. S. Narayan (1991) to evaluate three poverty alleviation strategies in India: (1) increasing subsidized food rations, (2) instituting a rural works program targeted at the poor, and (3) abolishing the fertilizer subsidy so as to augment aggregate government investment. Their simulation shows that of these options, a "rural works program has the greatest (positive) effect on the poor" (Parikh and Srinivasan, 1993: 408). Others have also argued that a rural works program is the most effective way of rapidly reducing rural poverty (Dandekar, 1994).

Rural development programs, such as land reform, must be tailored to the specific local realities of agricultural production. One important component of rural development is rural banking. With proper institutional support (most of it already in place but under threat of further banking deregulation), the significant rural savings raised by rural banks can be maintained and increased (Desai, 1993).

The Development Challenge

It is by no means uncommon for children to die for want of readily available and relatively inexpensive medical care in India. In 1989–95, the latest years for which data is available, nearly two-thirds of the under-five-years-of-age population in India showed signs of malnutrition (World Bank, 1995: 214, 240).

Poverty reduction is the obvious moral priority as well as the necessary foundation for economic development. It is generally recognized that an integrated approach to development is most successful and that measures of development achievements are best based on living standards rather than gross economic indicators. India's economic reforms have not made adequate outlays for human development. The economic reform process can not succeed by merely reducing fiscal and current account deficits and promoting growth. The productivity gains of adjustment must be plowed back into human and social development. Poverty alleviation and the strengthening of human capacity are high priorities. Economic development should proceed through reinvestment in agricultural infrastructure, implementation of land reform, and promotion of employment, education, and public health. However deeply entrenched in the social consciousness, compulsory childhood education is necessary (Weiner, 1991), as are programs for adult literacy. The government will need to expand its tax base not only to achieve fiscal targets but to finance essential public investment in literacy and education, community health (including medical care for pregnant women, mothers, and infants), sanitation, public distribution of food and essential commodities, irrigation, agricultural extension, and agricultural infrastructure, inputs, and other supports.

Social safety nets, by definition, are short-term arrangements for softening the impact of adjustment on the most vulnerable. "[I]t is increasingly recognized that poverty mitigation has to be defined into adjustment programs initially, not added as a tranquilizer later on" (Lipton and Ravallion, 1995: 2,563). Adjustment policies need to be tied to just social policies.

Jean Drèze and Amartya Sen (1995) have conducted an analysis of Indian development using a human capabilities framework, showing that the provision of basic social goods, such as literacy and health, has positive economic effects. Effective public policy designed to develop basic human capabilities (what Drèze and Sen refer to as social opportunities) through primary education and basic health care is an established international standard of development. While high levels of human development are good in themselves, higher standards of life also lead to higher economic productivity. Ensuring even basic human security, such as universal childhood education, would contribute enormously to correcting some of the economic dynamics that perpetuate areas of economic and social deprivation.

Further Reading

Ahluwalia, Isher Judge, *Industrial Growth in India: Stagnation since the Mid-Sixties*, New Delhi: Oxford University Press, 1985

An evaluation of recent discussions of the industrial decline in India in the 1960s and 1970s that looks critically at state policies in licensing and infrastructural investment. This study echoes other calls for liberalization in the 1980s.

Ahmad, Ehtisham, and Nicholas Stern, "Taxation for Developing Countries," in *Handbook of Development Economics,* volume 2, edited by Hollis Chenery and T. N. Srinivasan, Amsterdam: North-Holland, 1989

Asian Development Bank, *Key Indicators of Developing Asian and Pacific Countries*, Economics and Development Resource Center (27), 1996

Bardhan, Pranab K., *The Political Economy of Development in India*, Oxford and New York: Basil Blackwell, 1984

A major analysis of the political economy of India's economic policy making from a critical left perspective.

Brass, Paul R., *The Politics of India since Independence*, 2nd edition, Cambridge and New York: Cambridge University Press, 1994

Examines the decline of the Congress, the recent rise of regional and especially confessional political parties, caste, language, regional conflicts, and obstacles to economic development to argue that the Indian political system is in the midst of a deep crisis.

Burki, Shahid Javed, "Developing Countries' Debt: A South Asian Outlook," in *Journal of International Affairs* 38, no. 1 (1984)

Chakravarty, Sukhamoy, "On the Question of Home Market and the Prospect for Indian Growth," in *Economic and Political Weekly* 14, nos. 30–32 (1979)

Dandekar, V. M., *The Indian Economy, 1947–92*, New Delhi and London: Sage Publications, 1991–95

Desai, Bhupat M., "Rural Banking: Misdirected Policy Changes," in *Economic and Political Weekly* 28, no. 52 (1993)

Dhawan, B. D., and S. S. Yadav, "Public Investment in Indian Agriculture," in *Economic and Political Weekly* 32, no.14 (1997)

Drèze, Jean, and Amartya Sen, *India: Economic Development and Social Opportunity,* Delhi: Oxford University Press, 1995

Argues, through international and intrastate comparisons, that public action in the fields of nutrition, health, education, and other areas are not only vital to human development and social opportunity but are also the foundations for economic development.

Encarnation, Dennis J., *Dislodging Multinationals: India's Strategy in Comparative Perspective*, Ithaca, N.Y.: Cornell University Press, 1989

Analyzes the changing relations among multinationals, the state, and local enterprises in India in the first four decades of its independence. In a comparative study, the author demonstrates how Indian enterprises and states greatly overcame the bargaining imbalance with MNCs.

Frank, André Gunder, "Emergence of Permanent Emergency in India," in *Economic and Political Weekly* 11, no.11 (1977)

Ghosh, Jayathi, "India's Structural Adjustment: An Assessment in Comparative Asian Context," in *Economic and Political Weekly* 32, nos. 20–21 (1997)

Government of India, *Economic Survey 1996–97,* New Delhi: Government of India, Ministry of Finance, 1997

Hart, Henry, *Indira Gandhi's India: A Political System Reappraised*, Boulder, Colo.: Westview Press, 1976

Hazell, P. B. R., C. Ramasamy, and P. K. Aiyasamy, *The Green Revolution Reconsidered: The Impact of High-Yielding Rice Varieties in South India*, Baltimore, Md.: Johns Hopkins University Press, 1991

Hirschman, Albert O., *The Rhetoric of Reaction: Perversity, Futility, Jeopardy*, Cambridge, Mass.: Belknap Press, 1991

Jalan, Bimal, editor, *The Indian Economy: Problems and Prospects*, New Delhi and New York: Viking, 1992

Includes essays by accomplished economists and policy makers.

Kakwani, N., and K. Subbarao, "Rural Poverty in India, 1973–87," in *Including the Poor,* edited by Michael Lipton and Jacques van der Gaag, Washington, D.C.: World Bank, 1993

Kohli, Atul, *Democracy and Discontent: India's Growing Crisis of Governability,* Cambridge and New York: Cambridge University Press, 1990

An examination of the district, state, and national levels that argues that powerful state elites are undermining democratic and consensus-building institutions in India.

——, *The State and Poverty in India: The Politics of Reform,* Cambridge: Cambridge University Press, 1987

An examination of three states–Uttar Pradesh, Karnataka, and West Bengal–that represent varying political regimes; argues that "a well organized, left-of-center regime" is best able to alleviate rural poverty.

Lipton, Michael, and Martin Ravallion, "Poverty and Policy," in *Handbook of Development Economics,* volume 3, edited by Jere Behrman and T. N. Srinivasan, Amsterdam: North-Holland, 1995

Lucas, Robert E. B., and Gustav Papanek, editors, *The Indian Economy: Recent Development and Future Prospects*, Boulder, Colo.: Westview Press, 1988

Maddison, Angus, *Class Structure and Economic Growth: India and Pakistan since the Moghuls*, London: Allen and Unwin, 1971, and New York: Norton, 1972

Maddison's study, now dated, still provides a clear comparison of economic development in India and Pakistan.

Mencher, Joan, "The Lessons and Non-Lessons of Kerala," in *Economic and Political Weekly* 15, nos. 41–43 (1980)

Mitra, Ashok, *Terms of Trade and Class Relations: An Essay in Political Economy,* London: Frank Cass, 1979

Mundle, Sudipto, and M. Govinda Rao, "Issues in Fiscal Policy," in *Indian Economy: Problems and Prospects,* edited by Bimal Jalan, New Delhi: Viking, 1992

Narasimham, M., et al., "Main Report of the Committee on the Financial System," in *The Financial System: Report by M. Narasimham,* New Delhi: Nabhi Publications, 1992

Narayan, N. S. S., Kirit Parikh, and T. N. Srinivasan, *Agriculture, Growth, and Redistribution of Income: Policy Analysis with a General Equilibrium Model of India*, Amsterdam and New York: North Holland, 1991

Nayyar, Deepak, "Industrial Development in India: Some Reflections on Growth and Stagnation," in *Economic and Political Weekly* 13, nos. 31–33 (1978)

Papola, T. S., "The Question of Unemployment," in *The Indian Economy: Problems and Prospects*, edited by Bimal Jalan, New Delhi and New York: Viking, 1992

Parikh, Kirit, and T. N. Srinivasan, "Poverty Alleviation Policies in India," in *Including the Poor*, edited by Michael Lipton and Jacques van der Gaag, Washington, D.C.: World Bank, 1993

Rudolph, Lloyd I., and Susanne Hoeber Rudolph, *In Pursuit of Lakshmi: The Political Economy of the Indian State*, Chicago: University of Chicago Press, 1987

Remains the most thorough analysis of the political economy of development of India. Considers Indian state formation and the role of the state in politics, especially class politics, and the judiciary; the Indian National Congress and its challengers at the center and in the states; changing political regimes and economic performance; and the influence of various demand groups, such as industrial labor, students, and agrarian producers.

Sen, Amartya, "Fertility and Coercion," in *University of Chicago Law Review* 63, no. 3 (1996)

——, *Poverty and Famines: An Essay on Entitlement and Deprivation*, Oxford and New York: Oxford University Press, 1981

Shankar, Kripa, "Revamped Public Distribution System: Who Benefits and How Much?" in *Economic and Political Weekly* 32, no. 13 (1997)

Swami, Dalip, and Ashok Gulati, "From Prosperity to Retrogression: Indian Cultivation during the 1970s," in *Economic and Political Weekly* 21, nos. 25–26 (1986)

Thakurdas, Purshotamdas, J. R. D. Tata, and Ghanasyamadasa Birala, *A Brief Memorandum Outlining a Plan of Economic Development for India*, Harmondsworth and New York: Penguin, 1944

Toye, J. F. J., "Economic Trends and Policies in India during the Emergency," in *World Development* 5, no. 4 (1977)

Weiner, Myron, *The Child and the State in India*, Princeton, N.J.: Princeton University Press, 1991

World Bank, *World Development Report*, Washington, D.C.: World Bank, 1995

World Bank, *World Development Report*, Washington, D.C.: World Bank, 1996

Christopher Candland is a lecturer and research fellow in the Department of Political Science at the University of California, Berkeley. He is preparing a book on global labor transitions, with particular reference to India and Pakistan.

Table 2.1: India and Other Developing Economies
Economic and social indicators (%, except life expectancy)

	1980–90			*1990–94*		
	India	*low-income economies*	*middle-income economies*	*India*	*low-income economies*	*middle-income economies*
GDP growth	5.8	2.0	2.2	3.8	1.4	0.2
Agricultural growth	3.1	2.3	2.7	2.9	1.5	0.9
Industrial growth	7.1	1.7	3.2	3.2	-0.7	1.3
Services growth	6.9	3.4	3.1	4.6	2.1	3.7
Gross domestic investment	6.5	-0.4	-0.6	1.2	-1.8	2.1
Annual population growth	2.1	2.1	1.8	1.8	1.8	1.5
Life expectancy	58	54	66	62	56	67
Female literacy	29	44	69	38	55	73

Sources: GDP and sectoral growth rates from World Bank, 1996: 208, except 1980–90 middle-income economies, from World Bank, 1990:192, which is for 1980–88. Gross domestic investment from World Bank, 1996: 208. Annual population growth from World Bank, 1996: 194. Life expectancy, in years at birth, for 1988 from World Bank, 1990:178, and for 1994 from World Bank, 1996: 188. Female literacy, for ages 14 and above, for 1985 and 1995 fom World Bank, 1990: 178, and World Bank, 1996: 200, except for middle-income economies for 1990, from World Bank, 1993: 238. Figures from low-incomeeconomies exclude China.

Table 2.2: India and China
Economic and social indicators (%, except life expectancy)

	India 1980–90	*China 1980–90*	*India 1990–94*	*China 1990–94*
GDP growth	5.8	10.2	3.8	12.9
Agricultural growth	3.1	5.9	2.9	4.1
Industrial growth	7.1	11.1	3.2	18.8
Services growth	6.9	13.6	4.6	9.9
Gross domestic investment	6.5	11.0	1.2	15.4
Net reproduction rate	1.7	1.1	1.5	0.9
Life expectancy	58	59	62	69
Female school enrollment	20	37	38	51
Female literacy	29	55	38	73

Sources: GDP and sectoral growth rates from World Bank, 1996:208. Gross domestic investment from World Bank, 1996: 208. Net reproduction rate, for 1980–85 and 1990–95, from Asia Development Bank, 1996: 3. Life expectancy, in years at birth, for 1988 and 1994 from World Bank, 1990:178, and World Bank, 1996:188. Female school enrollment, for secondary schools in 1980 and 1993, from World Bank, 1996: 200. Female literacy, for ages 14 and above, for 1985 and 1995, from World Bank, 1990:178, and World Bank, 1996: 200.

Table 2.3: Kerala and India
Economic and social indicators (%, except life expectancy)

	period	*India*	*Kerala*
National/state product growth	1990–94 (annual)	3.8	15.2
Foodgrain production growth	1991–95 (term)	8.56	-10.0
Industrial employment growth	1970–80 (annual)	4.79	-0.58
Population in poverty	1977–78	48.3	48.4
	1983–84	37.4	26.8
Education expenditure	1980–81	2.89	5.81
	1988–89	3.65	6.52
Female secondary school enrollment	1995–96	54.9	101.5
Female literacy	1991	39.29	86.17
Decade population growth	1981–91	23.5	14.0
Life expectancy	1989–93	59.4	72.0

Sources: National/state product, in annual averages, calculate from Government of India, 1997: S11. Foodgrain production growth over the four year interval calculated from Government of India, 1997: S20–S21. Industrial employment, for registered manufacturing enterprises, from Oomen, 1993: 176. Population living below the poverty line from United Nations, 1992: 23. Education expenditure, as a percentage of national/state domestic product, from Oomen, 1993: 131. Female school enrollment, for classes six through eight as a percentage of the school age population, from Government of India, 1997: 187. Female literacy, for all ages, from Government of Kerala Directorate of Census Operations, *Census of India 1991: Kerala Final Population Totals,* Delhi: Controller of Publications, 1993: 9. Decade population growth from Government of Kerala Directorate of Census Operations, *Census of India 1991: Kerala Final Population Totals,* Delhi: Controller of Publications, 1993: 8. Life expectancy, in years at birth, from Government of India, 1997: 185.

Economic Policy

Chapter Three

The Political Economy of India's Economic Performance since 1947

Baldev Raj Nayar

Most current assessments of India's economic performance regard it to be weak, poor, and unimpressive, whether considered in its own terms or against the stated national targets, and especially when it is compared to the economies of the other regions in Asia. Many consider it to be disappointing. Bhagwati (1993: 17–25) calls it not only inadequate but also dismal, while Dhar (in Lucas and Papanek, 1988: 3–21) refers to India's "low-growth syndrome." In the late 1970s and 1980s, a whole literature developed on India's economic stagnation and industrial retrogression (Nayyar, 1994; Jha, 1980). So impressed (or depressed) was economist Raj Krishna (1988) by the long-term constancy of the "on any reckoning . . . unsatisfactory" average annual growth rate of 3.5%–believing it to be largely immune to changes in investment and policy–that he characterized it as the "Hindu rate of growth," implying some element of predestination in it. India's economic performance seems especially egregious when viewed against the population growth rate of 2% or more.

Is this an accurate assessment of the performance of the Indian economy? Dhar maintains that "it is not possible to draw up a balance sheet of its achievements and failures which would be acceptable to all." Still, the large size of the country, with respect to territory and population as well as its economic and military potential, mandates coming to grips with the different aspects of its performance. A first step toward a sharper understanding of India's economy is to note that, even acknowledging the poor nature of the economy's performance, the growth rate has not been uniform during the period under review, which stretches over half a century. Indeed, there has been considerable volatility in the annual performance, which is understandable in view of the fact that India is an underdeveloped economy and has not yet acquired adequate control over the forces of nature. India's economy has seen considerable turbulence over the period, a period punctuated by several major crises: there has been at least one major crisis in every successive decade. These crises have often served as an incentive for policy change, whether of the liberalizing or centralizing kind. It is necessary, therefore, to break down the half century since independence into several periods.

Empirically, for the period as a whole from the beginning of planning in 1951–52 to 1995–96, the overall average annual growth rate of the net national product at constant prices–here briefly defined as national income–has been 4.1% (see Tables 3.1 and 3.2). Of course, this overall figure does not provide us any indication of the performance of the different sectors of the economy, such as agriculture and industry, but it can be safely assumed that their performance is largely correlated, though the magnitudes may be different, with the overall economic performance (see Tables 3.3 and 3.4).

Although economic history is a continuous line and there is some hazard in artificially dividing it up, for analytical purposes the years since independence can be usefully broken down into three major periods: (1) 1951–52 to 1964–65–modest growth at 3.9%; (2) 1965–66 to 1974–75–low or poor growth at 2.4%; and (3) 1975–76 to 1995–96–respectable but not high growth, though rising, at 5.5% (see Table 3.2). Of course, even when divided thus, the periods are long and, depending on the purpose, they can be further broken down into smaller segments of time. For the present, it is sufficient to note that performance in the latest period seems to have slain the notion of the "Hindu rate of growth." Indeed, if one takes only the years from 1992–93 to 1995–96, the growth rate has been even higher, at 6.2%; however, that growth rate masks some very serious problems that need to be addressed.

Analytical Framework

Even given the higher rate of growth in the most recent period, India's performance compares poorly with that of the East Asian and Southeast Asian economies (see Table 3.5). However, such comparison among countries, while certainly important, must also take into account the particular constraints and opportunities of a country. A specific country may well not have the options open to it that are available to others, and any assessment of its performance must pay heed to that fact. In addition, any evaluation of performance must take into consideration the objectives that policy makers have for the national economy. This is essential, for the objectives may not place high value on growth per se. In the specific case of India, three key objectives in its economic planning are apparent:

1. A high economic growth rate with a view to raising the per capita incomes of the population and to providing greater employment
2. National self-reliance, which as an objective distinguishes India from many less-developed countries (LDCs)
3. Some measure of social justice along with a strong thrust for state ownership of the principal means of production.

The aim of the second objective, self-reliance, perhaps stands in contradiction to the first objective and may well have proven counterproductive for higher growth. It is precisely for that reason that it is usually not highly valued by economists. At one level, the second objective has implied that economic development be managed largely within the means of the country, or that at least dependence on foreign aid be minimized and eliminated as soon as possible. At another level, it has implied development of a relatively autarkic economy that can largely meet all its needs from production within the country without much interaction with the outside world, even if it is uneconomic to do so. Unlike other countries, which have measured their success by how well they have done in increasing their exports, India has measured its success by how well it has managed to limit its imports. Interestingly, in a chapter of his book *The Widening Gulf* (1978) entitled "Has India Failed?" Selig Harrison pointed out some two decades ago that critics of India had, from the perspective of conventional economics, faulted its policies for the inefficiency they generated. However, he himself saw India's economic strategy as part of the larger enterprise of nation-building and state-building. Without disputing the critics' economic argument, he maintained that "the Indian development experience cannot be meaningfully judged in a narrowly economic context. For the implicit rationale underlying Indian policies has been a nationalist rationale, a readiness to bear inordinate costs, if necessary, to maximize the independent character of the industrialization achieved" (326). Perhaps India erred in its overenthusiasm for policies of national self-reliance, but that has to be seen in the light of its colonial past and its established nationalism.

The third objective is part of a commitment to socialism, even if a truncated or distorted one. State ownership, too, has been counterproductive because of its inefficiency and its failure to generate savings for investment.

With respect to these key aims, India has

been similar to China, regardless of the contrast in their political systems, though India's implementation has not been as vigorous. Interestingly, the strategies of these two countries have often moved in a parallel rhythm, though at times with a certain time lag.

The preferred goal in the social sciences is not only to describe an event or phenomenon but also to seek an explanation for it. With respect to the economic performance of India, a summary assessment has been provided above for the long stretch of time since independence and also for the different periods within it. It seems necessary now to go beyond this general description, and search for an explanation for the variance in performance that has been demonstrated among the periods.

In the search for an explanation, four key variables seem particularly important. They are (1) geography, (2) the state, (3) the international system, and (4) society. These variables primarily lie outside the sphere of economics, but their inclusion here is a testimony to the fact that the explanation for economic performance is often profoundly noneconomic. Economic performance is no doubt in an immediate sense related to economic policy, but why some policy is adopted is not always a function of economics alone. As a German proverb says, whether there is meat in the kitchen is not determined in the kitchen. In laying out and elaborating on these variables, one can also gain some insight into a few of the major constraints that apply in India's case.

1. *Geography.* Here, the important elements for consideration are the size of the country, its location, its natural resources, and its climatic features.

A large-size country–which India is, both in terms of territory and, especially, population–is more likely to be oriented toward having a foreign policy of independence from the major powers and toward developing an economy that goes with such a policy of independence. Accordingly, apart from the natural tendency of a large-size country to be oriented to its internal market rather than foreign markets, its leadership is also likely to be self-consciously more inward oriented and less inclined toward exports. For a large-size country, a self-reliant economy is therefore a natural temptation, though other considerations may still move it in a different direction. In India, this tendency is reinforced by the Indian subcontinent's geographic isolation from the Eurasian landmass as well as by its ethnic and civilizational distinctiveness.

India suffers from a lack of large oil resources. On the other hand, it has abundant reserves of iron ore, which make for the belief that the country has a comparative advantage in steel production.

Climate can have a major influence on economic performance, depending on whether the country is subject to frequent droughts, floods, or both, which happens to be the case with India. The economic fortunes of India–as well as the fate of the budget and the political survival of the leadership–are on an annual basis hostage to the monsoons, as any textbook on Indian economics is likely to point out.

2. *The state.* This is a rather large category, but it includes such elements as the nature of state institutions, the nature of state power (state autonomy and state capacity), the class or class coalition in power, and the ideological and policy preferences of the leadership in power (dirigiste or market models, inward or outward orientation, sectoral emphasis on industry or agriculture, state or private ownership of the means of production).

Although Western scholarship is often reluctant to admit it, the adoption of democratic institutions, as in the case of India and in contrast to China and most LDCs, closes off some options in economic policy that are available to authoritarian or totalitarian regimes. Democracies must take into account the demands and tax-bearing capacity of the population in the development of policy. If abundant resources are not available through oil surpluses or foreign aid, there are severe constraints within a democracy in mobilizing savings for a high growth rate. Also built into a democratic regime, at least to a large extent, is the much-noted tendency of populism, and along with it the associated features of fiscal deficits to cover subsidies as well as, contrarily, the low tolerance for inflation.

3. *The international system.* Included here are not only conjunctural elements–such as global expansionary and contractionary phases in world trade and sudden shocks as in the case of oil prices–but also structural elements–such as the distribution of power among the major states and the configuration of allies and adversaries among them. It is noteworthy that the countries that have prospered in East Asia are chiefly those that have had strategic alliances with the United States; to this day, the United States has a major troop presence and military bases in Korea and Japan. The United States has also provided a strategic umbrella over the Association of Southeast Asian Nations (ASEAN) after its defeat in Vietnam. Even China has been a strategic partner of the United States since 1971. If a country is outside such a strategic framework, economic performance may well be constrained in terms of foreign aid and access to foreign markets, particularly if that country pursues an independent foreign policy that the United States regards as unfriendly.

4. *Society.* Several elements are of importance here. To begin with, there is the long colonial experience of South Asia under the British, which, along with the nationalism that it provoked, leads to the questioning of free trade and capitalism as a policy, for it was precisely through that mechanism that India was exploited for over 200 years. In addition, the nature of the colonial experience itself is related to the possibilities of growth, for British colonialism was exploitative and not at all development-oriented, unlike the colonial experience of Korea and Taiwan under the Japanese.

Other elements include social heterogeneity, levels of equality, and the nature of interest groups. For example, India is about the most heterogeneous society in the world; the contrast with China is striking, for China is a rather homogenous society, at least linguistically. National policy makers in India have therefore to be sensitive to the implications of policy for national integration; at times, high growth rates have to be sacrificed for the sake of satisfying regional and group aspirations. Equally, India is a very inegalitarian society, carrying the burden of the centuries-old caste system. In addition, a third to half of the population lives below the poverty line, which in the context of a democracy makes socialism almost a natural ideology for the country. Such poverty also makes intelligible the low tolerance for inflation among the population and therefore among the policy makers.

In terms of social classes and their relationships to the state, one influential view regards the new middle classes of the urban areas and the rich and middle peasantry of the rural areas–rather than the capitalist class–as constituting the ruling coalition in India (Nayar, 1989: 46–48, 112–18). It is this coalition that inherited power from the colonial authorities and continues to exercise it, though the internal balance has over the years shifted so that the rich and middle peasantry has replaced the new middle classes in terms of domination within the coalition. The capitalist class no doubt has certain strengths, but it is this nature of the class coalition in power that explains the ease with which policies unfavorable to the capitalist class could be adopted in India.

There can be the valid objection to this analytical framework that the four variables taken together constitute an exhaustive set and therefore, since any explanation would be correct by definition, the framework can not make for a satisfactory contribution. That is correct. On the other hand, it would be foolhardy to look for a single grand explanation for economic performance over such a long period as a half century in a single variable. Instead, the aim here is somewhat more modest, and that is to develop a tentative hierarchy within this set of variables. In other words, the aim is to see which variables have priority in explaining economic performance in given periods or given situations.

First Period, 1947–65: Modest Growth

In 1947, a genuinely nationalist elite took over power in India after a half century of active struggle against colonialism, which saw many mass movements, uprisings, and repression, even as many leaders and cadres spent innumerable years in jail. Its core values centered around anticolonial nationalism, an assertion of India's independent entity in the world, and a commitment to democracy and the economic regeneration of India. At the same time, the elite inherited an economic system that had been intensely exploited and structurally distorted as a subservient appendage of the British Empire. The first half of the twentieth century had been a period of utter economic stagnation; the issue in the scholarly debate over the impact of colonialism during the period has been between those who conclude that per capita income was only stagnant and others who maintain that it witnessed decline (Chandra, in Jalan, 1992: 1–32).

Since British colonialism was based on capitalism with free trade, its economic consequences did not exactly constitute a resounding recommendation for continued adherence to that system. Over and above the impact of colonialism, the economy had been exhausted by World War II even though India had not been directly in the theater of war. The war saw extreme inflation and tremendous shortages of food and basic consumer goods. Several million died in the terrible famine in Bengal during the war. To preempt resources for the war, the colonial authorities imposed controls and rationing; ironically, the wartime experience with controls subsequently facilitated the installation of a dirigiste economic regime in India in the name of socialism. On top of the ravages of colonialism and war came the disruption and turmoil of the partition of the subcontinent, which–apart from creating the problem of rehabilitation of 6 million refugees–abruptly tore into two parts what had been a single integrated economy.

The immediate tasks at independence were to survive politically, to stave off famine through food imports, and to stabilize the economy. India experimented with removing rationing and controls, but the result–a situation of shortages–was an economic disaster; steep inflation led to the reimposition of controls. The experience inculcated a profound distrust of the market and a perception of business as profiteering. A Planning Commission was established in 1950 to develop the basically agricultural economy with perhaps the lowest per capita income in the world. A series of five-year plans followed in the subsequent five decades. However, the First Five-Year Plan (1951–56) was a modest effort; it was less a plan with a rigorous approach than an aggregation of projects already under way or readily available on the shelf. The plan nonetheless proved successful; against the target of only 12% growth over the plan period, the actual growth was 18%, amounting to an annual growth rate of 3.6% (see Table 3.2). The primary reason for the success was, in actual fact, good monsoons, but it led to the false belief that the agricultural problem had been solved. Besides, the result made for excessive confidence among the planners, who seemed convinced that they could successfully launch India into a sustained economic takeoff through planning.

Real planning began with the Second Five-Year Plan (1956–61), which attempted to make a dramatic break with the inherited economic system. The second plan was modeled after the forced-draft industrialization effort of the Soviet Union under Stalin, even as it was incongruously and hazardously joined with an infant democracy. Its immediate conceptual origins lay in what has come to be known as the Mahalanobis model, after the economic adviser P. C. Mahalanobis, but its intellectual antecedents really go back to Nehru's own thinking in the 1930s and 1940s (Nayar, 1972: 129–34; Nayar, 1989: 143–68). It would not be farfetched to say that the Mahalanobis model merely provided an elegant theoretical scaffolding for an ideologically derived economic agenda. While the conceptual case for it was made on a seemingly sound economic rationale, its real origins belonged to Nehru's ideology of socialism, which Nehru at the same time combined with a deep and profound commitment to democracy in an effort to create a third way between capitalism and communism.

With the second plan, India adopted an inward-oriented, import-substitution-industrialization (ISI) strategy with a thrust for basic or heavy and capital goods industries primarily under state ownership. The plan aimed at an annual growth rate of 5% and an additional employment for 11 million people, but more importantly, these objectives were to be achieved through a strategy that would make India into a rather autarkic economy, with its own machinery producing at capacity, and that would launch India on a transition to an industrial society with a socialist system. In the underlying aim of national self-reliance, there was acute sensitivity to the importance that heavy industries had in developing a base for an independent defense industry later, especially in the context of the military alliance of the United States with Pakistan in 1954, which India correctly perceived–and some American statesmen acknowledged (Harrison, 1978: 267–68; Nayar, 1990b)–to be aimed against it. In that sense, the origins of the second plan strategy were profoundly geopolitical. The national consensus on heavy industry was broader than that for state ownership of it.

Most of the heavy industries and many others were to be state owned in order to develop what was referred to as the "socialistic pattern of society." The public or state sector was intended to grow rapidly, particularly relative to the private sector, with the expectation that the private sector would gradually but eventually fade away while the state acquired "the commanding heights" and became hegemonic over the economy. At the time of independence, India had a sizable indigenous–though regionally concentrated–capitalist class, which was one of the most sophisticated and most advanced in the developing world. However, the political and intellectual elites, who had grown up on the ideological thinking of Nehru, held a strong animus against this class. Rather than allowing the capitalist class to develop and expand its economic base, the political leadership aimed to contain and constrict it, endeavoring instead to add economic power to its existing political power. Accordingly, the leadership placed the new complex of heavy and capital goods industries under state ownership. However, it failed to evolve an adequate policy to ensure performance, apparently convinced that mere investment and state ownership would automatically translate into results. In this posture, the leadership was in line with contemporaneous economic thinking, which was marked by a lack of concern over the implications of the type of ownership for performance. Beyond ownership, the Indian state developed a rigid system of regulation and controls (which became more and more restrictive as time passed) over the private sector in order to take the economy in the direction it preferred.

While the second plan strategy concentrated on heavy industry, it neglected agriculture, apparently in the premature belief that agriculture no longer constituted a problem. At the same time, the plan relied on household industry for consumer goods and set itself against the factory production of consumer goods, no doubt because that would have resulted in expanding the private sector. In addition to (1) inward orientation, (2) heavy industry, and (3) state ownership of the principal means of production as the key pillars of the strategy, another pillar of Nehru's project of socialism at the time was (4) agrarian reform. However, by 1959, Nehru had met with failure on this score because of the opposition of the landed classes within his own party. As a consequence, Nehru's socialism became a truncated one, confined to the industrial sector while agricultural production remained organized on the basis of private property in a vast, mixed economy.

Within 18 months, the strategy of the second plan ran into serious trouble with a major balance of payments crisis. The immediate crisis was relieved by an emergency package of aid from the Western powers, including the United States, which additionally helped with PL480 food aid on a long-term basis. Although falling short of its targets, the second plan with a growth rate of 4.0% was nonetheless taken to be a reasonable success. Interestingly, despite the constraints placed on the private sector during the second plan and in spite of the privileged treatment accorded to the public sector, the private sector demonstrated enormous enthusiasm for growth by expanding at a rate higher than projected by the economic planners. Such a development really indicated, against all suggestions to the contrary, the inherent strong capabilities of the private sector. However, the response of the planners to this display of business enterprise was not commendatory endorsement or encouragement. Rather, it was a determination to apply more rigorously the network of constraints in the interest of preventing what the planners saw as an increased concentration of economic power.

The second plan was followed by the Third Five-Year Plan (1961–66), which persisted with the same overall economic strategy. Though the third plan fell far short of its targets, the period of the two plans taken together was considered one of success in terms of establishing a strong and broad industrial base. It is a testimony to the dramatically different perceptions of an earlier era that, even as he was cognizant of the problems the country faced, Higgins (1968: 653–78) included India among the "three success stories," the other two being Japan and Mexico; by way of contrast, he referred to "Indonesia: the chronic drop-out." The ISI strategy provided a tremendous boost to local industry, and a vast change took place in the structure of industry over the ten-year period of the two plans. Industrial production doubled over the period, growing at an annual rate of 6.9% during the second plan and at 10.2% during the third plan on top of the 8.1% in the first plan (Sandesara, 1992: 22). More striking was the change in the structure of industry. Over the period 1950–51 to 1965–66, while the value added in the manufacturing industry increased by about 90% over the three plans with respect to consumer goods, it increased by almost seven times for intermediate goods and by ten times for machinery. As a result, the picture of an industrial structure with an almost total concentration on consumer goods at the beginning of planning stood drastically altered (see Table 3.6). This industrial restructuring, pushed through at considerable sacrifice, laid the basis for the subsequent acquisition of industrial and technological capabilities in the defense arena, especially with respect to missiles. Until the last year of the third plan, agriculture did not emerge as a serious problem because of a reasonable rate of growth; this was more the result of the expansion of cultivated areas than an increase in yields. However, population was increasing at a higher rate than was envisioned at the beginning of planning and would soon make for a major food crisis.

Analysis. It is clear that in the economic performance of the period and the strategy underlying it, the state was the key variable. It commanded considerable autonomy in relation to society, the economy, and the international system. There was a strong consensus among the political leadership about the strategy, even if it was initially a forced consensus. At the same time, there was substantial capacity in the administration to carry out the strategy even though there were serious weaknesses in the implementation of the plans. Nehru was the political hero of the era, while the Congress Party itself was the bearer of institutional charisma inherited from the nationalist movement. As a result, the party was able to win massive electoral majorities, which enabled it to implement the strategy, even if it was ideologically driven. The period stands out as one in which the state was the independent variable, molding the economy and society according to its preferred design.

The international system proved to be supportive in this period. Even though the regime's agenda was ideologically driven, it was largely consistent with mainstream economic thinking of the time, which endorsed ISI because of the belief that export prospects were poor for the developing countries.

Moreover, the competition between the two power blocs enabled India, as one of the founders and leaders of the nonalignment movement, to benefit by way of aid from both; India was able to combine aid from the Soviet Union for building heavy industry, such as steel mills, with food aid from the United States, which was eager at the time to get rid of its accumulating agricultural surpluses. Society also proved supportive, even if by happenstance, in the sense that there prevailed during the period a low level of social mobilization, and as a consequence, demands on the political system were also low. Society seemed to be largely acquiescent, enabling the regime to carry out a strategy that otherwise more appropriately belonged to authoritarian systems in that it imposed sacrifices even as it postponed gratifications.

Second Period, 1966–74: Slow Growth

The second period opened with a tremendous exogenous natural shock in the form of an unprecedented drought in 1965, as a result of which food production fell by over 20%. A second shock followed in 1966 with another drought of similar severity. Although representing the forces of nature, these shocks called into question the underlying assumption of the second plan that agriculture did not require priority attention. However, the Indian attitude toward the strategy remained ambivalent, for the 1960s also saw another set of exogenous shocks that tended to reinforce what was seen as the essential correctness of the strategy in the minds of many. First, there was the short war with China in 1962, which brought on major inflationary trends as India hurriedly tried to build up its hitherto neglected defenses. Second, there was the war in 1965 with Pakistan, a technologically superior adversary because of the advanced military equipment provided to it by the United States. As the United States cut off economic aid to India, apart from any military aid, in the wake of this war, Indians felt confirmed in the innate soundness of Nehru's strategy of national self-reliance.

In addition to drought and war, India's political system was jolted by the death of Nehru in 1964, and the country entered a period of intense uncertainty and instability. Shastri, who succeeded Nehru, died within 20 months in office, after briefly emerging as a popular hero because of his determined leadership in the 1965 war with Pakistan. He was followed in office by Nehru's daughter, Indira Gandhi. Untried and untested, Indira Gandhi took over power in the midst of an enormous food and economic crisis. Meanwhile, the Congress Party saw rapid decline and internal conflict, and it eventually split.

Shastri was not enamored of the Mahalanobis strategy, which stood discredited with the food crisis confronting the nation; besides, the strategy had brought other perverse results, such as greater dependence on outside powers for food and funds. Shastri initiated important policy changes in agriculture that were later pushed through by Indira Gandhi and that eventually developed into the Green Revolution. The impulses for the Green Revolution were fundamentally nationalist; it constituted, as it were, the agricultural counterpart to heavy industry in the strategy for national self-reliance.

Shastri had also initiated some tentative steps toward liberalization. Initially, Indira Gandhi accepted the necessity to continue on that course. However, the endeavor proved abortive because it fell victim to internal conflict. Meanwhile, devaluation, which had been adopted under foreign pressure, ended in failure as there was a recurrence of drought in 1966 and the United States and other Western powers did not keep their promise to provide additional aid. Importantly, aid weariness had set in by this time because of the developing détente between the United States and the Soviet Union after the Cuban missile crisis. There was less competition in providing aid; indeed, there developed almost a condominium between the two superpowers over the management of the security of South Asia.

Apart from attempting to mobilize food supplies from around the world (and more specifically the United States), the state under Indira Gandhi had to respond to the high inflation that the droughts triggered. It did so by adopting a sharp and drastic contractionary policy. The ax fell most heavily on public investment; as a result, the infrastructure suffered serious neglect with damaging consequences for subsequent growth. At the same time, with the balance of payments under grave strain and the country lacking foreign exchange, industry worked at extremely low capacity in the absence of imports of components and intermediate goods. Import compression of the period resulted in India falling behind technologically. The economy entered into a powerful recession in the absence of countermeasures, which were avoided for fear of accelerating and prolonging inflation.

It was a sad and sorry period, one of unrelieved economic agony and largely barren of new initiatives except for the Green Revolution. The economic crisis was of enormous proportions. It brought planning to a halt; the Fourth Five-Year Plan was jettisoned in favor of a "plan holiday" with only annual plans for three years. The crisis also made for great misery and food riots. The elite's task of governance was made difficult by the continued hostility of Pakistan and China–which encouraged the local Marxist-Leninist guerrillas to rise up against the regime in a series of revolts across India–and the unfriendliness of the United States. India's very existence seemed at stake; the prognoses of disintegration associated with the notion of "after Nehru what" seemed to be coming true. When elections were held in 1967, the London *Times* correspondent Neville Maxwell declared that these would be India's last elections. While this assessment could be considered to reflect an arrogance springing from a combination of Marxist and imperialist attitudes, there was a basis for it in India's own sense of impending doom. This was the first major economic crisis in the country's history; later, India would become used to recurrent crises, but at this time there was no experience in overcoming crisis without Nehru.

As if the economic fallout from the exogenous shocks and the ensuing contractionary policy was not enough, the period saw the emergence and intensification of a power struggle in the Congress Party, which eventually led to the watershed event of the party's breakup. In the midst of the untold misery caused by food shortages and inflation in the wake of the wars and droughts, the Congress Party faced an uphill battle in the 1967 elections. As the masses revolted, society determined the fate of the state. In the process, the historic consensus within the leadership was destroyed and the state's autonomy eroded. A highly mobilized electorate lashed out against the Congress Party, and the party suffered heavy losses. The party did return to power at the center but with only a slim majority, where the defection of only 25 members could bring about the downfall of the government. At the same time, the party was ousted from power in eight of the states, making for increased political instability.

Meanwhile, the party became internally polarized between the left and right wings. Believing that the electorate wanted to move leftward, Indira Gandhi swung to radicalism in an attempt to rally mass support and nationalized the major banks. As the party split in 1969, she was able to stay on in power only with the support of the Communist Party of India (CPI) and some regional parties. Populism now came to dominate the political process. A severe attack was mounted against the private sector; big business was subjected to new controls on concentration of economic power through the MRTP (Monopolies and Restrictive Trade Practices Act) regime while the licensing regime was made even more restrictive. When Indira Gandhi called new elections in 1971, she made "poverty removal" the main plank of her Congress Party's platform. Her massive victory in the elections led to an even more radical leftward shift, culminating in a further nationalization spree. Supported by former communists in her government, she nationalized the coal industry, the copper industry, a large part of the textile industry, and general insurance; the public sector thus saw substantial expansion under her tutelage, beyond the capacity of the state to manage efficiently. Additionally, foreign corporations were made subject to stricter controls under the FERA (Foreign Exchange Regulation Act) regime, which required most to hold no more than 40% of equity in local ventures.

These radical measures inevitably led to a slowdown in investment in the private sector while the public sector was still reeling from the cuts in public investment during the mid-1960s. However, a whole set of new exogenous shocks generated another economic crisis that was then translated into a political crisis.

First, the military crackdown by Pakistan in its eastern wing in 1971 sent 10 million refugees into India; amounting to almost half of Australia's population, the refugees put an enormous burden on India's finances and social peace. When war ensued between India and Pakistan, China and the United States supported Pakistan, with the United States sending a naval task force against India in a show of support to an ally and to establish its credibility with China. The United States cut off economic aid to India, and India responded with renunciation of concessional food imports from the United States. This latter measure soon made management of the country's food economy very difficult.

Second, the oil price shock of 1973 dealt another severe blow to India's balance of payments.

Third, even though the Green Revolution constitutes a success story, it had not yet matured. After a series of bumper crops in the late 1960s, the revolution faltered. The early 1970s saw a failure in rains, with the result that, for four successive years following 1970–71, agricultural production was lower–as was per capita income–than in that year alone (see Tables 3.1 and 3.4). Correspondingly, the growth in industrial production slowed down to only about 3.0% in those four years (see Table 3.3). The only thing that flourished was inflation, which was nearly 75% over those same years. India was again in the midst of an economic crisis after having hardly recovered from the one experienced during the mid-1960s. The economic situation was no doubt compounded by ideologically driven policies, but the whole concatenation of events stemming from geography and the international system could not have made for a different outcome even if the leadership had been less constrained by ideology and factional conflict. Nonetheless, the economic stagnation of the period inspired a considerable literature that assumed the end of economic growth in India after the brief period of easy import substitution between 1956 and 1965; among the left-oriented economists, a basic cause for the stagnation was seen as inadequate domestic demand in the absence of agrarian reform (Nayyar, 1994).

The zenith of radicalism in economic policy was reached with the nationalization of the wholesale wheat trade in 1973, when the country was already in the throes of a major economic crisis. The move proved to be an absolute disaster, aggravating food scarcity. The public agony with food and other shortages expressed itself in riots, violence, widespread industrial unrest, and massive agitation against the government, more specifically against Indira Gandhi personally. It drove home the failure of nationalization of the wholesale wheat trade, and the decision was reversed. It was a turning point, and it resulted in the halt to further large-scale nationalization.

Because of the economic agony, the political scene witnessed an explosion of turmoil. A railway strike in 1974–which threatened not only the economy but was also perceived to be a dress rehearsal for a major political upheaval against the government–was crushed by the government with great ferocity. At that juncture, India changed economic course on its own, adopting first a severe deflationary package to cope with inflation and then turning to a growth-oriented economic orthodoxy through a hesitant and minor relaxing of the licensing guidelines in order to expand production in some parts of the industrial system. Inflation was soon brought under control. Subsequently, higher public investment was resumed. These various measures laid the basis for the renewal of economic growth in the next period. However, in the context of the cumulative economic deterioration and the consequent political turmoil, the threat to her own political survival led Indira Gandhi to declare a state of emergency.

Analysis. The period as a whole had witnessed the piling up of crises, domestic and international as well as economic and political, and it would be difficult to underline a single cause as an explanation for the problems of the period with any degree of certainty. To a certain extent, the explanation is shrouded in counterfactuals. One stance can be to take the economic crisis of the period to have been the natural result of the state policy of economic autarky of the earlier period. Indeed, there is a view that the roots of India's subsequent economic problems lie in the Mahalanobis model. But would the consequences have been of the same severity if the economic system had not received the exogenous shocks, both natural (droughts) and others originating in the international system (wars and oil price hikes)?

It has often been said, with the benefit of hindsight, that it was around this time that the state ought to have shifted to an export-oriented policy (Mohan, in Jalan, 1992: 85–115; see also Dandekar, in ibid.: 33–84). However, even if India had so decided, it is debatable whether the United States would have opened its markets to India at that time as it had done for Japan, Korea, and Taiwan. On the other hand, it was the occurrence of the shocks, especially the systemic shocks, that led to the contemporaneous conviction that the heavy industry ISI strategy was essentially correct, and that it was unfortunate that it had been disrupted so early in its career. Indeed, if China (1962, 1965, 1971), Pakistan (1965, 1971) and the United States (1965, 1971) were severally or jointly opposed to India, then the strategy seemed to have continued value. It is for that reason, no doubt, that the strategy evoked allegiance and remained in place, even though constrained by the inadequacy of resources.

Of course, the agricultural situation needed redress, and the state responded adequately on that score through the Green Revolution, which assured a long-term agricultural growth rate of over 2.5% and the building up of food stocks (Krishna, 1988). Undoubtedly, state policy erred in clamping down hard in relation to inflation and cutting down public investment, especially in infrastructure, but it was guided by the not unjustified fear of public reaction and, given the institutional structure, electoral revenge. The state's room for maneuver was thus very limited, given the totality of the situation, what with the aroused society, the economic crises, and the inhospitable international environment. Indeed, state power itself had been damaged, for state autonomy cracked as societal forces assumed the upper hand and elites fractionalized while state capacity deteriorated under the economic and administrative burdens increasingly taken on by the state. The state was no longer now the dominant element, but rather a prisoner of society and the international system.

On balance, it would seem that the exogenous shocks, both natural and systemic, set the initial crisis of the period in motion, in the process accelerating the mobilization of society, which in turn not only limited the scope of maneuverability of policy makers but also damaged state capacity, with the whole process further aggravated by still more exogenous shocks, systemic and natural. Nature, international system, and society were the key determinants; their impact attenuated state autonomy and eroded state capacity even as it brought about economic deterioration.

Third Period, 1975–97: Respectable Growth

The better performance of the third period would at first glance seem, at least in part, counterintuitive since the period witnessed enormous political instability and turmoil as well as exogenous shocks and economic crises. To begin with, there was the national emergency declared by Indira Gandhi in 1975. This was followed by her overthrow in the 1977 elections and the accession to power by a squabbling coalition known as the Janata Party. That coalition finally broke down in 1979 while India was reeling under the impact of an unprecedented drought and the second oil price shock. In the elections that followed, Indira Gandhi returned to power in 1980 feeling vindicated by the electorate. Not only did she have to face the economic crisis from the oil price shock and the drought but also the burden of higher defense spending in reaction to the renewal of US military aid to Pakistan in the wake of the Soviet intervention in Afghanistan. She had to approach the IMF (International Monetary Fund) for a major loan to cope with the problems emergent from the oil price shock.

There was no respite from political turmoil, however, which had its own independent sources. Indira Gandhi was dogged by the problem of national unity in Assam and the Punjab, which finally culminated in her assassination in 1984. Her son, Rajiv Gandhi (a former pilot), who succeeded her, at first won a landslide victory in the elections out of sympathy for his mother's martyrdom. Being new to politics, however, he was unable to maintain that level of support. In 1989, he was defeated in the elections, largely on the highly charged issue of a disputed religious site.

Rajiv Gandhi was followed in power by another squabbling coalition, which resorted to populist measures both in economics (liquidating loans to small and medium farmers) and in politics (radical affirmative policies on the basis of caste). A breakaway faction of the coalition took over power in 1990, but within less than a year, that, too, collapsed in the midst of an economic crisis, which was aggravated by the international conflict over Iraq's invasion of Kuwait. In the ensuing elections in mid-1991, the Congress Party suffered a major blow with the assassination of Rajiv Gandhi during the electoral campaign; it failed to win a majority and had to surmount the economic crisis as a minority government.

Since the third period covers a long time span, which also saw a great deal of political change–almost the alternation of the major coalitions in power–the discussion of performance here is organized largely around the successive governments and their economic programs. It would seem that the good performance during the second half of the 1970s after the earlier period of stagnation sprang from a combination of a sharp turn in economic policy and the fortuitous circumstance of what had been from another perspective a major blow to the economy–the oil price shock of 1973. Initially, the shock had, together with the droughts of the early 1970s, fueled inflation. But Indira Gandhi's deflationary package of mid-1974 soon brought the economy to an even keel. Then, India witnessed its own version of the recycling of petrodollars. The oil price hike resulted in a construction and consumption boom in the oil-exporting states of the Persian Gulf area, which had the unintended consequence of pulling in exports from India as well as contractors and migrant labor. The remittances from the migrant labor were a great boon for India's balance of payments. As a consequence, India did not have to clamp down on imports in reaction to the oil price shock.

By this time, the state in India also had the advantage of learning experience as a result of having witnessed the damaging consequences of the contractionary policy in the mid-1960s. Another element in the easing of the economic problem related to the fruits of successful import substitution in oil and foodgrains, which enabled India to reduce considerably its imports of these commodities. India's balance of payments was also helped by the expansion of exports as a result of a de facto devaluation through the mechanism of linking the rupee with the sterling in its period of weakness in the 1970s. India began to benefit as well from the emergent trend by this time of higher savings rates, especially in the household sector (private households and small-scale enterprises) (see Table 3.7).

Over and above these elements, the easing of controls through the several successive small installments of liberalization undoubtedly helped also. Indira Gandhi had started the process in 1974. Some further doses of liberalization were introduced by the Janata government during its brief tenure in office. However, the Janata government with its factional problems had neglected the infrastructure. Economic deterioration following the severe drought in 1979 formed the background of the collapse of the Janata government. The cumulative impact of the long neglect of the infrastructure from the 1960s plus the oil price shock of 1979 set the agenda for Indira Gandhi's new government in 1980. As she returned to power, Indira Gandhi was no longer enamored of socialism; instead, she was concerned about the failure of the public sector to provide economic surpluses expected of it for investment and over its propensity for dependence on budgetary support to cover its losses. At times, 50% of the gross fixed capital formation was in the public sector, but that sector's share in national savings was small; instead, the sector simply captured private savings for the purpose (see Table 3.7). The concern about the continued failure of the public sector to generate savings for investment marked the start of the elite's disenchantment with the public sector. Indira Gandhi then began to push liberalization a little further, but did it stealthily, without admitting that the old policies were no longer suitable.

After her death, Rajiv Gandhi accelerated the process of liberalization even though India faced no economic crisis, for Indira Gandhi had left behind a comfortable situation with respect to food stocks and foreign exchange reserves. His position was truly pathbreaking in the sense that he was the first prime minister, indeed anybody in high office, to openly espouse liberalization, at least at the beginning of his period in office. Socialism and the public sector were no longer sacred cows. The major impulse in his pushing liberalization was the increased perception–which of course reflected reality–of the marginalization of India on the international scene, a cumulative result of slow economic growth over the preceding two decades. He sought to reinterpret self-reliance, maintaining (as paraphrased by his economic advisor L. K. Jha) that "Self-reliance for a country like India cannot have the limited meaning of the country not being influenced one way or another by external economic forces. It should instead be measured in terms of India's contribution to the shaping of the international economic forces." In his own words, Rajiv Gandhi declared in the Seventh Five-Year Plan, "Self-reliance does not mean autarchy. It means the development of a strong independent national economy, dealing extensively with the world, but dealing with it on equal terms" (Nayar, 1990a: 46).

However, to meet the world on equal terms required, in Rajiv Gandhi's view, the restructuring of India's inefficient, technologically backward and high cost industrial economy. The route to restructuring was brought about in liberalization by way of deregulation and delicensing, at least in the domestic economy at first: technology upgradability through allowing imports of new technology; enhanced competition, at least internally; and reform of the public sector so as to provide surpluses for investment. Thus, for the first time, possibilities opened up of the eventual replacement of the statist model by the market model or at least in shifting the balance in favor of the latter in a mixed economy. Dismayed by the inefficiency of the public sector, Rajiv Gandhi sought to make the private sector the engine of growth. Ambitious as his vision was, however, Rajiv Gandhi did not as yet envisage a paradigm shift and an overthrow of the old system; rather, he aimed for incremental change within the framework of that system. Even so, he had to pay a heavy price for his acceleration of liberalization, for it led to unpopularity with the electorate, which perceived liberalization to be a policy for the benefit of the rich even as other vested interests that had been privileged by the old system, such as public sector labor, mounted a noisy campaign against Gandhi.

It would be comforting to attribute the acceleration of the growth rate in the 1980s simply to the progressive doses of liberalization introduced and the greater role accorded to the market, thus proving the superiority of the market model. They were, no doubt, a factor in improving productivity and enhancing the growth rate; for example, total factor productivity in manufacturing during the first half of the decade of the 1980s increased at a rate of 3.4% as against the annual decline of 0.3% in the period from 1965–66 to 1979–80 (Ahluwalia, 1991: 191). However, a major factor in the higher growth rate was that the state followed an expansionary policy, relying for investment–in a shift from its traditional fiscal and monetary conservatism–on large fiscal deficits and external commercial borrowings in the absence of adequate concessional aid. While Latin America was coming to terms with its devastating debt problem in the 1980s, India was in the process of quickly buying her way into it, banking on her good credit rating at the time. A future crisis was thus being quietly put in place through fiscal profligacy. The resulting higher growth rate was therefore in the long term unsustainable. On the other hand, a different perspective on the gathering crisis can justifiably be that it was better to have deliberately introduced dynamism, even if risky, into the economy than to have accepted a low-level equilibrium, which was not necessarily risk free.

There was also another economic price paid for liberalization. Since liberalization was unpopular, Rajiv Gandhi had to supplement it by major increases in allocations to social sectors and to state governments to attenuate their opposition. To compensate for the resulting higher outgo, his government then tried to mobilize resources by further adding to the fiscal deficit and increasing customs duties, which had the direct and indirect consequence of making Indian goods uncompetitive in the world markets and depressing exports.

The developing crisis finally burst when the Gulf War aggravated the already serious balance of payments problem that was developing in 1990. The origins of the crisis, however, lay in fiscal profligacy rather than any exogenous shocks. As investors lost confidence, there was a flight of capital. By early 1991, the government was frantically searching for ways to avoid defaulting on its debt payments, including the mortgaging of India's gold holdings. Foreign exchange reserves could hardly cover two weeks of imports, while inflation was nearing 15%. The crisis came at a time of governmental instability, and when the government fell, new elections returned the Congress Party to power as a minority government.

Impelled by the gravity of the crisis, the new government boldly seized the opportunity to undertake tough measures as well as launch India on a paradigm shift to a market model. On the face of it, the new economic policy marked a break with the earlier model of self-reliance launched with the second plan. The government's actions–delicensing over a vast expanse of industry and dismantlement of controls, lowering and rationalizing tariffs, and throwing the door open to foreign investment–constituted a historic retreat from the Nehru model.

These measures were no doubt inspired by the intent to preempt the imposition of "conditionalities" by the IMF, but they also sprang from a genuine conviction among key actors (Prime Minister Narasimha Rao and Finance Minister Manmohan Singh) that they were the right measures to adopt in light of the new situation of economic globalization and the consequent requirement to integrate the Indian economy with the world economy. The preference for the market was openly voiced as was preference for integration with the global economy. As in Kuhn's model of the scientific revolution, the glaring success by this time of a significant number of outward-oriented economies–not just Japan as a single anomaly–and the collapse of planned economies in the Soviet bloc had in a way already settled the issue in favor of a paradigm shift to the market, making it an almost universal orthodoxy among policy makers and economists alike. In India, there was particular concern, though not often expressed, about the country's place in the world. As Finance Minister Manmohan Singh (1992) explained in terms reminiscent of Rajiv Gandhi's perception,

> If our economy is not equipped to absorb, to assimilate, and to adapt this technical change which is taking place, all over the

> world, I think we will be marginalized. Many developing economies are already being marginalized in the new global economic system that has emerged. And it is only by successfully absorbing, assimilating, and adapting modern technological change that developing countries can acquire a minimum amount of bargaining power and influence in the management of the global interdependence. And, therefore, preparedness for handling problems of structural change is essential. This is the broad justification for economic policy initiatives taken by our government in recent months.

The stabilization package to overcome the crisis, through devaluation and some reduction in the fiscal deficit, was managed with skill, avoiding throwing the economy into a deep and savage recession as is usually associated with stabilization elsewhere. There was a quick recovery: growth rates picked up with respect to industry and exports; foreign exchange reserves bounced back to comfortable levels; and foreign investment increased significantly relative to the past. The economy seemed positioned on a higher growth path. The process was aided by a series of good monsoons, which were no doubt a factor in the better economic performance of the first half of the 1990s.

The liberalization process, however, soon stalled because of a banking scam in 1992, massive violence associated with religious conflict in 1992 and 1993, and the approaching elections in 1996. The Congress Party was defeated in the elections, but the defeat was apparently not related to liberalization. A new multiparty United Front coalition came to power in 1996, which included not only several regional parties but also the two major communist parties. The position of the coalition on liberalization seemed polarized; Prime Minister H. D. Deve Gowda and Finance Minister P. Chidambaram strongly affirmed support for it, declaring liberalization to be irreversible and pledging to carry it forward, while the communist parties applied brakes to the process, especially with respect to privatization of the public sector.

Analysis. The period of accelerated growth covers over two decades, and undoubtedly many factors were at work. Clearly, however, state policy must command first place in the hierarchy of variables. No doubt the state added to its problems through fiscal profligacy, but it also navigated policy fairly skillfully, albeit gradually and haltingly, in the midst of shocks emerging from geography, society, and the international system, even as it proved responsive to societal pressures. While India did not match the authoritarian states in the speed with which it moved on economic policy reform to accelerate economic growth, it did move. The state was creative in utilizing crises to change the course of policy, not in full measure but sufficiently within the constraints of a democratic system. As well, it made use of the resources in society and the international system in support of its aims, even as it was sensitive to demands emergent from these sources, in order to maintain support–not always successfully–for itself and its reform program. The process is made sharply visible in the reform of the early 1990s.

The crisis in 1991 was in the final analysis related to the fiscal profligacy of an earlier administration, but it was aggravated by international events in the Persian Gulf and the war that followed. As a minority government, the Congress Party did not have a mandate for reform as such, but it made use of the crisis not merely to bring about stabilization but to initiate the restructuring of the economy in a direction entirely different from its earlier trajectory. In this, it benefited from the changed international environment in favor of the market model. However, even while utilizing IMF support to rescue the economy, it did not adopt the IMF-preferred "big bang" or "shock therapy" strategy of wholesale reform at one blow. Rather, it attempted to enact reform through stages, starting with industry, then trade and foreign investment, and later partial financial sector reform, leaving privatization and labor market reform for some other propitious moment. At the same time, it made no drastic reductions in the fiscal deficit, for fear of recession and unemployment, even as it continued with subsidies for food, fuel, fertilizer, and power so as not to alienate powerful groups in society, particularly the farmers and the urban middle classes. The fiscal deficit simply reflected the power of society and organized groups within it. In this light, the calls for exertion of "political will" in economic reform stand downgraded, as they usually come from those who do not have to face elections. No doubt, the continuing fiscal deficit makes probable another economic crisis, but that seems to be a price imposed by democracy.

Economic reform also creates new pressures to slow it down even as it develops new sources of support. Foreign investment, both portfolio and direct, has no doubt been a factor in the accelerated growth of the mid-1990s. On the other hand, it has given rise to a new economic nationalism because of the fear that local entrepreneurship will be swamped by foreign corporations. In a sense, the problem is rooted in the socialist strategy of the mid-1950s, which blocked the growth of private sector giants that could have competed on the world scene while preempting the "commanding heights" for public sector giants that have largely proven inefficient and therefore unable to compete, and thus have become ripe pickings for foreign takeovers if privatization is pushed through. Apart from the dread of a Western-inspired consumerism, there is the feeling that the playing field is presently stacked against Indian companies, given the high cost of local capital compared to the awesome financial power of foreign corporations and the accumulated burden of past protective labor policies.

Summary and Conclusions

Over the long term, India's economic performance has been modest. However, there is considerable variation in performance over the years since independence, and three periods can be distinguished with different patterns of growth. A first period of modest growth of some 15 years was followed by ten years of slow growth, which was succeeded by some two decades of respectable growth. In trying to explain the variance in performance, it has been found that in the first period, the state attempted, with an ambitious, indeed audacious, economic strategy, to restructure and remold society and to largely detach itself from the world economy within a short span of time. A sophisticated and diverse industrial sector was the result, though achieved at considerable sacrifice. Consequently, a structural break occurred with the pattern of economic stagnation in the first half of the century under colonial rule.

Compounded by shocks from nature and the international system, other economic and social consequences of that ambitious strategy culminated in a revolt on the part of society at the same time that the state found the major Western powers uncooperative, indeed adversarial. The challenges were overwhelming for the state during the second period and largely immobilized the state except in the area of agriculture, where the crisis was the severest. The overall result was slow growth, but the period nonetheless saw the inauguration of the Green Revolution. That revolution provided highly prized food security for the country.

Finally, the third period saw the working out, through trial and error over the years, of a new accommodation between the state on the one hand and society and the international system on the other–an interactive process in which the state played a leading role even as it increasingly allowed society and the market a greater role, mindful of its own limited capacity, and adopted a more constructive engagement with the international system. The pressures from the internal and external environment in this period were, in Toynbeean terms, strong enough to provide an incentive for policy reform but not so strong as to almost overwhelm the state, as in the second period.

The interactive process is a continuing one. The year 1991 marked a watershed with a sharp break in policy from inward orientation to outward orientation and from the state to the market. The four years that followed the break witnessed a pattern of high growth, which, if sustained, may define a new period. To assure the promise of a period with a sustained high rate of growth demands persistence in economic policy reform to meet several remaining challenges: reducing the fiscal deficit–a very painful task–expanding exports, drastically improving and upgrading the infrastructure, shedding the inefficient enterprises in the public sector, rationalizing the highly protected small-scale sector, and reforming the labor markets. Sustaining a high rate of growth is essential to providing greater employment and removing poverty, an area in which India's record has been miserable (see Table 3.5). It is also essential to reducing the cumulative marginalization of the country in the world economy as manifest in the drastic decline in its share of world exports (see Table 3.8), for marginalization–reflective of slower growth comparatively–generates power imbalances that can be highly consequential for national survival.

Further Reading

Ahluwalia, Isher Judge, *Industrial Growth in India, Stagnation since the Mid-Sixties,* New Delhi: Oxford University Press, 1985

A pathbreaking study that calls into question the notion of a generalized industrial recession in Indian industry during the 1960s and 1970s and suggests that it is more applicable to the capital goods industry, which had been overemphasized in the Second Five-Year Plan. Ahluwalia attributes the industrial slowdown to the failure to make adequate investments in infrastructure and to the restrictive policy regime. Ahluwalia also blames state policy for slow growth, among other things, and therefore calls for liberalization of controls.

——, *Productivity Growth in Indian Manufacturing,* Delhi and New York: Oxford University Press, 1991

A quantitative study that shows the decline in productivity in manufacturing in the 1960s and 1970s and the rise in productivity in the first half of the 1980s.

Bardhan, Pranab K., *The Political Economy of Development in India,* Oxford and New York: Basil Blackwell, 1984

Joins Olson's notion of "distributional coalitions" with the Marxist understanding of classes to argue that India's slow economic growth is a result of a stalemated coalition of industrial capitalists, rich farmers, and state professionals.

Bhagwati, Jagdish N., *India in Transition: Freeing the Economy,* Oxford and New York: Oxford University Press, 1993

The eminent trade theorist and long-standing critic of India's approach to planning provides a highly intelligible survey of developments leading up to the liberalization of 1991.

Byres, Terence J., editor, *The State and Development Planning in India,* Delhi and New York: Oxford University Press, 1994

An important collection with different viewpoints, primarily of the progressive variety.

Chakravarty, Sukhamoy, *Development Planning: The Indian Experience,* Oxford and New York: Oxford University Press, 1987

Distinguished economist and eminent planner, the late Chakravarty examines the theoretical underpinnings of Indian planning with a definitive insider's view. Though acutely aware of the shortcomings of Indian planning, particularly with respect to agrarian reform and poverty removal, he regards the Indian model as both justified and moderately successful.

Frankel, Francine R., *India's Political Economy, 1947–1977: The Gradual Revolution,* Princeton, N.J.: Princeton University Press, 1978

A pioneering study and a standard text for many years.

Harrison, Selig S., *The Widening Gulf: Asian Nationalism and American Policy,* New York: Free Press, 1978

A veteran observer takes a critical look at the American encounter with nationalism in Asia.

Higgins, Benjamin, *Economic Development: Problems, Principles, and Policies,* New York: Norton, 1968

The important textbook in the 1960s and 1970s.

Jalan, Bimal, *India's Economic Crisis: The Way Ahead,* Delhi: Oxford University Press, 1991

An important adviser on economic policy inquires into what went wrong with Indian economic planning.

Jalan, Bimal, editor, *The Indian Economy: Problems and Prospects,* New Delhi and New York: Viking, 1992

An important collection of articles.

Jha, Prem Shankar, *India: A Political Economy of Stagnation,* Delhi: Oxford University Press, 1980

An eminent journalist argues that India's economic stagnation after the mid-1960s was related to the rise to dominance of the intermediate class, consisting of the self-employed such as small-scale industrialists, traders, and rich peasants.

Joshi, Vijay, and I. M. D. Little, *India: Macroeconomics and Political Economy 1964–1991,* Washington, D.C.: World Bank, 1994

An authoritative, comprehensive account of macroeconomic policy.

Krishna, Raj, "Ideology and Economic Policy," in *Indian Economic Review* 23, no. 1 (1988)

Establishes the roots of India's institutional and policy regime in the ideological orientation of Fabian socialism of the founding fathers, and links India's poor performance to it.

Lucas, Robert E. B., and Gustav F. Papanek, editors, *The Indian Economy: Recent Development and Future Prospects,* Boulder, Colo.: Westview Press, 1988

A good compilation on liberalization under Prime Minister Rajiv Gandhi. The article cited in the text is by P. N. Dhar, formerly principal secretary to Prime Minister Indira Gandhi. There is also an excellent review by Montek S. Ahluwalia, former economic adviser to Prime Minister Rajiv Gandhi.

Nayar, Baldev Raj, *The Modernization Imperative and Indian Planning,* New Delhi: Vikas, 1972

Demonstrates the importance of national power considerations in the evolution of the economic strategy of the Second Five-Year Plan.

——, *India's Mixed Economy: The Role of Ideology and Interest in Its Development,* Bombay: Popular Prakashan, and London: Sangam, 1989

A comprehensive analysis of the state and public sector that establishes the ideological origins of the public sector.

——, *The Political Economy of India's Public Sector: Policy and Performance,* Bombay: Popular Prakashan, 1990(a)

Provides paired comparisons between private and public sector enterprises in the steel and aluminum industries. Also discusses the politics of liberalization under Rajiv Gandhi.

——, *Superpower Domination and Military Aid: US Military Aid to Pakistan,* New Delhi: Manohar, 1990(b)

Utilizes declassified US documents on foreign relations in support of the proposition that the United States as the global power has pursued a policy of containment in relation to India as a regional power.

Nayyar, Deepak, editor, *Industrial Growth and Stagnation: The Debate in India,* Bombay and New York: Oxford University Press, 1994

A collection of important articles of varied viewpoints from the influential left-wing *Economic and Political Weekly* on the industrial stagnation during the 1960s and 1970s. Includes, among others, the often-cited article by S. L. Shetty. The most comprehensive discussion is by C. Rangarajan.

Rudolph, Lloyd I., and Susanne Hoeber Rudolph. *In Pursuit of Lakshmi,* Chicago: University of Chicago Press, 1987

A magisterial work on political economy by two scholars with long experience, marked both by empathy and critical reflection.

Sandesara, J. C., *Industrial Policy and Planning, 1947–91: Tendencies, Interpretations, and Issues,* New Delhi: Sage, 1992

Based on quantitative data, the study provides a balanced assessment of the debate on industrial growth and stagnation. The author considers Indian industrial growth to be satisfactory and highlights the role of growth alone in meeting social objectives without special targeting.

Singh, Manmohan, "Keynote Address," in *Social Dimensions of Structural Adjustment in India: Papers and Proceedings of a Tripartite Workshop Held in New Delhi, December 10–11, 1991,* New Delhi: International Labour Organization, Asian Regional Team for Employment Promotion, 1992

Sundrum, R. M., *Growth and Income Distribution in India: Policy and Performance since Independence,* New Delhi and Beverly Hills, Calif.: Sage, 1986

A sound and authoritative account.

Baldev Raj Nayar is emeritus professor of political science at McGill University, Montreal. His most recent publications include *The State and Market in India's Shipping* (1996) and *The State and International Aviation in India* (1994).

Table 3.1: Annual Growth Rates of Net Domestic Product (at 1980–81 Prices)

Year	*Rate*	*Year*	*Rate*	*Year*	*Rate*
1951–52	2.4	1966–67	0.5	1981–82	5.8
1952–53	2.8	1967–68	8.2	1982–83	2.2
1953–54	6.3	1968–69	2.5	1983–84	8.1
1954–55	4.0	1969–70	6.7	1984–85	3.4
1955–56	2.5	1970–71	5.2	1985–86	3.9
1956–57	5.5	1971–72	0.6	1986–87	3.8
1957–58	-1.8	1972–73	-0.8	1987–88	3.8
1958–59	7.6	1973–74	4.9	1988–89	10.7
1959–60	1.7	1974–75	1.3	1989–90	7.0
1960–61	7.0	1975–76	9.5	1990–91	5.1
1961–62	2.7	1976–77	0.9	1991–92	-0.1
1962–63	1.7	1977–78	7.7	1992–93	5.1
1963–64	5.0	1978–79	5.6	1993–94	5.9
1964-65	7.4	1979–80	-6.0	1994–95	6.8
1965–66	-4.7	1980–81	7.5	1995–96	6.9

Source: Government of India, *Economic Survey 1996–97,* New Delhi: Government of India, Ministry of Finance, 1997, S-4

Table 3.2: Average Annual Growth Rates of Net National Product by Categories

Category		*Rate*	*Category*		*Rate*
Plans			*Growth Periods*		
First	(1951–56)	3.6	Modest	(1951–65)	3.9
Second	(1956–61)	4.0	Low	(1965–75)	2.4
Third	(1961–66)	2.4	Respectable	(1975–96)	5.5
Annual	(1966–69)	3.7			
Fourth	(1969–74)	3.3			
Fifth	(1974–79)	5.0	*Regulation*		
Annual	(1979–80)	-6.0	Dirigiste	(1951–85)	3.6
Sixth	(1980–85)	5.4	Liberal I	(1985–92)	4.9
Seventh	(1985–90)	5.8	Liberal II	(1992–96)	6.2
Annual	(1990–92)	2.5			

Table 3.3: Annual Growth Rates of Industrial Production (Base 1980–81)

Year	*Rate*	*Year*	*Rate*	*Year*	*Rate*
1951–52	-	1966–67	-0.4	1981–82	9.3
1952–53	3.8	1967–68	-1.3	1982–83	3.2
1953–54	1.8	1968–69	6.4	1983–84	6.7
1954–55	6.7	1969–70	7.1	1984–85	8.6
1955–56	8.4	1970–71	4.9	1985–86	8.7
1956–57	8.5	1971–72	4.4	1986–87	9.1
1957–58	3.6	1972–73	6.0	1987–88	7.3
1958–59	1.7	1973–74	0.5	1988–89	8.7
1959–60	8.8	1974–75	1.8	1989–90	8.6
1960–61	12.1	1975–76	5.4	1990–91	8.2
1961–62	9.2	1976–77	12.1	1991–92	0.6
1962–63	9.7	1977–78	3.4	1992–93	2.3
1963–64	8.1	1978–79	6.9	1993–94	6.0
1964-65	8.6	1979–80	1.1	1994–95	9.4
1965–66	9.3	1980–81	0.8	1995–96	11.7

Source: For the years until 1980–81, J. C. Sandesara, *Industrial Policy and Planning, 1947–91: Tendencies, Interpretations, and Issues,* New Delhi: Sage: 1993, 118; for the subsequent years, Government of India, *Economic Survey 1996–97,* New Delhi: Government of India, Ministry of Finance, 1997: 115

Table 3.4: Annual Growth Rates of Real GDP in Agriculture, Forestry and Logging, Fishing, Mining and Quarrying (at 1980–81 Prices)

Year	*Rate*	*Year*	*Rate*	*Year*	*Rate*
1951–52	1.7	1966–67	-1.3	1981–82	6.2
1952–53	3.1	1967–68	14.5	1982–83	-0.7
1953–54	7.6	1968–69	-0.1	1983–84	10.4
1954–55	3.0	1969–70	6.4	1984–85	0.0
1955–56	-0.8	1970–71	6.6	1985–86	0.5
1956–57	5.4	1971–72	-1.7	1986–87	-1.0
1957–58	-4.3	1972–73	-4.7	1987–88	0.5
1958–59	9.9	1973–74	7.0	1988–89	16.3
1959–60	-0.9	1974–75	-1.3	1989–90	2.0
1960–61	6.9	1975–76	12.9	1990–91	4.2
1961–62	0.2	1976–77	-5.5	1991–92	-2.0
1962–63	-1.6	1977–78	9.8	1992–93	5.8
1963–64	2.4	1978–79	2.3	1993–94	3.5
1964-65	9.0	1979–80	-12.3	1994–95	4.8
1965–66	-10.4	1980–81	12.9	1995–96	0.3

Source: Governmnt of India, *Economic Survey 1996-97,* New Delhi: Government of India, Ministry of Finance, 1997: S-10.

Table 3.5: Comparative Per Capita Income, Life Expectancy, and Illiteracy

Country	*GNP per capita US$ 1994*	*PPP* estimates of GNP per capita 1994 (International $)*		*Life expectancy at birth (years) 1994*	*Adult illiteracy (% of population) 1995*
		Current	Share of US (%)		
Bangladesh	220	1,330	5.1	57	62
India	320	1,280	4.9	62	48
Pakistan	430	2,130	8.2	60	62
China	530	2,510	9.7	69	19
Sri Lanka	640	3,160	12.2	72	10
Indonesia	880	3,600	13.9	63	16
Philippines	950	2,740	10.6	65	5
Malaysia	3,480	8,440	32.6	71	17
South Korea	8,260	10,330	39.9	71	-
United States	25,880	25,880	100.0	77	-

Source: World Bank, *World Development Report 1996,* New York: Oxford University Press, 1996: 188–89
*Purchasing power parity

Table 3.6: Value Added in Manufacturing Industry (Rs. Million at 1960–61 Prices)

Industry	*1950–51*	*1960–61*	*1965–66*
Consumer Goods	2,607	4,237	4,876
Intermediate Goods	895	3,461	6,202
Machinery	309	1,513	3,159
Others	31	69	103

Source: Government of India, *Fourth Five-Year Plan: A Draft Outline,* New Delhi: Government of India, Planning Commission, 1966: 10

Table 3.7: Domestic Savings and Gross Fixed Capital Formation (As % of GDP at Current Market Prices)

	Private Sector			
Year	*Household*	*Corporate*	*Public sector*	*Total*
		Gross Domestic Savings		
1950–51	7.7	1.0	1.8	10.4
1955–56	11.0	1.3	1.7	13.9
1960–61	8.4	1.7	2.6	12.7
1965–66	9.9	1.5	3.1	14.5
1970–71	11.3	1.5	2.9	15.7
1975–76	13.4	1.3	4.2	19.0
1980–81	16.1	1.7	3.4	21.2
1985–86	14.6	2.0	3.2	19.8
1990–91	20.5	2.8	1.0	24.3
1995–96	19.5	4.1	1.9	25.6
		Gross Fixed Capital Formation		
1950–51	6.9		2.4	9.3
1955–56	7.3		5.2	12.5
1960–61	6.8		6.5	13.3
1965-66	8.0		7.8	15.8
1970–71	9.1		5.5	14.6
1975–76	9.8		7.1	16.9
1980–81	10.7		8.6	19.3
1985–86	10.2		10.5	20.7
1990–91	13.8		9.4	23.2
1995–96	16.3		8.3	24.6

Source: Government of India, *Economic Survey 1996–97*, New Delhi: Government of India, Ministry of Finance, 1996: S-8. Household also includes small-scale enterprises.

Table 3.8: India's Share of World Exports

Year	%	*Year*	%
1948–49	2.23	1970–71	0.63
1950–51	1.88	1975–76	0.49
1955–56	1.35	1980–81	0.43
1960–61	1.03	1985–86	0.47
1965–66	0.81	1990–91	0.53

Source: CMIE, in Siddheswar Prasad and Jagdish Prasad, editors, *New Economic Policy: Reforms and Development,* New Delhi: Mittal Publications, 1993: 55

Chapter Four

Planned Development and the Search for Self-Reliance

George Rosen

The roots of India's economic development policies go back to well before 1947 and independence. Leaders of the independence movement, major Indian policy makers, and leading private businessmen within India had been thinking about appropriate policies for an independent India in the early 1930s and had published their ideas. M. Visvesvaraya, a leading official in the princely state of Mysore, proposed an economic plan for India in 1934; the Indian National Congress set up a National Planning Committee with Jawaharlal Nehru as chairman in 1938 to prepare a plan for development; in 1944, both the Indian Federation of Labor and a group of India's leading businessmen, including J. R. D. Tata and G. D. Birla, presented two separate economic plans for an independent India; and one of Mohandas Gandhi's leading disciples also published a Gandhian economic plan in 1944. In that same year, the viceroy's government set up a Department of Planning and Development, headed by a leading Indian industrialist, and this agency prepared a postwar program of industrial development. While the government did not accept that plan and disbanded the department, an advisory board was set up that recommended a Planning Commission in the postwar government. The British government during the war years had also set up an elaborate wartime control system to guide the Indian economy toward the support of the British war effort.

The various Indian plans advocated as goals for India's economic development a significant improvement in living standards of the people and the elimination of the worst extremes of poverty. The Visvesvaraya, National Congress, and businessmen's Bombay plans all emphasized the development of large-scale, capital good industry, using the Soviet Union as the model of the means to achieve those goals as well as a high degree of economic self-sufficiency and independence to accompany the political independence. The Indian Federation of Labor urged concentration on the agricultural sector for rapid benefits for the mass of the population that was dependent on agriculture. The Gandhian proposal favored the encouragement of small-scale, cottage industry and decentralization of the economy and of economic policy making. All the proposals, except the Gandhian, favored major government involvement in economic development with extensive state support, ownership, and control of the proposed large-scale capital goods industries. Nehru and the Congress Party's Planning Committee were strongly influenced by the democratic socialism of the British Fabians, as well as the apparent Soviet success in achieving rapid industrial growth and political power. In addition, the Great Depression of the 1930s and the decline in foreign trade resulted in disillusionment with capitalism and the urge toward self-sufficiency. There was also a fear of "economic imperialism" replacing political imperialism if India encouraged foreign investment and undue trade dependency.

With Nehru as prime minister after independence, the ideas of the National Planning Committee of the Congress Party were adopted as the basis of the economic policy of the new government. The new Constitution, adopted in 1950, set forth certain Directive Principles for economic policy. One of the goals was a "just society," in which inequalities of income and in opportunities for personal advancement would be minimized, and there would not be an undue concentration of wealth and control over the means of production. The rights to work, to education, and to assistance in the event of unemployment and illness were set forth. These were not laws but principles of policy; a Planning Commission was also established in 1950 to prepare the program to implement those principles. Prime Minister Nehru was the first chairman of the commission, which issued the First Five-Year Plan in 1951.

The First Five-Year Plan stressed industrial development, a high rate of capital investment in industry, and the future extension of government ownership of industry as well as direct control over private industrial production and investment and of foreign trade and foreign investment. But these goals were more indicative of intention rather than implementation; much of the period during the First Five-Year Plan was devoted to preparation of the Second Five-Year Plan, issued in 1956, which in effect laid out India's development strategy for the next three decades. This plan was prepared by P. C. Mahalanobis, India's leading statistician and a classmate of Nehru's in England and a close friend thereafter. It was also prepared after extensive consultations with and visits to India by many of the world's leading economists from both market and socialist economies.

In both the Second and Third Five-Year Plans, investment in capital goods and basic industry was to be the main instrument of development. This new large-scale industry was to be government owned, reflecting the "socialist pattern of society" in which government owned the "commanding heights." The stress on investment in goods production would be both the basis for further industrialization and for freeing India from foreign dependence for key machinery and equipment. "Import substitution" was a major policy tool both to save foreign exchange and increase economic independence over the longer run. Foreign investment and foreign trade were to be tightly controlled during this development period for the same reason. Consumer goods production was given low priority for a variety of reasons. While consumer demand would increase as a result of the massive investment program, capital goods production was seen as the key to the future. There was also a fear of undue "consumerism," reflecting Gandhian thinking. Thus consumer goods production was allotted to the private sector; small-scale and handicraft industries were to be protected from competition from both large Indian firms and foreign competition, and it was hoped that the existing consumer goods industries and future small-scale enterprises would meet the expected increase in demand for such goods. Consumer goods industries would also be more labor intensive than the capital and basic goods industry, thus providing greater industrial employment, but they were essentially considered marginal to the development process. One cost of this marginality of the export and consumer goods sectors was that the modernization of India's large-scale textile industry was sacrificed to protect small-scale and handicraft production. As a result, India lost its trade advantage in this field. While at the end of the war, it had one of Asia's few remaining factory textile industries, other Asian countries rapidly started new industries or rebuilt former ones at more advanced technological levels and gradually replaced Indian exports of those goods.

Accompanying the rapid expansion of the public sector in the capital and basic goods industrial sector, an elaborate network of controls over the private sector was introduced to limit entry of new firms or expansion of existing ones in the various low priority fields. Many of these controls were extensions of the British wartime controls over production, but now they were used to siphon investment to the high priority industrial areas to control use of scarce foreign exchange while limiting foreign investment to prevent "monopolistic" practices of private firms and prevent encroachment of large firms on fields reserved for smaller firms. Agricultural development was given lower priority both in terms of investment allocation and in areas of policy reform. This emphasis reflected five-year plan thought on priorities, but it also arose from the major position of the state governments, rather than the central government, in the area of agricultural policy making.

This development program has had serious problems and has been strongly criticized. Nevertheless, it did have significant overall positive effects despite the fact that it did not fulfill the hopes of its early advocates, and it has been modified. During the final 80 years of British rule, per capita incomes were estimated to have increased by 0.6% per year from 1868–1930 and then remained roughly stagnant. After independence, from 1950–80, per capita incomes are estimated to have risen at a rate of 1.5%–2.0% per year. This was a product of increasing total output at an annual rate of 3.5% and population increases of about 2.0% annually: "[In] round terms 360 million Indians produced US$50 billion of goods and services [in 1950]; in 1995 925 million Indians produced $250 billion" (Adams, 1996: 151). Thus, there are two and a half times as many Indians today, and on average, each consumes twice as much. With this rise in incomes, the percentage of the population below a very low poverty level "fell from over half the population in the mid-fifties to about a third in the late eighties." But with the far larger population, the absolute number below that poverty line may have increased in that same 30 years. (The poverty figures are highly approximate and controversial and there was an estimated range of 200 million and 360 million people below the poverty line in 1987/88). With this rising income and relatively reduced poverty, the quality of life of the population has also improved. Between 1950/55 and 1985/90, life expectancy at birth has risen from 31 years to 56 years, and the literacy rate from 18% to 52%. The Overall Physical Quality of Life Index almost doubled, rising from 30.0 to 55.0 from 1960 to 1990. This index, while rising relative to that of other countries in that period, is still below the world average (for poverty estimates, see Srinivasan, 1996: 211; Dandekar, "Forty Years After Independence," in Jalan, 1992: 42–43; Government of India, 1996: 169–70, Table 10.2; for the quality of life, see Vaidyanathan, 1995: 34–35, and 104, Table 6; Morris, 1996: appendices 1 and 4, esp. 82–83, 102–7).

This improvement was in part a result of the industrial growth that was planned and much of which occurred between 1950 and 1990. The index of industrial production rose by about 12 times in that period. Manufacturing output as a share of India's gross domestic product (GDP) doubled from 10% to 20%; the share of capital goods production and such basic goods as steel, cement, and nitrogen fertilizer rose from 20% of manufacturing output in 1951 to 60% in 1990, while that of consumer goods fell from 50% to 20% over the same period, despite massive absolute output increases. Not surprisingly, the output of coal, petroleum, and electric power all showed enormous increases. This growth in the output of the capital and basic goods sector and power inputs was a product of the massive increase in industrial investment that the government carried through. Annual levels of gross domestic capital formation rose from 10% of GDP in the early 1950s to 24% in the late 1980s; the fixed capital stock in the nonagricultural sector almost quadrupled in real terms from 1950 to the late 1960s (Vaidyanathan, 1995: 98).

At the same time that industrial output was increasing, the role of the public sector was also becoming much greater. The public sector's share of India's productive assets rose from 18% in 1950 to 45% in 1995. By the late 1980s, the public sector's share of the gross investment in manufacturing reached 50%, and it produced over 50% of the country's total industrial output compared to only 11% in 1960–61. Over two-thirds of the total employment in the organized industrial sector was also in government owned enterprises in the late 1980s (Vaidyanathan, 1995: 45; Dandekar, in Jalan, 1992: 57).

Despite these increases in industrial output and apparent improvement in the quality of life, the overall results were disappointing and well below hopes. India has lagged behind not only the newly industrializing economies of South Korea, Taiwan, Singapore, and Hong Kong in health and literacy standards but also behind the People's Republic of China (the one comparable country in terms of size) and its small neighbor, Sri Lanka. While India's industrial growth has been relatively rapid in historical terms, India remains a predominantly nonindustrial country in terms of employment. In the late 1980s, manufacturing employment was only about 11% of the total labor force; agriculture employed over 60% of the total employed workers, and there was a large and growing number of unemployed (T. S. Papola, "The Question of Unemployment," in Jalan, 1992: 306–10).

The achieved increase in industrial output was almost entirely for domestic use: India's share of world exports fell from 2.5% in 1947 to 0.5% in 1990. The contrast with South Korea's trade performance was striking, and this was reflected in total industrial output over time. In 1970, India had produced approximately US$8 billion of manufactured products compared to Korea's US$2 billion. By 1989, Korea was exporting US$43 billion of manufactured exports compared to India's US$8 billion, and in 1988, Korea's total manufacturing output of US$54 billion was well above India's US$44 billion (Vaidyanathan, 1995: 35–36; Srinivasan, 1996: 212).

The effects of India's tight controls over industrial production and investment, over the entry of new firms into industry and the production of new products, and over the growth of existing enterprises as well as of foreign trade and foreign investment created an environment that discouraged technological change and quality improvement in both the public and private sectors. Internal markets were protected against any competition, and both public and private firms were under no pressure to improve either technologically or financially. The degree of labor protection contributed to overstaffing and low labor productivity, and discouraged labor-intensive production in the organized sector. This was a major factor behind the disappointing industrial employment and the lack of comparative advantage in labor-intensive production. It had been hoped that the public sector industrial firms would be profit-producing enterprises contributing significantly to public financial resources for further industry and infrastructure investment. In fact, however, these enterprises were drains on the government finances and the economy; rates of return were low, and many were consistent loss makers requiring large public subsidies, thus contributing to inflationary pressures in the economy. Another major problem associated with the system of controls was the "rent-seeking" practices of private and government businessmen, as firms sought to use the controls to enhance monopoly positions and profits. This contributed to the corruption believed to be associated with the required permissions; permissions were used to exclude competitors, and there was a growth of what was believed to be a large black-market economy. The corruption also had significant political effects in weakening support for the Congress Party and possibly contributing to the rise of nonsecular parties.

The longer-term consequence of this regime was a declining annual growth rate of industrial production after 1965. From 1951 to 1965, the overall rate (including mining, quarrying, and electric power production as well as manufacturing) had been 7.7%, but it fell to 4.0% in the next 15 years. Within that decline, the growth of manufacturing output per se fell even more rapidly, from 7.8% in the first period to 3.6% from 1966 to 1980. The experience of declining output and trade induced both criticism from sources outside the government and calls for reviews of performance and suggestions for reform within the government.

In the mid-1960s, the apparent decline in exports and resulting foreign exchange problems led to an initial devaluation of the rupee in 1966 and suggestions for reduced controls. This was politically very sensitive, contributing to a split in the Congress Party. Unfortunately, the problems of the world economy in the late 1960s did not allow exports to increase in the short run, despite the devaluation. The World Bank supported the devaluation of the rupee with a promise of substantially more aid, but it did not fulfill its commitment. The net effect was that the reform was discredited, and continued reforms were considered politically threatening. In the agricultural field, however, the policy reforms introduced by then Minister of Agriculture C. Subramanian contributed to the Green Revolution and the rapidly rising wheat and rice outputs of the next decade.

The low industrial growth rates of the 1970s led Indira Gandhi to set up various committees to review policies with respect to trade, controls, and the role of the public sector. These committees all recommended reform of those policies, and Indira Gandhi took some tentative first steps before her assassination. Rajiv Gandhi introduced a much broader range of internal industrial reforms to reduce domestic controls in the mid-1980s. As J. C. Sandesara summarized these in a recent paper, the government began first by calling a halt to increased state intervention and then reducing controls with respect to licensing, trade, prices, etc., as well as opening up of areas earlier reserved for the public sector to the private sector. The changes in controls included raising the limit of investment for licensing, delicensing a number of industries, broad banding (allowing expanded production of related items freely), streamlining of licensing procedure, raising investment limits for "monopolistic" large firms, and reducing some of the other monopoly controls on larger firms. With these reforms and an expansionary fiscal policy that stimulated demand, the annual growth rate of overall industry output rose to 7.7% in the 1980s while the manufacturing output growth doubled to a rate of 7.3% (R. Mohan, "Industrial Policy and Controls," in Jalan, 1992: 102–3).

These reforms were within the internal control system established following independence, and their extension to the foreign trade and investment areas was minor. Nevertheless, exports of manufactured goods more than doubled from US$5 billion in 1985/86, at the start of Rajiv Gandhi's government, to US$12 billion in 1989/90. Following the defeat of Rajiv Gandhi in 1989, the internal political instability that followed, the Gulf War, the decline in remittances, and higher oil prices, India's foreign exchange and trade positions worsened significantly and inflation accelerated. The World Bank was appealed to, new elections were held, the Congress Party once again won power following the June 1991 election, and the new prime minister, P. V. N. Rao, and finance minister, Manmohan Singh, accelerated reforms sharply and broadened their extent. Industrial licensing was largely abolished, except for 15 products, the number of industries reserved for the public sector was reduced to six, government approval for expansion or new undertakings by large firms was ended, and the approvals of foreign investment and foreign technology agreements were greatly eased. Reforms were also greatly expanded in the foreign trade area. These included rupee devaluation and full convertibility, abolition of import licensing and many import restrictions, tariff reductions both in products taxed and in rates, and reduced excise duties permitting lower prices.

There was a decline in the rate of industrial growth from 1991–93 as a result of a deflationary policy to reduce a 12% inflation rate in 1990–91. But after 1993, there was a major pickup, with growth rates from 1994–96 exceeding those of the 1980s. Within that growth, there was a significant shift toward the production of consumer goods, which had been given the lowest priority from 1950 to 1990. The large absolute number of middle and upper income Indians in both urban and rural areas (the figure ranges between 100–200 million) created a large potential demand for such products as TV sets, automobiles, bicycles, washing machines, and refrigerators, among others. With the easing of the restrictions on their production, both actual demand for them and their production increased significantly. As the output of final products increased, the output of the mining and electricity sectors, which produce the power for manufacturing, increased at about the same rate as manufacturing output. With the foreign trade reforms, exports rose from US$12 billion in 1989–90 to US$21 billion in 1994–95, and within the larger total, the exports of chemical products, machinery, and metal manufactures rose from only 1 billion dollars in 1985–86 to US$6 billion in 1994–95 (Government of India, 1996).

Despite this accelerated growth of the past decade, India still has far to go both in terms of its own economic goals and in comparison to many of its neighbors. The employment benefits of the greater industrial output remain limited; the economy remains largely agricultural in terms of employment, and while the incidence of poverty may have declined, it is still great in absolute numbers and as a proportion of the population. In terms of the 11 Asian countries that have introduced economic reforms since the early 1980s, when measured by growth rates in manufacturing, overall industry, and GDP, India ranked eighth in manufacturing and GDP, and sixth in industry as a whole. The People's Republic of China, of comparable size, had much higher growth rates; Korea, Thailand, and Pakistan had significantly higher rates (figures from J. C. Sandesara, "Industrial Reforms in India," unpublished paper).

The Congress Party lost the 1996 elections in part because of the sense of mounting corruption during its government and the belief that the benefits of the reforms had not spread widely enough in terms of improved well being. The replacement government represents a coalition of a large number of parties, national and regional, which may limit its ability to govern as well as its life. But there does seem to be general agreement as to the desirability of continued economic reform among the parties, although there are differences on the extent and areas of reform. Such questions as the future of the public sector enterprises, further opening of the economy to foreign trade and investment and the appropriate fields, and the continued subsidization of various sectors of the economy are unresolved. There is agreement as to the need for improved infrastructure in a wide variety of areas–power, road transport, telecommunications–but the financing of such investments and the relation to macro-policy have not been settled. But these issues are on the agenda, and some movement is probable in the next few years. India is a democracy, which is a significant element in its quality of life, but it also complicates problem-solving while making less likely the gross errors possible in a dictatorship.

Further Reading

Adams, J., "Reforming India's Economy in an Era of Global Change," in *Current History* 95, no. 600 (1996)

A good short summary of India's economic achievements and problems since independence.

Ahluwalia, Isher Judge, *Productivity and Growth in Indian Manufacturing,* Delhi and New York: Oxford University Press, 1991

One of the best analyses of India's industrial performance since independence.

Bardhan, Pranab K., *The Political Economy of Development in India,* Oxford and New York: Basil Blackwell, 1984

A major analysis of the political economy of India's economic policy making.

Bhagwati, Jagdish N., *India in Transition: Freeing the Economy,* Oxford and New York: Oxford University Press, 1993

A brief but excellent book on the economic reforms of the early 1990s.

Bhagwati, Jagdish N., and Padma Desai, *India: Planning for Industrialization: Industrialization and Trade Policies since 1951,* London and New York: Oxford University Press, 1970

This is the first book to point out the economic problems of India's planning and control system.

Chakravarty, Sukhamoy, *Selected Economic Writings, Part III: Development Strategies in India and Abroad,* Delhi, Oxford, New York: Oxford University Press, 1993

The collected essays on the topic of India by one of India's most distinguished economists, now deceased, who played a major role in making economic policy under Indira and Rajiv Gandhi.

Drèze, Jean, and Amartya Sen, *India, Economic Development, and Social Opportunities,* Delhi: Oxford University Press, 1995

A major review of the broader social consequences of India's economic development.

Government of India, *Economic Survey 1995–96,* New Delhi: Government of India, Ministry of Finance, 1996

The annual economic surveys published by the Government of India are the best source of official data and analysis for the economy and its problems.

Government of India, *Second Five-Year Plan,* New Delhi: Government of India, Planning Commission, 1956

The Second Five-Year Plan laid the foundation for India's development policies until the 1980s.

Jalan, Bimal, editor, *The Indian Economy: Problems and Prospects,* New Delhi and New York: Viking, 1992

A very good collection of essays on current economic issues.

Joshi, Vijay, and I. M. D. Little, *India's Economic Reforms, 1991-2001,* Oxford: Oxford University Press, 1996

An excellent overall review of the issues of economic reform in India in the 1990s.

Kumar, Dharma, editor, *Cambridge Economic History of India,* Cambridge and New York: Cambridge University Press, 1983

An encyclopedia review of Indian economic history. Volume 1 covers the pre-British period.

Kurien, C. T., *Global Capitalism and the Indian Economy,* London: Sangam, 1994

A very good review of economic and social issues arising form the reform process.

Lal, D., *The Hindu Equilibrium,* volume 1, *Cultural Stability and Economic Stagnation,* Oxford: Oxford University Press, 1988

One of the best economic histories of the Indian economy and its current problems.

Lewis, John Prior, *India's Political Economy: Governance and Reform,* Delhi: Oxford University Press, 1995

A study of Indian policy making, and especially foreign aid and development in India, by one of the most knowledgeable American political economists.

Little, I. M. D., "India's Economic Reforms, 1991–1995," in *Journal of Asian Economics* 7, no. 2 (1996)

An excellent brief summary of Indian economic reform achievements and issues of the 1990s.

Morris, Morris David, *Measuring the Changing Condition of the World's Poor: The Physical Quality of Life Index, 1960–1990,* Providence, R.I.: Thomas J. Watson Institute for International Studies, Brown University, 1996

A pathbreaking statistical review of world social development since 1960 and India's achievements in that area.

Rosen, George, *Contrasting Styles of Industrial Reform: China and India in the 1980s,* Chicago: University of Chicago Press, and Bombay: Voray, 1992

A review of India's industrial reform goals and achievements in the 1980s and a comparison with those of China.

Rudolph, Lloyd I., and Susanne Hoeber Rudolph, *In Pursuit of Lakshmi: The Political Economy of the Indian State,* Chicago: University of Chicago Press, 1987

One of the best analyses of the political economy of India's development since independence.

Sandesara, J. C., *Industrial Policy and Planning, 1947–91: Tendencies, Interpretations, and Issues,* New Delhi: Sage, 1992

One of the major reviews of India's industrial policies and achievements since independence.

Srinivasan, T. N., "Economic Liberalization and Economic Development," in *Journal of Asian Economics* 7, no. 2 (1996)

An excellent short summary of Indian economic policies and achievements since independence and of reforms of the 1990s.

Srinivasan, T. N., with contributions from J. Y. Lin and Y. W. Sung, *Agriculture and Trade in China and India,* San Francisco: ICS Press, 1994

A major review of India's agricultural policies and growth, and its trade policies and problems since independence and reform; these are also contrasted with China's achievements.

Tomlinson, B. R., *The Economy of Modern India 1860–1970,* Cambridge and New York: Cambridge University Press, 1993

One of the best of the economic histories of India from later British rule to the early decades of independence.

Vaidyanathan, A., *The Indian Economy: Crisis, Response, and Prospects,* New Delhi: Orient Longman, and London: Sangam, 1995

An excellent short review of India's current economic problems and policies, both overall and in the agricultural area.

George Rosen is emeritus professor of economics at the University of Illinois, Chicago. He served as chief economist of the Asian Development Bank, Manila, from 1966 to 1971, and he is currently the book review editor of *Economic Development and Cultural Change.* He has published numerous articles and books, including *Economic Development in Asia* (1996) and *Contrasting Styles of Industrial Reform: China and India in the 1980s* (1992).

Chapter Five

The Western Aid Community and Post-colonial India: On Development's Continuities with Colonialism

Daniel Klingensmith

[I]t would not be proper for you to say that you have obtained Home Rule if you have merely expelled the English.

–M. K. Gandhi, *Hind Swaraj* or *Indian Home Rule*

Two Critiques of Development in India

Over the past 15 years or so, the grand Nehruvian project of "developing" India, the project of restructuring and reforming Indian civil society by means of state-led interventions in economic and social relations, has faced increasingly sharp criticism from various sources inside and outside the country. The indictments most audible in the Western media and in the discipline of economics are centered around the question of whether or not the goal of development has been well served by the means adopted by the state after independence. Another strand of criticism, however, is based on a questioning of the fundamental character of the nationalist development project and the degree to which it actually lives up to its own claim to constitute a break with colonial domination and political economy. In this essay, I will delineate the basic contours of this second critique and discuss the emulation and appropriation of Western development strategies and models by India's nationalist movement and the postindependence state, and I will argue that while there is much that is useful in this second perspective, its claim that development in some senses is a continuation of liberal colonialism needs to be supplemented by a more careful consideration of the question of Indian and Western agency and collaboration.

Let me begin with a few words on what is sometimes called the "mainstream" debate in order to highlight the distinction between it and the different way of looking at development that I want to discuss in greater detail. During the 1990s, the neoliberal position on the early postindependence state's development strategies has come to be widely familiar, at least in its broad outlines. According to that position, the development strategy commonly attributed to Jawaharlal Nehru, P. C. Mahalanobis, and their colleagues was highly inefficient and therefore fatally flawed. It steered capital toward unproductive enterprises and raised high barriers to entry into and exit from the market. In other words, it disrupted the free flow of capital to enterprises that could produce the sorts of things India is thought (in retrospect) to have needed. In the planned economy, little importance was attached to "free trade" along the lines of comparative advantage; the emphasis instead was on the autarkic nurturing of a domestic industrial sector for the sake of economic independence. In the usual neoclassical formulation, this sector (composed of both large public and private undertakings) all too often produced heavy industrial and consumer goods that were more expensive or of lower quality than those that might otherwise have been bought from abroad. Furthermore, as the state occupied "the commanding heights of the economy," its functionaries assumed a great deal of power, enabling them to manipulate bureaucratic barriers to entry in ways that benefited them and their kind personally–a state of affairs Morarji Desai famously described as "permit-license raj." In short, the Nehruvian strategy allowed a too-cozy relationship between those already in the market and the state, to the detriment of those not included in this alliance. India (it is claimed) would perhaps have done better to have allowed more trade and competition with, and investment from, Western economies.

There is a counterargument in this debate, common in India if not among Western economists, that the semisocialist, planning-oriented state and its autarkic strategy, whether or not they continue to be suitable in an age of free-flowing, globalized private capital, were necessary for several decades to build the India that exists today: a country that, while poor, is militarily secure, is more or less self-sufficient in food, and has a substantial industrial base with a strong high-tech sector. If India becomes a major power in a global capitalist economy, as many of the disillusioned political heirs of Nehru and Mahalanobis hope it will, its success will arguably rest on foundations laid by a generation much more skeptical of the market and international free trade.

Both sides of this debate, especially the first, are comparatively well known to those in the West and in India who are interested in global trade and investment, public policy, or foreign affairs. What has received less attention is a self-consciously alternative critique that has taken form in the last 20 years. In brief, this alternative analysis has claimed that the Nehruvian approach, despite its insistence on autarky, was in fact based on too strong a connection to European modernity, and that it helped constitute the basis for a putatively independent nation's continued, quasi-colonial dependence on the West (*West* and *European* in this context being terms that include the United States and Soviet Union, the chief sources of aid and advice).

In this formulation, the modernist, and modernizing, project of development (statist or not) has privileged the Western foreign aid community's understandings and models of development, imposed Western-inspired projects on India's peasants, tribals, and urban poor, and more generally subjected the nation to the Western telos of "progress." Modern, more high-tech, and more expensive production technologies, with little relevance to the needs of the majority of the country's population, have effaced "traditional," lower-tech, more easily reproduced techniques of production, jeopardizing the livelihood of millions. The infrastructural projects required by and prized in the Nehruvian and market approaches (power plants, hydroelectric dams and their reservoirs, factories, etc.) have cost the nation dearly in terms of foreign exchange and ultimately domestic resources, as well as in opportunity costs. Moreover, they have displaced an enormous number of people, often with inadequate or no compensation: according to one estimate, as many as 25 million people had been uprooted by development projects by 1994 (Seema Paul, "Home Is Where the Project Is," in *Telegraph* [Calcutta], Sunday, 26 February 1995: 19). All too often, Claude Alvares insists, "'development' is a label for plunder and violence" (Alvares, 1992: 1), because it has required that people, and the land, forests and other resources they have used to sustain themselves, be arrogated, at times quite forcibly, to the purposes of the state, the national and international market, and the Westernized, urbanized classes that seek to make India a modern nation.

This alternative critique is sometimes labeled leftist, but it is not from the Marxist left, which historically has often shared many of the basic assumptions of both foreign advisers and state officials about the nature of progress and even about how to realize it. In India, at least as it has been published in English, this critique is rather from what might be called the neo-Gandhian left, especially from Indian social scientists who might originally have had a Marxist or liberal training but who have been influenced by Gandhi and by the failures of Marxism and liberalism in India. It has thus far had little currency in academic economics, though it is much more widely accepted (and embraced) by representatives of other social science disciplines, in the West as well as in India. Several such authors have approached the issue in ways suggested by the work of Michel Foucault and other poststructuralist, cultural analysts. Whether they have a direct debt to Foucaultian perspectives or not, what is common to almost all of these critics is an interest in development discourse as a subject of analysis as well as development practice. In other words, whereas both the defenders of the Nehruvian state and its neoliberal critics focus on the question of whether or not that state found an optimal application of available means toward desired ends and take the cultural and conceptual foundations of this attempt more or less as unproblematic, the various authors who have formulated this alternative critique are particularly concerned with how economic strategies emerged from the way development was understood, illustrated, talked about, and debated–from particular points of view, implying and implied by particular political positions.

It would be a mistake to claim that any one author in this emerging discourse is entirely representative of what is a great diversity of views. For present purposes, however, the work of Vandana Shiva on the Green Revolution is reasonably illustrative. In brief, Shiva argues that the technology of the Green Revolution in the Punjab has been highly inappropriate, socially, environmentally, and even economically. The advances it has brought in food production are illusory, she maintains, because they are based on grain varieties that require much-increased amounts of fertilizer, an unsustainably high amount of irrigation, and the neglect of a nutritionally and agriculturally better-balanced variety of crops in favor of maximized production of grain alone. Furthermore, after initial spurts in output, more and more fertilizer and water have been required to ensure the same production. The inputs required by high-yield varieties of grain, moreover, are relatively expensive and have to be purchased on the market rather than reproduced locally. Hence, the Green Revolution has favored cultivators who were already well-off enough to afford its inputs; smaller cultivators in the Punjab, according to Shiva, have found themselves less and less able to compete, and many have lost their land (Shiva, 1989).

According to Shiva, while Indian agricultural planners after 1947 initially sought to refurbish an indigenous agriculture that was both environmentally sustainable and avoided dependence on foreign technology and inputs,

> another vision of agricultural development was taking shape in American foundations and aid agencies. This vision was based not on cooperation with nature, but on its conquest. . . . It was based not on self-reliance, but on dependence. It was based not on diversity but uniformity. Advisors and experts came from America to shift India's agriculture research and agricultural policy from an indigenous and ecological model to an exogenous and high input one, finding, of course, partners in sections of the elite, because the new model suited their political priorities and interests. (Shiva, 1989: 29).

The Green Revolution is dealt with in detail elsewhere in this volume (see Chapter Eight), and in any case the point here is not the accuracy or validity of Shiva's critique but that it illustrates some themes common to a larger body of literature. Shiva represents an indigenous model of agriculture as "appropriate technology," technology more environmentally and socially sound. Shiva stresses that the model which was ultimately adopted was, significantly, of foreign provenance, promoted by foreign foundations, governments, and the World Bank for their own material and political ends. Finally, in her view, these foreign agencies were aided by the self-interested, uncritical participation of the postcolonial elite.

Similar critiques have been made of forestry policy in the Himalaya (Guha, 1989), the dairy cooperative program in Gujarat (Alvares, 1992), and smallpox vaccination programs (F. Apffel Marglin, in Marglin and Marglin, 1990), among others. In these instances, it should be noted, there is less concern with direct Western intervention, but all are concerned to point out that the conceptual sources of these programs are in Western (often British colonial or American) approaches to economics, reform, and nature. A common point of departure for all these studies is the assumption that development knowledge and models come hand in hand with particular power relations. An approach to economics implies, and is implied by, a politics, and the politics of postcolonial development, from this perspective, resemble the politics of the colonial period in at least one crucial respect: the dominant voices in setting policy agendas and in guiding the worldview of India's political elite are voices from outside, from the West, which sends advisers who (it is charged) legitimate anti-people policies. From this vantage point, the foreign advisers provided by the Ford Foundation, the Rockefeller Foundation, US Agency for International Development (USAID) and others (including, a decade or so later, the USSR) are seen as, in effect, the new missionaries in India, carrying forward the ongoing project of saving India by making it more like the West, carrying forward a colonial project that has outlived the Raj itself. Not all those who have contributed to this critique use the word *colonial* to describe development politics, but implicitly, and often explicitly, they maintain that the legal-diplomatic decolonization of 1947 has not brought "a decolonization of the mind" (S. Apffel Marglin, in Marglin and Marglin, 1990: 1).

In the rest of this essay, I address foreign influences on the Nehruvian development project from this perspective and consider in what sense this influence can be usefully described as colonial. I also attempt, however, to suggest a caveat in our usage of the word *colonial,* that in using the term, we ought not to edit out the agency and intellectual independence of the postcolonial elite that embraced modernist development.

Development's Foundations, the Nationalist State, and the West

In order to distinguish India's situation from that of some other postcolonial states, it ought to be noted that one specific mode of foreign domination has not particularly marked the political economy of India, and is not at issue here. Relatively speaking, India's economy and development since 1947 have not been "dependent," in the sense that Marxist dependency theorists have brought to debates on neocolonialism and relations between the "first" and "third" worlds. That is to say, unlike many former European formal and informal colonies, India has not experienced much of the dark side of integration into the global market. Its economy thus far has been largely independent of the advanced capitalist economies of Europe and America: its upturns and downturns are not purely a function of booms and busts in the developed world, and it is not dependent on exports of raw materials or cash crops to more industrialized countries in exchange for finished consumer goods and capital. While it has had to import industrial capital and technical advice, India has become progressively more able to produce both heavy equipment and technical cadres itself. In short, India's political economy is not "colonial" in the sense that still describes some African political economies. It has not been set up by foreign advisers, its most important sectors are not linked directly to the interests of foreign capital, and it is not bolstered by occasional foreign military and political intervention.

The reasons for this independence are not hard to find. Unlike many African former colonies, at independence, India was geographically large and fairly well endowed with important natural resources. It already had a substantial infrastructural and industrial base and a fairly sophisticated technical and scientific sector, at least compared to much of the colonized world. It was governed by a stable, well-defined, and professionalized bureaucracy. It was, obviously, too large to be easily pressured militarily. The state was able to increase and capitalize on these assets to ensure the country's economic as well as its political independence (Rudolph and Rudolph, 1987; Tomlinson, 1993). In addition, although the government needed foreign technical specialists after independence, for the most part there was little need for a substantial foreign managerial presence. Immediately after 1947, there remained a substantial British presence in public administration and private firms, but this presence declined precipitously in the state, and rapidly even in the private sector (Tomlinson, 1993: 144). Furthermore, while economic independence was explicitly the goal of the independent state, even the government of India under the Raj had been forced, by the long-term contradictions of its situation in the country and in the British Empire, to insulate India partially from both the British and the world economies. Moreover, due in part to the enormous strains placed on the economy by the Second World War, the Raj had taken the first steps toward a planned economy for India, an economy partly shielded from the play of even internal market forces (Tomlinson, 1993). Thus, the nationalists' approach to economic management had already been prefigured, to a limited extent, by late colonial policies. This approach was also very much in harmony with mainstream, midcentury European and American economic thinking.

The new state's approach to India's economy was geared toward maintaining national autonomy and, somewhat less directly, to increasing the standard of living and the opportunities available to citizens. The basic outlines of the Congress government's policies had been formulated in the 1930s and 1940s, partly by Nehru and the Congress "left," and partly by others. There was broad agreement among the intellectual and political pioneers of economic management in India about what the state's role in the economy ought to be. What Sir M. Visvesvaraya (divan of Mysore and author of an early plea, *Planned Economy for India*), Sir Ardeshir Dalal (the first Member for Planning on the Viceroy's Council in the 1940s), Meghnad Saha (editor of *Science and Culture* and a strong proponent of Soviet-style economic planning), the Congress's National Planning Committee of 1938–40 (on which Nehru and Saha served), and the Indian industrialists behind the "Bombay Plan" of 1943 all had in common was agreement on three important general points. First, the development of India could not be left to the forces of the market, which were thought to have produced India's poverty: the state must take the lead in planning a rationalized, efficient economy that could provide prosperity to all Indians (and, implicitly, thereby eliminate the worst abuses of Indian society). Second, the state must as a corollary be prepared to protect national industries from harmful foreign competition and to intervene to create infrastructure and resources, much of which would be beyond the capabilities of Indian private industry. Third, the key to the development planned by the state was industrialization, with a special privileging of heavy industry and large-scale infrastructure: this would make possible both the defense of independent India and later increases in consumption.

Although at that point, there was little in the way of pro–free market protest from organized, private capital, there was already a strong critique of this economic management approach from Gandhi and his followers. Gandhi's criticism of what he called "machine civilization" is complex, but it is worth some attention here: as I have noted, it has informed the alternative critique of development substantially. Abstracted from its context, it might even appear to have been an indigenous economic philosophy, circumvented by outside influences. In a nutshell, Gandhi (who was himself much influenced by nineteenth-century Western critics of industrialism) charged that under modern industrial conditions, machines and capital ended up dominating people, forcing them to conform to the logic of industrialized production rather than vice versa, and that industrialization necessarily meant the adoption of labor-saving machinery, which was inappropriate in an economy with such a pronounced labor surplus already (Gandhi, *Indian Self-Rule,* Ahmadabad: Navajivan Press, 1939; *Gandhi in India in His Own Words,* edited by Martin Green, Hanover and London: University Press of New England/Tufts University, 1987). Gandhi's alternative was of course famously symbolized by *khadi* (homespun cloth) and the *charka* (the hand-powered spinning wheel). These were meant to emphasize development in a different direction, toward small-scale, low-technology, geographically dispersed industry as a means of ensuring maximum employment, the self-sufficiency of both individuals and villages, and a low but socially, morally, and environmentally sustainable standard of living. This has sometimes been called "Gandhian economics."

In the Gandhian view, for modern industrial civilization to function, people would necessarily have to be greedy rather than self-restrained and would have to seek to gratify an endless series of desires rather than mastering desire and remaining content with sufficiency. This was a recognition, not unlike that of Keynes, of the need for increased consumption to counter the global economic depression, but Gandhi refused to acknowledge that, even in a short or middle term, the economic principle that "fair is foul and foul is fair" (in Keynes's famous words) could be socially or morally anything other than corrosive. Alternatively, Gandhi's critique of industrial capitalism can be compared to that of Marx, but unlike his lieutenant and political heir, Jawaharlal Nehru, he lacked Marx's faith that complex technology could someday be socialized, allowing people to work only as much as they wanted. India could never be truly free, genuinely and literally self-governing, according to Gandhi, until its citizens cultivated personal as well as political self-government. If it sought development toward material wealth, along Western and Soviet lines (toward a state of affairs that he believed to be unhealthy even for Western and Soviet peoples), India would always be striving after a material standard inappropriate to it. The state would always be demanding sacrifices from India's poor and rural population so that the middle and upper classes of the cities could ape Western norms, live in Western-style comfort, continue to rely in a direct or indirect sense on the West. "Resisting" a pernicious Western political and economic domination by, in effect, reproducing it in a nationalized form was hardly an improvement: industrial civilization was pernicious for reasons beyond its foreign origin.

However much Nehru, his colleagues in the first government, and other nationalists inside and outside the Congress may have respected Gandhi, it generally wasn't for his economics. Most of those involved in laying the foundations of planning and development in India saw Indian self-rule's long-term survival as a function of India's ability to defend itself and to become modern like the West while keeping itself insulated from Western material domination. They, too, wanted to end the colonial economic domination of India, but not along the antimodernist path Gandhi proposed. In the modernist-nationalist view, India remained "backward" (i.e., it had not made adequate "progress" toward modernity) precisely because colonialism was a flawed vehicle for modernity, whereas Gandhi implicitly argued that colonialism's modernizing project had already been all too successful. The self-conception and justification of the modernist majority of India's nationalist elite was heavily dependent on the notion that they could and would provide the benefits of modernity to India's backward poor that the British, for all their talk of progress, had failed to bring (Chatterjee, 1993). The state and the central classes that staffed it were committed to improving the lives of India's poor, who were assumed to be wretched and ignorant in large part due to a lack of basic amenities and infrastructure (so wretched and ignorant, in fact, that they could have only a limited role in defining the specifics of their own transformation). More, not less, industrial modernity was needed according to this view.

In all of this, India's secularizing, modernist elite was by no means unique. The thinking of these nationalist technocrats-in-waiting closely parallels that of contemporary Soviet, Western European, and American constituencies for planning. For these, of course, the immediate issue was not the failure of colonialism to bring economic progress but rather the failure of the free market to bring progress and stability. In as much as many left-leaning liberal observers in the West believed that the failures of the free market had generated Europe's expansionist tendencies, and hence its colonialism, while at the same time undermining its domestic social stability, an intellectual space opened in the 1930s for a consensus between Western liberal democrats and African and Asian anticolonial nationalists on the question of the state's relationship to capitalism. Liberal internationalists increasingly agreed that a rationalized, state-planned or state-regulated economy was the antidote to social instability, international conflict, and colonialism, all of which were generated by a combination of technological backwardness and the irrationality and chaos of the free market. By 1945, two years before India's independence, this consensus had become deeply influential in the West, India, and elsewhere. This was, after all, a generation that was living through a prolonged, violent, and global economic and political crisis, a crisis in which the conceptual foundations of the international market and colonialism both appeared increasingly untenable. "Development," as an intellectual field and policy problem, was born of this consensus, which profoundly influenced it until very recently.

There are two points of special importance here. First, it should be noticed that, for all their invocation of Soviet planning techniques, Nehru, Mahalanobis, and their colleagues were just as much, if not more, in dialogue with liberal Western European and American proponents of planning and state-led intervention. With these they shared the basic assumption that a rational, apolitical, and authoritative planning state was, on the one hand, viable, and on the other, readily compatible with a liberal, democratic polity and even a good deal of private enterprise. One can find Indian planning's conceptual sources as much in British Fabian Socialism and the American New Deal as in a watered-down, lukewarm Stalinism. Second, it would be misleading to say that the political-economic approach of the Indian government after independence was "imposed" or directly upheld from outside. The new state's economic visionaries had known more or less where they wanted to take India since at least the early 1940s. Their views had undoubtedly been formed in dialogue with Western sources, but the hegemony of these views had emerged in the contexts of the elite-led nationalist movement and the larger crises of the mid-century world rather than as a result of the crude play of economic exigencies or political interventions from outside.

This is not to say that foreign advisers and capital did not play an important role in postcolonial India, especially in the first two decades after 1947. The Ford Foundation and the World Bank, which in different ways were centrally important to the foreign aid and development community that sprang up in Delhi in the 1950s, were of course foreign institutions, the first being an American foundation, the second an international agency over which the United States exercised a dominant influence. Not coincidentally, both of these, and other, American-influenced institutions in India, were disproportionately staffed by former New Dealers, the sort of American liberals most readily in sympathy with the state-led development mission of the Indian state. Moreover, especially in the late 1940s and early 1950s, a number of prominent former New Dealers trooped to India to take development-oriented tours, dispense advice, and report back to the

American public: David Lilienthal, Eleanor Roosevelt, William O. Douglas, and others inspected and advised the new state's efforts in these years and then wrote positive accounts of "awakening" India for American readers and policy makers. They supplemented a small but important American community with longer-term commitments to the country, among whom some of the most notable and successful (again, New Dealers) were Douglas Ensminger, the Ford Foundation's chief representative, Albert Mayer, designer of the Community Development Programme, and of course Chester Bowles and John Kenneth Galbraith, the most effective American ambassadors of the 1950s and 1960s. There were, of course, development advisers from other countries as well; after 1955, an increasing number of them were from the Soviet Union. These advisers were important in setting up projects of all kinds and occasionally (especially in the case of engineers at India's dam and factory projects) manning them in their first few years. The Ford Foundation alone was involved in, among other things, establishing institutions to train Indian economists, urban planning for Calcutta, the formulation of agricultural policy and institutions both before and during the Green Revolution, and in Mayer's Community Development Programme, one of the state's most important early ventures (Rosen, 1985).

These advisers, and the institutions they represented, were used by the Indian political establishment to provide both technical means toward, and legitimation of, ends that had already been determined in outline. The circumspection with which Ensminger and the Ford Foundation had to proceed is telling: as George Rosen has noted, Ensminger owed his influence to his ability to foresee which of the prospective development schemes and economic insights offered by foundation representatives would be acceptable, in the context of Indian politics, and to his tactful awareness of nationalist ambivalence about the presence of foreign advisers (Rosen, 1985: 53–56). Western advisers did not and could not dictate programs and strategies to nationalist Indian leaders.

All this is to characterize more precisely the connection between the Western aid community and the Indian state. I have tried to show here that the hegemonic understanding of development in postindependence India is by no means a simple function of "dependency" in the political and economic sense that has marked many other former colonies and Third World colonies. It would be equally inaccurate to suggest that the Indian nationalist movement had formulated a philosophy on economic issues, Gandhian or otherwise, which was later circumvented by Western capital or aid agencies. The modernist, statist approach of the Western-educated elite that dominated the development thinking was articulated before independence in the larger context of an internationalist liberalism in crisis. This approach was paramount in nationalist thinking before 1947 and dominated policy without serious challenge afterward. While Western advisers were important in translating it into practical steps, Nehru and his colleagues on the whole were already in agreement about where they wanted to go.

Much of this is readily admitted by those critics of development who see it as a reformulation of colonialism. At their best, however, the great insight these critics offer is an insistence that the truly important issue is not finally structural or political dependence, not the literal dictation of India's political-economic course by the West, but rather a fundamental conceptual and cultural dependence on Western political-economic thought, norms, and models, an internalization of Western discourses of economics and politics rather than domination by Western pressure and interests. In this view, the postcolonial state's enthusiasm for building a modern economy and society along Western lines represents the ultimate triumph of one particular strand of colonialism: the liberal colonial civilizing mission of redeeming "backward, lost" India by making it over in the image of European modernity. Building on Gandhi's criticism of mainstream, nationalist economics, many theorists implicitly see development as the culmination of liberal colonialism because it is a modernizing project conceived by "colonized minds," by Indian modernists who affirm that project's validity independent of Western prompting and who have established a state dedicated to equaling or even someday outstripping their former colonial masters in the march of progress. Western cultural and intellectual hegemony continues to define the standards of that progress, although in the modernist imagination (in the West as in India), these standards are claimed to be universally valid and "Western" only in terms of historical origin.

This point of view has been described as an "external" critique (by Tariq Banuri, "Development and the Politics of Knowledge," in Marglin and Marglin, 1990), as opposed to "internal" critiques, neoliberal and otherwise, that accepts the basic assumptions of development, while focusing on the question of the adequacy of particular means toward that end. Both points of view proceed, obviously, from implicit ethical and normative judgments on the desirability and moral feasibility of political modernism. The external critique I have described is compelling in many ways, but it makes an unwarranted assumption about the intellectual passivity of Indian modernists. In the next section, I illustrate this point by briefly discussing just how America exported and India emulated one of the first widely-copied specific models for development: the New Deal's federally-legislated but bureaucratically autonomous TVA (Tennessee Valley Authority), especially in the building of "India's TVA," the DVC (Damodar Valley Corporation) in Bihar and West Bengal, established in 1948, a few months after independence.

Building "India's TVA"

In principle, one could illustrate many different themes in the history of development by recourse to one or another of a great variety of different strategies, models, programs, and projects that were tried or considered. What I want to highlight here is that the question of conceptual dependence, of India's alleged attempt to copy blindly or to mimic Western, especially American, approaches to development, was never quite so simple as that. We need to question more carefully both the Western presentation of development models, supposedly based on the West's historical experience, and Indian acceptance of these models. The case of India's emulation of TVA, at the DVC and elsewhere, is interesting precisely because it is marked by details that do not quite fit the conventional descriptions of what this relationship of tutelage was about as given by American advisers, the Indian development community, or anti-development critics.

There are additional reasons to consider this particular example, beyond the fact that it suggests the complexities underlying tutelage and emulation. The new state spent an enormous amount on "river valley development" after independence, in particular on building massive dams–as much as 15% of all national expenditures up to the early 1980s, according to one estimate (Ganguly Thukral, 1992: 9). The new dam projects–the DVC, Bhakra-Nangal, Hirakud, and later many more–were the most visible prestige projects of the Nehruvian state, its "temples of the new age," as Nehru described them on more than one occasion. Exactly because they were once so central in the state's physical and political construction of modernity, they are now in turn focal points in the critique of modernism in India. This is especially true of the multi-dam plan for the Narmada river, the most recent TVA-descendent in India. In the decade after independence, however, such projects were the pride of the state and were ceaselessly compared to TVA's dams. The TVA was the first development model to be well known (and even popular) among India's colonially trained professional classes. It became an early standard for the state's attempt to transform society by transforming and harnessing nature for the purposes of heavy industrialization.

At first glance, the Indian emulation of TVA might appear as an instance in which an approach was directly abstracted from the industrialized West, possibly even for the purpose of reordering and remaking the Third World to fit the needs of Western capital. Indeed, TVA was an influential paradigm for natural resource development in the 1950s, and many other "new nations" in addition to India hired TVA personnel to recreate in their countries the miracle supposedly wrought by the New Deal in the American South. In several cases, the end products of these emulative schemes did indeed work to tie Third World polities and economies, not only to Western sources of legitimation, but to Western economies as well, in ways suggested by dependency theory. The Volta river project in Ghana, for example, chiefly aided the interests of the Kaiser Aluminum Corporation, which needed power to smelt aluminum, but it did not provide enough extra power to enable significant, autonomous industrialization (Austen, 1986: 245) of the sort TVA was thought to have made possible in the South. Nevertheless, the DVC and other Indian projects cannot easily be described as having reshaped India to meet the needs of foreign capital. The power and irrigation water they supplied were a central part of the government's strategy of development in order to avoid dependency.

The replication of an American institution in India and elsewhere, however, in itself recalls colonialist precedents of a different kind. One highly important strand of British colonialism, from at least the time of Macauley's famous educational memorandum of the 1830s onward, was the attempt to recreate the institutional forms and cultural norms of emergent European modernity in India, to thereby "redeem" India by creating a class of "brown Englishmen" committed to Enlightenment notions of progress and their institutional manifestations. Whatever else this liberal colonialism may have accomplished, one can argue that the very idea of a progressive, modern Europe was to some extent dependent on the construction of its opposite, a backward, stagnant, and traditional realm, located in the "Orient," among other places (Edward Said, *Orientalism*, New York: Vintage Books, 1978). From this perspective, the reformist, educational, and infrastructural projects of the colonial era in India helped to constitute and validate the self-conceptualization of post-Enlightenment Europe as modern. In short, European institutions and technology in colonial Asia and Africa made an ideological statement as well as serving an economic function. This process did not end with decolonization. In this dimension, as in others, the postwar United States can be seen as carrying on some of the legacies of the European colonial era in its relationships with former European colonies.

Of course, the liberal colonial venture demanded the presence of European guides (teachers, doctors, engineers, missionaries, and others). Progress was both a moral commitment and a career–not only for Europeans in the colonies but for a later generation of Americans who took their places after decolonization. The TVA employees who went to India had obvious material stakes in their new jobs, but ideological and cultural stakes in them as well. William Voorduin, dispatched by TVA in 1944 in response to a request from the Government of India for advice in dealing with floods on the Damodar, planned a new agency that strongly resembled TVA and thereby reaffirmed the understandings of development that were central to TVA's conception and presentation of itself as an institution with potentially global significance. The DVC's first chief engineer, a former TVA man named Andrew Komora, considered himself a teacher and saw it as his difficult mission to train Indians in the use of modern techniques, to overcome the appalling backwardness of "the 'head-basket' idea" (i.e., the reliance on unskilled manual labor to build large projects). Komora designed a highly mechanized construction process for the DVC that avoided as much as possible using the abundant labor supply, demanding instead the importation of expensive, high-maintenance heavy equipment. This professedly value-neutral technical decision resulted in higher construction costs, but for Komora the idea of using "traditional" methods to construct a modern project made little sense. Development was as much an assertion of cultural and moral norms as a calculated, instrumental project.

It would be inaccurate to say, however, that Voorduin, Komora, and others merely imposed a simple blueprint, appropriate or not, on a passive Indian government and development community. Both American engineers and development publicists, and their Indian interlocutors, quite selectively appropriated certain themes and strategies from the different possibilities that TVA had to offer. The dominant, widely known image of TVA in the 1940s and 1950s was a representation of a more complex reality. Many of the early Indian development publicists who discovered the New Deal as a model for the democratic planning state interpreted that legacy in ways that distorted American history but served particular interests in India.

In the United States, the proper functions of the TVA had been hotly contested during the late 1930s. The details of this history are too complicated to recite here; suffice it to say that David Lilienthal, the TVA director who wanted the agency to be oriented toward increasing market opportunities, industrialization, and consumption, prevailed over Arthur Morgan, his rival at the agency who was interested, among other things, in using TVA to set up an economy based on artisanal industry and even limited local autarky. Lilienthal went on to write *TVA: Democracy on the March* (New York: Harper, 1944), a best-seller that made his agency world famous as a model for post-war reconstruction and peaceful, scientific, and above all democratic decentralized development. This book was a representation, and an over-simplification, of a complicated and perhaps less attractive economic and political history, but it was Lilienthal's version that Indian politicians, planners, and engineers embraced: Nehru and the nation's top engineers expressed their admiration for the book and made pilgrimages to Tennessee to learn more about the TVA's achievements. In later years, both Morgan and Lilienthal toured India, dispensing development advice. But while Lilienthal's three week visit was superficial compared to Morgan's year of extensive observation, Lilienthal's opinions received considerably more attention. Morgan was popular only with the marginalized Gandhians and much less so with mainstream planners, who were unimpressed with his critique of heavy industrialization. Lilienthal's vision was much closer to their own priorities and hopes. As a result, not all foreign advisers were given equal attention.

Even before Lilienthal had made the TVA famous, Meghnad Saha, an Indian physicist and development crusader, had been campaigning for a "TVA for the Damodar." Saha had written extensively on issues such as planning, flood control, and power production before discovering TVA in the early 1940s. As the key member of the committee of inquiry into the Damodar River's 1943 flood, he persuaded the government to take the first steps toward what would eventually become the DVC. Saha published several articles and editorials on the TVA in his laymen's journal, *Science and Culture,* partly based on Lilienthal's book and partly on his own peculiar ideas of what the New Deal was about. The important point is that Saha, who more than any other single person popularized the TVA among India's English-reading public and created a constituency for the DVC, imagined this technology in terms different from those used by either Lilienthal or Morgan.

Saha outdid Lilienthal in arguing that the real achievement of the TVA had been heavy industrialization. Moreover, he was almost completely uninterested in Lilienthal's rhetorical efforts to prove that TVA could reconcile scientific authority with popular democracy. To Saha, TVA–and the New Deal more generally–were worthy of India's attention precisely because they appeared as the triumph of a rationalist, scientific, apolitical and even authoritarian state over all petty local concerns and vested interests (Saha, 1986). Saha's understanding of Roosevelt and the New Deal as apolitical and purely technocratic may be startling to American readers. It makes sense, however, in the context of Saha's peculiar account of the TVA and its relationship to India: geographically and hydrologically, the Damodar was represented as nearly identical to the Tennessee (which it does not much resemble). By extension, India was made to look, potentially, like America–in his view, a land of science and progress, a nation that had arrived at modernity (Saha, volume 4, 1994). India, too, would arrive there if it would commit itself to scientific and industrial development ventures, in this particular case by copying what Saha interpreted as the New Deal's scientific approach to transforming and harnessing nature. Saha constructed an image of America, and of an institution that he thought represented its ethos, for his own ideological purposes.

The DVC that eventually resulted from Saha's public campaign, an institution supposedly copied from the TVA, as both agencies have often insisted, was in fact politically, administratively, and technically a very different affair. (This is quite apart from the question of whether any of the understandings of what the TVA was about were actually "appropriate" to India or to the Damodar Valley and its flood problem.) Both the Indians and Americans involved in the project throughout the 1940s and 1950s insisted, however, that this was a simple project of tutelage, of American teaching and Indian learning. That representation elided the substantial differences between the way each group (and each institution) functioned in its own political environment. It is true that Saha, Nehru, and other backers of the project took practical and theoretical advice from the TVA, the leaders of which were certainly eager to have their unique organization copied internationally (or at least to be able to tell the US Congress and public that it was being copied). But they were selective in what they took, reshaping and reinterpreting the American model for their own political purposes. One could argue that the TVA, as often as it was cited in Indian development circles of the 1940s and 1950s, was more a signifier in a discourse of development than an actual economic and technical blueprint. Ironically, Sudhir Sen, the DVC's first general manager and the official who probably took the TVA most seriously as a blueprint, left it in frustration in 1954.

Conclusion

What does this particular example suggest about colonialist reverberations in development discourse and practice? Is there some way in which the Nehruvian development project may be reasonably seen as an extension of certain facets of colonialism? Development in India has not exactly been "neocolonial" in the usual sense of the term, connoting the obvious forms of economic and political domination over formally independent states, but is it "postcolonial" (to use a word that has been variously employed, but generally to emphasize in particular the cultural, conceptual legacies of colonialism).

The answers to these questions rest to some extent on normative judgments. Whether or not one sees a connection between colonialism and the course of development in India since 1947 depends in part on one's approach to the legacies of the eighteenth-century Enlightenment, legacies that profoundly shape our ideas about economics and other social sciences, about the purely utilitarian character of technology, about the political neutrality, rationality, and self-evidence of "progress." If one agrees with Nehru (and Roosevelt, and any number of other liberal internationalist, or alternatively Marxist, statesmen and development pioneers of the mid-twentieth century) that technological development along modern lines is obviously beneficial, culturally neutral (except where it actually helps to eliminate "pathologies"), and universally desirable, then it can be difficult to believe that a politically independent, science-ordered state is not genuinely free, as long as other nations have not found ways to interfere politically or economically. To modernist nationalists in India, British "colonial modernity" may have been a perverted modernity, but it was at least on the right track, primarily requiring only reform via its "nationalization." From this point of view, there is nothing incongruous about India's looking West for technical inspiration after independence.

If, on the other hand, one agrees with Gandhi (and others who came to similar conclusions from different starting points) that what is called progress, in the Enlightenment sense, is not a self-evidently desirable thing, that it is not neutral but caught up in politics, that it makes possible new and more powerful forms of domination, and that these possibilities were fundamental to colonialism's effect on India's economy and culture, then the fact that independence means that the development project has been taken out of British hands and put into Indian hands may not really make much difference. To Gandhi, what made colonialism in India illegitimate at the deepest level was not the fact that it meant the exercise of power by Britons instead of Indians, but that it meant the exercise of kinds of power that were different from (and worse than) what had come before (Gandhi, 1939). Decolonization was of limited importance, then, because the transfer of power was not, after all, a renunciation of modern forms of power. In this sense, the development-oriented regimes that took over from colonial bureaucracies in many formerly colonized lands can be seen as having directly inherited the mission of liberal colonialism–even though that mission has of course been reformulated to suit the sensitivities of a world organized around formally independent, formally equal nation-states.

One cannot fully resolve the difference between these two different ways of seeing things. What I want to suggest, instead, is that while I believe we can meaningfully speak of a continued hegemony of ideas that originally came to India as an aspect of British imperialism–in that sense, it is accurate to think of development as continuing certain aspects of colonialism–we ought to be wary of speaking of development as symptomatic of "colonized consciousness," insofar as that implies a lack of intellectual agency on the part of those whose consciousness is supposedly colonized. Let me elaborate on this by pursuing some of the homologies and historical relationships between liberal colonialism and the transnational project of development that succeeded it.

The British colonial presence in India was never only about trade, investment, or the extraction of wealth. Britons and other Europeans in India pursued many different projects, some economic, some educational or religious, some purely personal. It is indeed impossible to characterize that presence in terms of any coherent master project, economic or cultural. Colonial economics, extractive or otherwise, was situated in a larger cultural and political context. The same can be said of development economics. Both before and after 1947, a frequent means of describing and legitimating economic interventions–whether infrastructural (like the DVC), industrial, agricultural, or transaction related–was by reference to their role in bringing secular progress and in permanent, this-worldly improvement. The prominence of this idea is comparatively recent, even in Western history.

In both cases, the mission of "improvement" empowered Westerners and certain Indians–those trained in colonial or postcolonial institutions. They were the ones who supervised the construction and management of the transformational institutions, the elaboration of the normalizing processes, that were someday supposed to result in a nation of fully modern Indian citizens. Progress, before and after independence, was thought to require a certain amount of trusteeship (or, if you prefer, authoritarianism), exercised by the modern and educated over the backward and uneducated, who might not (yet) fully know what was in their best interest.

Colonialism is generally used to describe a particular kind of domination of one culture over another, one political system over another, one economy over another. In notable ways, the domination I have described has colonial features. Both before and after 1947, there was (and continues to be) a sense in which Western institutions, norms, and people had authority in India, whether this authority had a direct political dimension or not. This was an authority that did not readily flow in the opposite direction. That is, Nehru, Mahalanobis, and their colleagues were in dialogue with sympathetic Westerners: they cannot be said to have had the fundamentals of their approach actively imposed upon them. However, although they had been trained according to modernist norms in colonial institutions, as putatively "new recruits" to modernity, they had much less direct influence on American, British, or Soviet economic and political thinkers than these had on them. Indians looked to the West, especially to the United States and the Soviet Union, for models and strategies of development, a venture that itself was heavily imbued with post-Enlightenment understandings of what constituted progress, freedom, and justice. As I have observed, it does make sense to speak of Western intellectual hegemony in Indian development practice.

The particular example of India's emulation of the TVA suggests an important caveat, however. Both Americans and Indians claimed that this was an example of Indians learning a modern technique of managing nature, one that happened to have been pioneered in America but that was universally valid. We have seen something of the particular American ideological interests in training Indians. We have also seen, however, that this relationship was something different from straightforward American "teaching" and Indian "learning." On the one hand, the TVA employees who educated Indian pupils abstracted their model of development from a much more complex history. Theirs was an idealized representation, one that brought a messier history in line with a more acceptable political morality, a more acceptable understanding of themselves. On the other hand, the Indians who pressed for TVA-like projects, on the Damodar and elsewhere, in turn did so on the basis of their own representations of what the TVA was and what it had done. These ideas were based partly on a very selective reading of earlier American representations and partly on their own abstraction of the "relevant" features of the New Deal and American history.

As the organizing principle of the Indian state and its economic efforts, development continued the liberal colonial project of transformation and improvement. This project of "emulation" was complex, however; it was marked by the making (and sometimes also breaking) of ideological alliances between Indian and Western political actors with different agendas and concerns. Indian development policy after 1947 was by no means a blank slate on which Western norms were written. Perhaps India's development-oriented elites were intellectually part of the modern, Western world, but they also had to make claims to and even perform for one another and for other domestic audiences. Their particular political and cultural position made any simple, blind emulation impossible. Indeed, one might say the same of emulative projects, personal and social, under colonialism more generally. *Colonialism* as an explanatory concept in general needs to encompass a greater awareness of agency on the part of the colonized.

These reflections suggest a final question. There has been some criticism of the move toward economic liberalization in India, both from those who are uncomfortable with economic modernism and from others who are comfortable with it but who do not trust privatization to make up for state socialism's defects. On both sides, the accusation that liberalization is "colonial," or at least risks subjecting India to neocolonial domination, is occasionally made. These positions are of course related to other, possibly well-founded fears about the globalization of capital and its effect on national autonomy. Whether or not "colonial" is indeed the right word to characterize liberalization policy, in any case one can wonder if there might be something more to liberalization than meets the eye, on the basis of the history of earlier development policy. If we acknowledge that there was some elision of historical reality in both American and Indian representations of the New Deal as a scientific, development-oriented state–if we acknowledge more generally that development models have been abstracted from historically messy contexts and cleaned up somewhat to serve ideological and political interests, then perhaps we have to question whether what goes by the name of liberalization in Indian political rhetoric is quite the same thing as what is meant by it in its American or European usages and whether any of these usages themselves adequately reflect the history of private capital in the West. Economic programs, like any other programs of the state or of civil society, are based on representations of a reality that indeed can only be apprehended and communicated by means of representations. And at the back of every representation–every model or critique of politics, economics, or anything else–is a point of view that both limits perception and empowers actors.

Further Reading

Alvares, Claude Alphonso, *Science, Development, and Violence: The Revolt against Modernity,* Delhi and Oxford: Oxford University Press, 1992

A polemical book but one with a great deal of information on hidden aspects of development projects in India.

Brass, Paul R., *The Politics of India since Independence,* Cambridge: Cambridge University Press, 1990

Provides a useful critical survey of Indian political economy since 1947 and explicitly addresses important scholarly debates on the subject.

Chakravarty, Sukhamoy, *Development Planning: The Indian Experience,* Oxford and New York: Oxford University Press, 1987

A good, readable overview from the point of view of a professional economist and onetime member of the Planning Commission.

Chatterjee, Partha, *Nationalist Thought and the Colonial World: A Derivative Discourse?* London: Zed Books for the United Nations University, and Minneapolis, Minn.: University of Minnesota Press, 1986

An interesting discussion of the relationship between development and nationalism.

Dos Santos, Theoantonio, "The Structure of Dependence," in *American Economic Review* 60, no. 2 (1970)

A terse statement of dependency theory.

Escobar, Arturo, "Discourse and Power in Development: Michel Foucault and the Relevance of His Work to the Third World," in *Alternatives* 10 (Winter 1984–85).

Discusses the Foucaultian understanding of the relationship between power and knowledge.

Fuchs, Martin, editor, *India and Modernity: Decentering Western Perspectives,* a special issue of *Thesis Eleven* 39 (1994)

A collection of valuable essays, particularly Ashis Nandy's "Culture, Voice, and Development: A Primer for the Unsuspecting" and Sudhir Chandra's "'The Language of Modern Ideas': Reflections on an Ethnological Parable." The second half of Chandra's article examines an important and telling 1945 exchange of letters between Gandhi and Nehru on the future course of development in India.

Ganguly Thukral, Enakshi, editor, *Big Dams, Displaced People: Rivers of Sorrow, Rivers of Change,* New Delhi and Newbury Park, Calif.: Sage, 1992

Criticizes water-resource development policy, mainly after independence.

Guha, Ramachandra, *The Unquiet Woods: Ecological Change and Peasant Resistance in the Himalaya,* Delhi and New York: Oxford University Press, 1989

Discusses the connections between colonial and postindependence forestry development policy.

Kothari, Rajni, *Growing Amnesia: An Essay on Poverty and the Human Consciousness,* New Delhi and New York: Viking, 1993

All of Kothari's works are useful; he has played an important part in the elaboration of the "alternative" critique. This one is deliberately less theoretical. It is useful on the unacknowledged commonalities and connections between state socialist Nehruvian development and the liberalization policies of recent years. Kothari is less sanguine than others about the Gandhian rural utopianism and anti-industrialism that informs many other recent Indian criticisms of development, and he is also perhaps more obviously interested in practical alternatives.

——, *Rethinking Development: In Search of Humane Alternatives,* Delhi: Ajanta Publications, 1990

Includes an extended description of other possibilities for a less authoritarian, less violent development politics in India.

——, *State against Democracy: In Search of Humane Governance,* Delhi: Ajanta Publications, 1988

Includes interesting essays on development and authoritarianism.

Ludden, David, "India's Development Regime," in *Colonialism and Culture,* edited by Nicholas Dirks, Ann Arbor: University of Michigan Press, 1992

Discusses the state and economic development in particular over the entire course of modern Indian history.

Marglin, Frederique Apffel, and Stephen Apffel Marglin, editors, *Dominating Knowledge: Development, Culture, and Resistance,* Oxford and New York: Oxford University Press, 1990

This has many interesting essays, including Tariq Banuri's "Development and the Politics of Knowledge: A Critical Interpretation of the Social Role of Modernization Theories in the Development of the Third World." It is Banuri who makes the distinction between "internal" and "external" critiques of development in the course of a useful survey of development literature.

Masani, Minocheher Rustom, *Our India,* London and New York: Oxford University Press, 1940

A primer on nationalist development for children. This may seem an unlikely selection for suggested reading, but *Our India* (*Hamara Hindustan* in its Hindi translation) was a bestseller in its Indian publication and shaped a whole generation of professional-class children growing up in the Nehru era; for that reason it is important to the history of economic ideas in India.

——, *Picture of a Plan,* Bombay and New York: Oxford University Press, 1945

A primer for adults on the so-called Bombay Plan of 1943. Masani later grew disillusioned with state-dominated development, embracing more free-market approaches.

Nandy, Ashis, *At the Edge of Psychology: Essays in Politics and Culture,* Delhi: Oxford University Press, 1980

An important book about political economy from a perspective outside the disciplines traditionally devoted to it.

——, *Traditions, Tyranny, and Utopias: Essays in the Politics of Awareness,* Delhi and New York: Oxford University Press, 1987

This book is very useful for exploring the relationship between colonialism and the dominant political culture (and hence the development culture) of postindependence India.

Nandy, Ashis, editor, *Science, Hegemony, and Violence: A Requiem for Modernity,* Delhi: Oxford University Press, and Tokyo: United Nations University, 1988

A collection of essays that speak to many of the same issues raised in the Marglin volume.

Rosen, George, *Western Economists and Eastern Societies: Agents of Change in South Asia, 1950–1970,* Baltimore, Md.: Johns Hopkins University Press, 1985

Useful for the history of ideas and the people who formulated them. Like Chakravarty, above, Rosen is less skeptical than others about the overall project of development while being by no means uncritical.

Rudolph, Lloyd I., and Susanne Hoeber Rudolph, *The Modernity of Tradition: Political Development in India,* Chicago: University of Chicago Press, 1967

An important earlier contribution to the social science critique of Western modernity in and for India.

——, *In Pursuit of Lakshmi: The Political Economy of the Indian State,* Chicago: University of Chicago Press, 1987

A good overview of Indian political economy and a consideration of what sort of models might best be used to interpret it.

Sachs, Wolfgang, *The Nation and Its Fragments: Colonial and Postcolonial Histories,* Princeton, N.J.: Princeton University Press, 1993

Includes interesting discussions on development and nationalism as well as an important recent discussion of Indian modernism.

Saha, Meghnad, *The Collected Works of Meghnad Saha,* volumes 2–4, Calcutta: Saha Institute of Nuclear Physics/Orient Longmans, 1986, 1993, 1994

Includes this influential author's contributions to public debates about planning, the Soviet Union, Britain, and the United States as models for India, "TVA"-style development for the Damodar, and many other topics, most of them in the form of short articles originally written for *The Modern Review* or his own journal *Science and Culture* from the 1920s to the 1950s.

Shiva, Vandana, *The Violence of the Green Revolution: Ecological Degradation and Political Conflict in Punjab,* Dehra Dun: Natraj, 1989

An important but controversial attack on Green Revolution technology.

Tomlinson, B. R., *The Economy of Modern India: 1860–1970,* Cambridge and New York: Cambridge University Press, 1993

Locates the postindependence political economy of India in the much larger economic history of India since the advent of colonial industrialization. This work is rich in historical details and presents a much more textured overview of themes raised in the Ludden article listed above.

Visvesvaraya, Sir M., *Planned Economy for India,* Bangalore: The Bangalore Press, 1934

An early statement by an engineer-statesman.

Daniel Klingensmith is a doctoral student in history at the University of Chicago. He is currently writing a dissertation on development and its relationship to colonialism in India.

Chapter Six

Growth with Justice: Understanding Poverty

Reeta Chowdhari Tremblay

Although the eradication of poverty and the promotion of social justice have been recurrent themes in the Indian polity for the last five decades, poverty remains persistent in India. The World Bank recently reported that out of more than 1 billion poor people living in the developing countries, 420 million are in India alone, of which 80% are rural poor. While official and academic estimates in India vary and are somewhat lower than those of the World Bank, there is a general agreement that more than a quarter of India's population lives below the poverty line. On the basis of the most recent National Sample Survey income figures, Minhas estimates that in 1991, out of the 360 million poor, 280 million were in rural areas (Minhas, 1991: 27–28). Although the Indian state's pursuance of a strategy of direct attack through poverty alleviation programs since 1970 has reduced the proportion of rural households in poverty from 54% in 1973 to 39% in 1991, agricultural labor households (which constitute about half of the rural poor), the scheduled castes and tribes, and backward regions have not benefited from either growth or antipoverty strategies. Recent structural adjustment economic policies have further contributed to the immiserization of these groups. In the three-year period from 1990 to 1992, the poverty ratio in the rural sector rose from 35% in 1990–91 to 42% (Hanumantha Rao, 1995: 47). Now female labor and female-headed households form a major part of this hard-core group of rural poor due to the increasing feminization of agriculture resulting from a larger participation of male laborers entering nonfarm occupations.

Since its inception, the perceived and declared goals of the Indian state have revolved around the pursuance of the difficult task of accelerated economic development and improved social welfare and equity. The so-called equity-growth approach, initially perceived as uniting complementary objectives, came to pose a policy dilemma, the resolution of which has been sought in recent years in a decoupling of the two objectives. While in the early 1950s, the accepted public philosophy in India favored agricultural growth via structural reorganization (i.e., growth attained by equitable land redistribution), the late 1960s and the early 1970s witnessed a shift in emphasis to equity through growth. The past two-and-a-half decades have seen yet another stage in the separation of the two goals. The failure of the 1950s growth-equity approach and the ushering in of the Green Revolution (growth first, equity later) resulted from several factors such as a stagnant agriculture, an excessive dependence on foreign aid in the pursuance of an ambitious industrial policy, the constraints of legitimacy underlying democratic politics on the implementation of land reforms, an embedded bureaucracy, and the natural calamity of drought. The final and most recent policy approach of separately attacking the growth and the equity issues emerged from the concern among social scientists regarding the inequitable distribution of the benefits of the new Green Revolution technology. Equity, in the present context, does not imply structural reorganization but has the much narrower focus of relieving the misery of the poorest of the poor through a program of productive asset creation and employment schemes. These poverty alleviation programs (PAPs) have had limited success, and the benefits of these programs have not trickled down to the asset-less poorest of the poor.

The following discussion, in its analysis of India's growth and equity strategies of the first and the third period, will suggest that the present separation of the growth and equity goals is only a short-term solution to entrenched poverty in India. Since the failure of the 1950s' integrated growth and equity policy, the Indian state has lacked the political will to resolve the problems of poverty and of socioeconomic structural imbalances through the implementation of a redistributive agenda. Instead, it will be pointed out, the Indian state has adopted a soft, nonideological alternative whereby it has hoped to reconcile its fundamental goals of accumulation, legitimacy, and the maintenance of social and political harmony. Fundamentally, the Indian state is committed to doing as much as it can to resolve the poverty problem but without rocking the boat. The limited success of the PAPs in the last two decades has led to the resurrection of the local institution of the panchayat, which in the 1950s had been the main instrument of planning and implementation of the simultaneous goals of redistributive reform and agricultural growth. However, the revival of the panchayati raj institutions within the framework of antipoverty programs, where growth and equity goals are decoupled and where its functions are limited to participatory planning and the implementation of various schemes, is likely to prove far from sufficient to significantly address the problem of rural poverty. Indeed, the poor can derive the maximum benefit from the poverty alleviation program if they sever their feudal dependency on the rural elite and if they are able to take advantage of social welfare reforms such as education and health. Thus once again, in a roundabout way, the Indian state's political and economic survival has come to be linked with a redistributive agenda.

The Failure of the Growth-Equity Approach: Neither Growth nor Justice

In its 1948 report on the economic program for independent India, the All India Congress Committee outlined the goals of the new state. Faced with the challenging tasks of nation building and economic development, the program was to create an egalitarian society in which a "quick and progressive rise in the standard of living should be the primary consideration governing all economic activities and relevant administrative measures of the central and provincial governments." In its pursuance of the twin goals of equity and growth, India's ruling party, the Indian National Congress, introduced a variety of rural reforms with the objective of bringing a large number of peasants in direct relation to the state through the abolition of intermediaries, the imposition of land ceilings, changes in the tenurial system, and cooperativization in credit and marketing. A vast bureaucratic structure, in the form of the Community Development Programme, was set up in order to educate the farmers, train them in political and economic democracy, and imbue them with a sense of community. The Joint Cooperative Farming Resolution of 1959, also known as the Nagpur Resolution of the Congress Party, was to provide a broader and more systematic perspective to previous policy efforts by laying down an integrated agricultural policy with simultaneous social, political, and economic goals. It is imperative to focus on this resolution for two reasons: (1) it is the most explicit and vocal statement of India's first prime minister's government outlining an integrated agrarian strategy with comprehensive redistributive implications; in short, the Indian state, through a network of panchayats and cooperatives, was to take on the task of rural reconstruction of a just and equitable society, and (2) this was the Congress Party's last attempt to reconcile productivity goals with those of equity, within the constraints of a liberal democracy and an economic strategy emphasizing the development of heavy industry. This was also the democratic state's last attempt, within the agricultural arena, to assert its autonomy vis-á-vis the dominant classes. Political leaders were to learn from this experience that coerciveness associated with redistributive politics has negative implications with regard to political support and hence, legitimacy for a democratic state, and the students of India's agrarian reforms were to learn that the country's political leaders lacked the will to enhance the reforms' capacity to implement the equity-growth approach. The state's failure to realize its agenda had much to do with its inability to deal with the embedded bureaucracy, the prevailing bureaucratic ideology and culture, and the coincidence of socioeconomic power with political power in the newly designed state structures. All this was compounded by a not a very well thought out economic policy logic on the part of the political leadership.

On 11 January 1959, the Congress Party, at its annual session at Nagpur, unanimously adopted a resolution that expressed in unequivocal terms the state's commitment to the implementation of both equity and growth goals in the agricultural sector within a democratic framework. It proposed that the state governments should establish land ceilings on existing and future land holdings and complete the abolition of the intermediaries within a one-year time period. Although the resolution endorsed joint cooperative farming as the future agrarian pattern (this step to be completed within a three-year period), two modes of production were suggested. While the small farmers were to pool their land voluntarily to create a cooperative farm (with their property rights intact), the new beneficiaries of the ceiling legislation, also participating in a cooperative farm, could not claim rights to the newly acquired land since the surplus land resulting from ceilings was to be vested in the panchayats.

The radical implications of the resolution were quite evident. Though it had explicitly stated that the state had no intention of imposing a ceiling on agricultural income (on the contrary, it hoped that it would rise with improved agricultural practices and additional employment), the imposition of ceilings on land holdings within a fixed time frame affirmed a certain normative position of the state. The existing concentration of ownership of agricultural holdings, the resulting social and political differences, and the enormous economic disparities between large landowners, small farmers, and the landless population were considered unacceptable. Through the imposition of ceilings, not only was unlimited accumulation of private property brought into question but also the legitimacy of those institutions and processes that produce and maintain a skewed distribution of land.

Similarly, in spite of the resolution's assurances of voluntariness and the maintenance of property rights, cooperative farming as a mode of production questioned the individual's control over his labor and his physical property. Cooperative farming can be defined as a voluntary association of small farmers who pool their land while retaining their individual property rights in order to enjoy the benefits associated with an integrated large farm enterprise. Decisions in such an enterprise are made by an elected management. Wages are determined according to a stipulated rate in which quality and quantity of work are taken into consideration. Moreover, the landowners are periodically rewarded through property dividends. Theoretically, the concept of cooperative farming appears to reconcile the goals of democratic rights with those of the socialistic society. In practice, however, one can discern certain difficulties both with the notion of retention of property rights and with the concept of voluntariness. Once the land has been pooled in a cooperative farm, the farmer has only legal control rather than physical control over his property. In distinguishing a cooperative farm from a collective one, A. M. Khusro and A. N. Agarwal, strong advocates of cooperative farming, present the advantages of the former in providing the opportunity to the member to leave the cooperative anytime with compensation for his assets. In other words, if the farmer decides to leave the joint farm, he will be paid compensation equivalent to his share of assets, which may include the enhanced value of his property since the formation of the producers' cooperative (Khusro and Agarwal, 1961: 56). One has to agree with Otto Schiller (who proposed individual farming along cooperative lines) that in a joint farm, as the farmer's claim to a visible unit in the field disappears, this absence of a real possession of land could be perceived as negating the fundamental democratic right to retain private property .

Agrarian reforms imply a certain degree of coerciveness on the part of the state. Wolf Ladejinsky observes in his discussion of land reforms in Taiwan and Japan that voluntariness does not lead to the resolution of the basic problems associated with the skewed distribution of land (Ladejinsky, 1977: 396). The Congress Agrarian Reforms Committee of 1949 foresaw the possibility of compulsion in convincing the farmers of nonoptimum size holdings to pool their land in a cooperative farm. Though the two Indian delegations sent to China underplayed the use of coercion in the formation of Chinese cooperatives, retrospective analysis by various authors on China and India suggests that compulsion, violence, and intimidation were associated with the Chinese reforms. While the Chinese had clearly understood the impact of the Soviet experience of forced collectivization without the support of the peasantry and therefore had adopted a cautious three-stage process for the establishment of collective farms throughout the nation, they had quite often to launch rectification campaigns for overzealous cadres (Bernstein, 1979: 73–105).

Though the Nagpur Resolution was not as blunt as the 1949 Congress Agrarian Reforms Committee report, voluntary action on the part of the peasants, as proposed in the resolution, cannot be interpreted as the laissez-faire attitude of a state "doing nothing to provoke such a voluntary action" (Khusro and Agarwal, 1961: 35). Khusro and Agarwal envisioned a positive role for the state in informing the peasants about the benefits associated with the cooperatives and in creating an environment conducive to the establishment of joint farms. While the internal organization of the producers' cooperative was left to its members, the government was expected to provide external economies through land reforms, ceilings, land redistribution, subsidies, tax exemptions, and various other services. Khusro and Agarwal conclude that the state targets "can be fixed only where voluntariness means voluntary action in the face of incentives and these incentives can come through an integration of cooperative farming with land reforms and other benefits" (Khusro and Agarwal, 1961: 35). Under such conditions, does the principle of voluntariness remain intact?

The use of force in the formation of cooperative farms and the threat to the institution of private property became the focal points of criticism and contributed largely to the conversion of the Nagpur Resolution into a nondecision. Although the resolution had been passed unanimously at Nagpur, mostly out of deference to Nehru, there was wide disagreement among the political leadership, both within and outside the Congress Party. On 4 June 1959, the Swatantra party was organized through the efforts of two groups, the Forum for Free Enterprise and the All-India Agriculturalist Federation, to bring all the elements of the right into one platform and provide organized opposition to the Congress Party's economic and agrarian policies.

Ceilings, redistribution of land, and cooperative farming had been persistent themes in the Indian agricultural policy arena. What was new in the Nagpur Resolution was that for the first time, the Indian political leadership set a time frame for the implementation of these reforms. The ceilings on land holdings were to be determined by the end of 1959. Over a three-year period, multipurpose service cooperatives were to be established. After that, joint cultivation was to begin. One can discern two reasons behind this urgency, one ideological and the other developmental.

Ceilings and redistribution were considered to be essential for the successful realization of cooperative farming as a mode of production. The homogenization in terms of ownership and the virtual absence of large inequalities were said to be conducive to making cooperatives operational. The question is, why did the political leadership consider it necessary to endorse cooperative farming instead of peasant proprietorship after redistribution of land and within a very short period of time? Ideologically, joint cooperative farming was viewed as a higher, socialist form of social and economic activity. It was perceived that smallholder agriculture, if left alone, gives rise to a politically threatening class of rich or capitalist farmers. Embourgeoisement of small farmers, it was felt, could be avoided by shortening the time between the completion of land reforms and the onset of socialistic transformation in the agricultural practices. A quick transition from the redistribution to the organization of cooperatives would ensure that there was too little time for a smallholder economy to develop and take hold. With the failure of the Nagpur Resolution, such a transition became impossible, and eventually a strong bullock capitalist economy became firmly entrenched in the Indian system.

However, during the 1950s, from the point of view of the political leadership, developmental rather than ideological reasons were more compelling for implementing the new agrarian strategy as quickly as possible. The new industrial strategy, emphasizing import substitution and claiming most of the state's resources, required increases in agricultural productivity without shifting resources away from the industrial sector by means of extracting increased revenues from the agrarian sector (whose proportionate burden was much lower than that of the urban sector), generating large agricultural surpluses, and increasing employment in the rural sector. Cooperative farming, along with a universal system of multipurpose service cooperatives and state trading in foodgrains, was perceived to be the best solution in meeting the state's demands.

A more efficient agricultural strategy had to be devised in order to meet the higher productivity targets, since the earlier policy had achieved only limited success. Encouraged by reports that China had been able to enhance its agricultural production by 15 to 30% in those villages where cooperative farms had been formed, a strong feeling prevailed in the Planning Commission that India could achieve a similar, perhaps even better result through cooperativization, instituted along democratic lines (Frankel, 1978: 124–25). Retrospective evaluations of China's agricultural performance subsequently were to reveal that the official Chinese statements had been highly exaggerated. In addition to these ideological and developmental factors behind the urgency to initiate an integrated agrarian policy, it should be noted that the results of the 1957 state and national elections also played a modest role by shifting the Congress Party's agenda to the left. The Congress Party's popular vote for the Lok Sabha improved from 45% in 1951 to 48% in 1957 (a result that was not matched until December 1984), and from 42% to 45% for the state assemblies. However, in the 1957 election, the Communist Party, which almost trebled its popular vote in the Lok Sabha from 3.3% to 8.9% and formed the first Communist Ministry in Kerala, appeared to emerge as the potential alternative to the Congress Party. In its postelection analysis of the results, the Congress leadership observed that growing economic discontent among the population required the party to move more to the left and to narrow the gap between promises and actions.

The Nagpur Resolution's exclusive endorsement of cooperative farming (as "the future agrarian pattern") implied that peasant farming was no longer considered rational and appropriate by the policy makers. If family farming as a mode of production was to be discontinued, the authors of the resolution neglected to address the issue of optimal size for a cooperative farm. There was no indication even of upper or lower bounds.

The economic rationale underlying the policy owed more to ideology than science, for the policy makers took a stand in favor of large-scale farming to improve productivity without any supporting data or analysis. Although ceilings were a hot issue in Indian planning circles in the 1950s, empirical questions were not sorted out. The debate continued mostly at a political level "largely in ignorance of economic and technological facts" (Khusro, 1966: 16). By and large, the political leadership addressed the problem of land reforms and reorganization based on redistributive rather than efficiency considerations. The inability of the state to clarify its economic logic led to a perception that proposed large-scale farming would be either along the lines of Western-style capitalist farming or of collectivist farming as practiced in the Soviet Union or Poland, where mechanization, especially the use of tractors, had been predominant during this period. Khusro and Agarwal view the rejection of cooperative farming on this ground as "guilt by association" (Khusro and Agarwal, 1961: 45).

The Nagpur Resolution proposed that the new political-bureaucratic rural structure, recommended by the Mehta Committee and accepted by the Indian government, should serve as the vehicle for the implementation of the equity and growth strategy. The political leadership's objectives were transparent in the statements of the Nagpur Resolution. They intended to take advantage of the proposed viable partnership, to be introduced under the new program (the "panchayati raj"), between the "newly empowered" rural masses, the bureaucrats, and the politicians in order to realize their goals. Underlying this proposal was the complex corporatistic strategy entailed in the state's giving the local panchayats a monopoly on the representational, economic, and social regulating functions over the rural population. Through careful regulation by the instrumentalities of the state, the democratically elected panchayats, and the local bureaucratic apparatus, the political leadership expected a rational and planned outcome. In this case, it was hoped that goals of productivity and redistribution would be realized in a short period of time with negligible resources and minimum disruption.

The concept of corporatism suggests an active role for the state in creating official or legally constituted organizations and in defining or redefining their powers. The Nagpur Resolution's major stipulation regarding joint cooperative farming as the future agrarian pattern was that it be accomplished through a new organizational structure based on a geographical division of duties. Each village, its territorial delineation determined historically, was to be represented by village panchayats and cooperatives. These two organizations were to be vested with exclusive and comprehensive socioeconomic and developmental powers. As the resolution stated, both these agencies were to "be the spearheads of all developmental activities in the village and, more especially, should encourage intensive farming with a view to raising the per acre yield of agricultural produce." While the panchayats were to be democratically elected, with some seats reserved for women and the scheduled castes, the membership of the cooperatives was to consist entirely of permanent members of the village.

The formal exchange relationships between the village panchayats and the Indian state were seen as the building blocks of the whole planning structure. The panchayats were to be used as bureaucratic instruments that would initially determine demand from below, essentially generate the national plan at the village level, and ultimately implement and monitor national policy goals. The significance of this "bureaucratic creativity" (Chalmers, 1985: 75) lay in the fact that while these organizations were a legal creation of the state, no decision of the state could be imposed. The state, through these organizational devices, was to give peasantry a push. However, it was hoped that it would thereby be able to effectively mobilize the rural masses regarding its political and economic agenda.

These attempts of the political leadership to introduce democracy at the grassroots level are not inconsistent with those aimed at strengthening the state. Within a corporatistic context, the strength of the state should be determined by the "increasing density of relationships that centre on the official apparatuses of the state" (Chalmers, 1985: 75). Corporatistic interactions between the groups and the state take the form of an exchange relationship, and it is the content of exchange that specifies and defines the relative power and the degree of mutual autonomy of the state and its officially sanctioned agencies and organizations. The Indian government's attempt to try more rather than less democracy via the panchayati raj in order to achieve real growth and genuine reform cannot been seen as weakening the state. Through its proliferation, the state perceived that it could mobilize the masses more efficiently toward the realization of its goals. Although the panchayats were popularly elected, their agenda had already been defined by the national leadership. Goals had been articulated and formulated, and the local-level organizations were intended to act as the agencies of the state in order to appropriate legitimacy for its actions from the masses. It was expected that the legitimacy of the state goals would be further enhanced when the new institutional forms were able to take over the collective and individual functions of the landlord class as patrons and village leaders.

From the policy makers' point of view, the introduction of the panchayati system was quite attractive. A traditional Indian institution could be revived, reformed, and rehabilitated in order to maintain continuity with the cultural and political heritage. But as a developmental agency representing the multiple interests of the villagers and regulating their social and economic life, the panchayats had no parallel in traditional India. A traditional symbol had been given a new meaning, a new identity. In initiating novel political and economic agendas, the tasks of the political leaders could only be facilitated through the revival of the nostalgic notion of a self-sufficient rural India, with panchayats as its political instrument. Moreover, through the revival of panchayats, the political leadership explicitly endorsed the notion of harmony associated with its traditional counterpart: in rural India, the unanimous voice of the panchayat had traditionally been perceived as the voice of God; it gave expression to the consensus of the traditional moral order.

In the rural reconstruction of modern India, however, a new moral order, no longer based on karma and dharma, was being created. The panchayats were to remain free from party politics, and primordial ties to castes and factions were to be replaced by loyalty to the community and ultimately to the nation. The rural institutions of panchayat and cooperatives were to train the peasantry in transcending their subjective parochial interests and in understanding their common interests. It was perceived that development of a community of interests, the absence of party mobilization, and unanimity among the panchayat members would foster the harmony essential for the realization of the developmental goals of the state.

Later experiences were to show that the panchayats did not perform the functions that the planners had envisioned for them. The panchayati raj institutions, local government, and village council were unable to get involved in land reform implementation, especially in the areas of the maintenance of land records, determination of surplus land, regulation of common land, and protection of tenants from illegal evictions. Though party competition was eliminated in the selection of the representatives, village and factional politics and the interests of the dominant castes could not be kept out. Moreover, panchayati raj was unable to remove the excessive bureaucratization associated with the earlier community development program. In spite of a large mandate given to the elected bodies in terms of supervising and directing development planning, the bureaucratic impact was pervasive in terms of defeating the goals underlying the new institutional setup. The bureaucracy does not constitute a neutral instrument to be used by the state in implementing its goals. It is an embedded bureaucracy in terms of its class constraints and the normative values associated with it. Also, the bureaucratic concept of the new democratic institutions as essentially playing the agency role obliterated the corporatistic view of the state with respect to a solid partnership between the rural masses, the bureaucrats, and the political leadership, and as Wolf Ladejinsky points out, reluctant civil servants and not-so-bold farmers are not a good combination to advance the cause of reform (1977: 401)

The Nagpur Resolution was an innovative effort on the part of the Indian state to pursue its multiple agenda within the agrarian sector. With the goals of justice, productivity, political stability, and maintenance of a liberal parliamentary democracy in its mind, it devised a complex organizational structure to pursue an integrated agricultural policy. The inability of the political leadership to legislate the strategy approved at the party level is attributable to the overriding concern of the state elites to preserve democratic principles and to the leadership's failure to defend the policy logic on economic grounds.

Following the adoption of the resolution, the element of compulsion in the formation of farming cooperatives and the negation of an individual's fundamental right to possess and dispose of private property became the major issues. Nehru's withdrawal of emphasis on universal cooperative farming can be related to the mounting pressure from the opposition parties on these two points. Prior to the adoption of the resolution, the Indian prime minister had very often assured both his party members and the public that the government would not compel the farmers to pool their land into cooperatives. However, given the need to increase productivity, he claimed that India had no choice but to create production cooperatives and that in pursuance of this goal, the members of the party should try to persuade the farmers of the advantages of this mode of agricultural production. In response to the criticisms that followed the adoption of the Nagpur Resolution, he asserted that cooperative farming was the only solution to the problem of productivity and that party workers should not "wait for legislation in this regard but to go to the villages to prepare the ground for the introduction of new changes" (*Times of India* [New Delhi], 11 January 1959). Although he insistently reaffirmed the government's and the party's policy of achieving redistributive goals through peaceful methods, with respect to the clarity of the party's policy statements, he noted the difference between the Nagpur Resolution and Congress Party's earlier statements. He observed that the Congress Party could no longer be vague in its pronouncements: "We have to state our objectives and programmes in a clear and unambiguous language" (*Times of India* [New Delhi], 31 January 1959). These pronouncements were interpreted by the critics of the new agrarian strategy as being antidemocratic and against the Indian tradition of a consensus-oriented society.

The Communist Party of India's endorsement of the new agrarian strategy did not help the Congress Party government in convincing its opponents that the new policy would be carried out along democratic lines. Moreover, China's crushing of the Tibetan uprising in 1958 and Chinese incursions into Indian territory made an economic model based on Chinese experiences unacceptable to the general populace. The first sign of retreat from the Nagpur Resolution appeared when no mention of cooperative farming was made in the president's address to the opening session of Parliament. During the debate on the address, in response to the opposition parties' attack on joint cooperative farming, Nehru assured the house that no act (concerning production cooperatives) was going to be passed by Parliament. The Congress Party itself began to emphasize two phases of the agrarian program. The first and more immediate phase was to consist of the formation of service cooperatives, on the successful completion of which cooperative farming was to follow. In March 1959, Nehru reassured the nation by stating that the setting up of service cooperatives was the only way to solve the country's problems. The Congress Working Committee made a conscious attempt to publicly emphasize service cooperatives.

The Indian political leadership's constant and consistent emphasis on voluntarism and respect for the right to own property resulted in the state having neither the capacity nor the will to implement the integrated agrarian strategy proposed in the Nagpur Resolution. The success of panchayats and the Community Development Programme was dependent on an effective implementation of ceilings and a quick transition to cooperative farming. Without the implementation of a complementary and integrated agrarian policy, the new organizational structure had come to be dominated by the local elites, who were interested in protecting their dominant position in rural India, and with the existing power relationships left unaltered, agrarian reforms were a remote possibility. Moreover, an embedded bureaucracy and the prevailing bureaucratic ethos made the implementation of reforms ineffective.

As mentioned earlier, agrarian reforms involve a certain degree of coercion. The Indian state's negation of this structural aspect of the agrarian redistributive policy and its insistence on the maintenance of democratic rights were, to a large extent, responsible for the failure of the new strategy. Consequently, over the years, the Indian state has unsuccessfully pursued various agrarian programs individually underlined in the resolution, such as ceilings, multipurpose and production cooperatives, and state trading whereby the goals of neither justice nor growth were realized. In the absence of a cohesive public policy, the attack on poverty met with disastrous results. Various agrarian reforms, if implemented, had major loopholes. By invoking their fundamental right to own private property, the wealthier rural sections managed to consolidate their socioeconomic positions through legal battles. Many absentee landowners deprived small tenants of their traditional rights of tilling by using land for self-cultivation. The major beneficiaries of the land reforms were the middle peasants and the bullock capitalists, who were to translate their economic power into political power by initially taking control of the regional government in the late 1960s and the central government in the mid-1970s.

Separating Equity and Growth Goals

Opposition to the Nagpur Resolution, acute food shortages, the emergence of the medium farmers as a demand group, Chinese infiltration of the Indian border (thus making a policy modeled after the Chinese agricultural experimentation politically less expedient), and lack of enthusiasm of the landed-elite-based regional governments in the implementation of land reforms–all these factors contributed to the adoption of the Green Revolution technology. However, the subsequent breakthroughs in the agricultural sector led to concern among social science analysts about the inequitable distribution of the benefits of the new technology. It was maintained that the lower rural sector was restrained from fully participating in and benefiting from agricultural growth, thereby widening the gap between the small, marginal farmers and landless laborers on the one hand and medium and large farmers on the other, the true beneficiaries of the Green Revolution (Brown, 1971; Desai, 1969; Gaikwad, 1977). Concerns began to be expressed regarding the substantial increase in marginal holdings of less than one acre (Rao and Deshpande, 1968). Moreover, immiserization of the poor was cited through economic and health statistics: some economists maintained that real per capita net domestic product (NDP) in the agricultural sector had remained more or less constant for over 30 years and that per capita availability of calories was 10.7% short of the recommended requirements of 2,300 calories per day (Rath, 1985).

Indian democracy was now faced with one of its most severe tests: how to maintain the pattern of accommodative politics where the goals of economic growth and alleviation of poverty should be reconciled without antagonizing the powerful bullock capitalists and politically unleashing the impoverished peasantry. In order to resolve the contradictions of the accommodative politics, which claimed to advance the interests of the poor while simultaneously accommodating the propertied elite, the Indian state embarked on a new course by decoupling the two goals, redistribution and growth, and resolving to deal with each separately. The emergent political predominance of the capitalist peasantry at both the central and the regional levels was to facilitate this decoupling whereby, in the official thinking, agricultural productivity and poverty would come to be ideologically perceived as two distinct issues. While agricultural growth was to be pursued and based on pricing policies and new technologies that had demonstrated their relevance in the past, the problem of rural poverty was now to be dealt with through policy instruments specific to the antipoverty program, such as productive asset creation and employment schemes (Dogra, 1986: 689). The earlier policy instruments such as land reform, used previously to bring about changes in land relations, were not to be perceived in this latest context as structural responses creating equitable relations but more simply as one of the many policy instruments for fighting poverty. Given the constraints of democratic politics, the political leadership moved from teleological commitment to nonideological solutions. As Vyas and Bhargava point out, "an important implication of this approach is that efforts at poverty alleviation were not conditioned by political labels" (Vyas and Bhargava, 1995: 2560).

From the political leadership's viewpoint, the appeal of decoupling the equity and growth goals was and is that they can, in principle, lead to the simultaneous accommodation of a variety of often diverging interests and thus, in the short run, resolve the contradiction between the two basic requirements of the state: legitimacy and accumulation. While the growth-oriented agricultural strategy satisfies the more entrepreneurial segment of the agrarian population, without threatening their rights and privileges, the antipoverty policies independently address the interests of the disadvantaged rural constituency. Both political and economic reasons have made it possible to generate a wide acceptance of this segmental view of rural development. Mathur notes that politically, such a strategy "provided visible methods of working for the rural poor and thus could pay political dividends," and economically, it did not "hamper the pursuit of policies of growth." Though the alleviation of poverty was to retain prominence in the political agenda of the Indian state, it "was not the core theme around which development strategy was conceived," and therefore, the "poverty alleviation schemes did not threaten either the strategy of development nor the structural relationships in the society" (Mathur, 1995: 2704).

This decoupling strategy of the past two decades has, however, been accompanied by an extremely cautious attempt at social engineering on the part of the Indian state, which I would like to designate as the generation of "equity via incrementalism." The ad hoc policy measures of distress relief in the fourth and the fifth plan were, in the sixth plan, translated into a package of economic reforms for the poorest of the poor. Encouraged by the results of the antipoverty measures and concerned about their inability to provide the maximum benefit to the ultrapoor, the designers of the seventh and the eighth plan moved toward the enfranchisement of the poor through the revitalization proposals of block developmental planning and the political organization of the poor constituency. However, the plan's program of decentralization through local bodies is confined to ensuring a proper selection of projects and beneficiaries for the antipoverty projects and to encouraging the rural poor to take up group-oriented activities through cooperatives and nongovernmental organizations (NGOs). The political leadership is ensuring that this earnest attempt to make the dependent rural households economically free agents will not disturb the class harmony of rural India. Though the political balance of forces within the country has made it exceedingly difficult for the Indian state to initiate a social engineering agenda, it has become compelling for the state to sponsor the goal of a politically active and organized peasantry to ensure that the state's objective of accumulation and its need to maintain social harmony are fulfilled. Consequently, a soft option has been adopted. Any socioeconomic restructuring is to be realized incrementally within the context of separate and distinct policy prescriptions for growth and equity.

Poverty Alleviation Programs (PAPs)

The 1970s witnessed the emergence of certain ad hoc policies aimed toward the small and the marginal farmers, such as small farmer development agencies (SFDA) and marginal farmers and agricultural laborers (MFAL), to provide them with productive assets and inputs. In the Sixth Five-Year Plan, these were subsequently merged in the Integrated Rural Development Programme (IRDP). The original concept of the IRDP, formulated in the draft sixth plan, as a total developmental plan for the block based on local resources, was shelved for a more practical individual beneficiary-oriented program. The IRDP was initially introduced in 2,300 blocks, and in 1980, it was extended to the entire country. Similarly, the Food for Work program of 1977, aimed toward the creation of employment of agricultural laborers by utilizing the accumulated surplus food stock, was restructured as the National Rural Employment Programme (NREP) in 1980. Other small-scale, beneficiary-oriented programs that were created between the fourth and the seventh plans were the Drought Prone Area Programme (DPAP) and the Desert Development Programme (DDP), the Rural Landless Employment Guarantee Programme (RLEGP) and the program for Training of Rural Youth for Self-Employment (TRYSEM), Development of Women and Children in Rural Areas (DWCRA) and Jawahar Rozgar Yojana (JRY–Jawahar Lal Nehru wage employment plan). During and after the seventh plan, most of these programs were integrated into the IRDP.

The objectives of these antipoverty programs have been to enable poor families eventually to attain income levels that will bring them over the poverty line. Such a policy logic can be distinguished, on the one hand, from the structuralist argument that fundamental changes such as land reform in the relations between classes resulting from the nature and mode of production are a precondition for poverty eradication and, on the other hand, from the neoclassical approach that believes in enlarging the resource base and productivity as a precondition to improving the level of welfare. Instead, what is stated is that while productivity enhancement is independently pursued, rural poverty can be attacked by endowing the poor with the productive assets and inputs that will result in their self-employment. Where this cannot be achieved, government-sponsored wage employment schemes are to supplement the self-employment programs. This emphasis reflects the structure of rural employment in India: according to the Thirty-Second National Sample Survey, 62.5% of the rural working force, in the agricultural and nonagricultural sectors combined, is self-employed.

Unlike the 1950s' attack on institutional poverty, the beneficiaries of which were underprivileged socioeconomic classes, the antipoverty program targets a narrower clientele, the "poorest of the poor." The initial *antyodaya* (uplifting of the poor) approach entailed the identification of the neediest among those below the poverty line, defined in the sixth plan as either a total annual household income of Rs. 3,500 or an annual per capita income of Rs. 700 and revised upward in the seventh plan to Rs. 6,400 per annum per household. The *garibi hatao* campaign, through an increase in the productive potential of the rural economy, was perceived by the political leadership as a direct means of targeting benefits to the "special publics," that is, groups of individuals in special need. Unlike the past rural development programs, which were oriented toward the community as a whole, the IRDP, the antipoverty strategy pursued since the sixth plan, directly targets the poor constituency and provides them assistance in the form of income-earning assets. In addition to the state's concentration on a narrow clientele, the goals of its war on poverty appear to be limited to the economic emancipation of the family. The social needs of the latter, such as education, health, housing, nutrition, and family welfare (e.g., rural electrification and water supply, rural roads) are to be taken care of by the governmental departments outside the district Integrated Development Agencies under the Minimum Needs Programme. In contrast with the earlier solutions to rural development, equity has come to be defined narrowly so that the goal of human development implied in the state's welfare function takes second place to the immediate and short-term goal of income generation by the targeted beneficiaries. The sixth plan clearly stated that IRDP was conceived as an antipoverty program. Because its beneficiaries would also need support from social services such as health, education, and housing, it became necessary to link, to the extent possible, the prospective beneficiaries under the IRDP to these social services, particularly programs such as applied nutrition, compulsory primary education, adult education, family welfare, children and women's welfare, and activities. The prospective beneficiaries having been identified, these lists should be made available to the departments concerned in order for them to follow these persons with respect to the services handled by them. Consistent with this theme, both the seventh and the eighth plans placed a special emphasis on training of unemployed women. During the seventh plan, DWCRA, through a group strategy, targeted the formation of 15,000 women's groups with the objective of motivating them to extricate themselves from traditional social taboos so that they can take advantage of the IRDP and other poverty-alleviation programs. Similarly, in 1986, the Indian Parliament approved the National Policy on Education, which reiterated that "education will be treated as a crucial area of investment for national development and survival" (Vashist, 1991: 168).

Despite the limited objectives of the antipoverty programs, there appears to be a general consensus, even among their staunchest critics, that these measures have had some degree of success. Disagreement, however, prevails regarding the depth of this success and the policy logic of the planners (Bandyopadhyay, 1988: A-84; Bagchee, 1987: 139–47). Leading the pack of the structuralists, Dantwala points out that a direct attack on poverty without a radical restructuring of the society is self-defeating and counterproductive (Dantwala and Souze, 1986). Kurian similarly suggests that these target-oriented programs through their privatization drive to generate a new crop of penny capitalists, have threatened the traditional economic and social structure without replacing it with a new framework. He emphasizes, "growing privatisation of resources and the commercialization of economic activity have almost completely marginalised the weaker sections who find increasingly that they have to buy things which they formerly used to receive in the form of traditional claims. . . . [Moreover] while antipoverty programmes are launched by the dozens, social and economic processes which continuously deplete the resources from which the poor derive their incomes and employment and make their traditional skills redundant are either not noticed or ignored and effective action to stop such process is rarely taken" (Kurian, 1989: A-14). In the same spirit, Bandyopadhyay observes that the IRDP, without any particular plan for collective economic action, can only have limited success, for the beneficiary, subjected to the inherent insecurity of impersonal and ruthless market forces, "is likely to lose both ways as a seller of product and as a buyer of inputs. Market forces would toss him about, since he would not be in a position to bargain individually" (Bandyopadhyay, 1988: A-84). However, these structuralist critics of the antipoverty programs recognize their successes, however limited. The IRDP has significant "public appeal" and has caught "the imagination of the poor." They admit that a fairly large number of assets have reached the very poor via a program that aims at a direct assistance in the form of income-earning assets.

The thirty-eighth round of the National Sample Survey (NSS) conducted during the year 1983 estimated a steep decline of poverty, by 11 percentage points, since 1977 (the thirty-second round of the NSS). It placed poverty figures for the rural poor at 39.9% or 222 million. Kurian cautions that these figures of decline should not necessarily be viewed as a "welcome trend" in the decline of poverty. He suggests that there is a strong correlation between a good crop year and the figures on poverty. The thirty-eighth round coincided with the extremely good crop year of 1983–84. For him, any sustained reduction in poverty can only be a product of a structure of economic growth "which is socially equitable and an agricultural growth which is regionally balanced" (Kurian, 1989: A-15).

The seventh plan, in a self-congratulatory note, touts the success of its two-pronged policy: "There is now evidence to suggest that the process of economic growth and the anti-poverty programmes have made a significant trend in the problem of poverty." With its additional emphasis on decentralized planning and state-sponsored voluntary organizations of the poor, it targeted the reduction of rural poverty to 28% by 1990 and to 18% by 1995. Minhas has estimated that in 1991, the proportion of rural households in poverty has been reduced from 54.3% in 1973 to 39% in 1991. Vyas and Bhargava observed in 1995, after the completion of a research project on poverty in nine Indian states, that in the last two decades, there has been a continuous reduction of poverty in practically every state in India, and this decline is a result of the Indian state's deliberate and targeted attempts to alleviate poverty. However, there are "sharp intra-state variations in the incidence of poverty as well as the extent of its decline over time" (Vyas and Bhargava, 1995: 2563).

Major evaluation studies of the self-employment and wage-employment measures of the antipoverty program have identified their specific shortcomings as either administrative failures or weaknesses in the planning procedures. The four all-India level IRDP studies have singled out two problem areas: wrong identification of the beneficiaries and the selection of activities without regard for the skills of the beneficiaries and the capacities of the infrastructural support system. The NREP and RLEGP studies, conducted only at regional levels, have observed that these programs have been successful in providing only short-term employment. In response to these administrative deficiencies and given the rate of success of the antipoverty measures where the local *gram sabhas* participated in the selection of beneficiaries, both the seventh and the eighth plans proposed organizational changes in the planning and implementing process.

Following the 1985 recommendations of the Planning Commission's Committee to Review the Existing Administrative Arrangements for Rural Development and Poverty Alleviation, the seventh plan proposed the reactivation of the panchayati raj institutions. It proposed that the *zilla parishad* should be made the principal body for the management of all developmental programs and that other panchayati raj institutions should be given the responsibility for planning, implementing, and monitoring the rural development programs. To this effect, the plan recommended the restructuring of the administrative machinery at the district level. A new post, that of District Development Commissioner (DDC), that would coordinate all the development activities of the district would be instituted. In the hierarchy of the district administration, the DDC would occupy a higher position than that of the District Commissioner, thus establishing the "primacy of developmental administration over maintenance administration" (Bagchee, 1987: 145). Moreover, concerned with the regional variations in the poverty structure (the eastern and central regions have the highest incidence of poverty), the plan also proposed that the targets should be fixed locally, preferably at the block level, instead of being determined at a uniform level throughout the country. The eighth plan reiterated the need for integrating the various antipoverty programs within an administrative structure of decentralized planning. In December 1992, the Indian Parliament passed the seventy-third constitutional amendment, whereby the panchayats were recognized as the agents of rural transformation. However, as Mathur notes, "the planning exercise at the local level continues to be confined to compilation of schemes in specified sectors according to the guidelines issued from above," whose ability to be implemented is decided on mostly political or administrative grounds (Mathur, 1995: 2706). While the central government is reluctant to decentralize planning on the basis that a lack of uniformity and standard programming may contribute to ineffectiveness and inefficiency, the local staff, on the other hand, have "little training or incentive to perform better" (Mathur, 1995: 2707).

Land reforms, the importance of which had been religiously referred to in the fourth through the sixth plans but which occupied an ambiguous place in the strategy of rural development, have been integrated into the antipoverty program under the seventh plan. Unlike the sixth plan, which made no mention of agrarian reforms in the strategy and methodology for accelerated rural development and which confined discussion of these reforms to a critique of their indifferent implementation, the seventh and the eighth plans include them with other policy instruments of income generation for "special publics." With the goal of the antipoverty program to endow the indigent with income-generating assets, in the opinion of the planners, "redistribution of land could provide a permanent asset base for a large number of rural landless poor for taking up land-based and other supplementary activities" (Bandyopadhyay, 1986). This renewed but limited emphasis on land reforms does not represent, as some structuralists would have it, their consecration as a core activity of the antipoverty package. Instead, land reform, the cornerstone of the first period's equity-oriented policy, has been dusted off by the political leadership and taken down from the shelf to which it had been consigned during the growth-oriented period. It has been reactivated, but with the lower status of just one more weapon in a larger arsenal, in line with the present nonideological goals of the state.

Since the failure of the 1950s' integrated growth-equity approach, the Indian state has lacked the political will to address the issue of economic and social disparities through the implementation of a redistributive agenda. Instead, it has adopted a soft alternative and has opted for the slow, incremental process of attacking poverty through poverty alleviation programs that would neither threaten the strategy of development nor the structural relationships in society. Through a decoupling of the goals of equity and growth, it has hoped to reconcile its fundamental requirements of accumulation, legitimacy, social justice, and the maintenance of social and political harmony. As the above discussion points out, such a strategy where equity is narrowly defined has only marginal success. Unless the Indian state addresses the issues of societal environment and the removal of structural constraints that inhibit the poorest of the poor from taking advantage of the antipoverty policies, poverty will remain entrenched in Indian society. The Indian state has no option but once again to resurrect the redistributive thrust of its developmental goals.

Further Reading

Bagchee, Sandeep, "Poverty Alleviation Programmes in the Seventh plan: An Appraisal," in *Economic and Political Weekly* 22, no. 4 (1987)

Bandyopadhyay, D., "Direct Intervention Programmes for Poverty Alleviation: An Appraisal," in *Economic and Political Weekly* 23, no. 26 (1988)

——, "Land Reforms in India: An Analysis," in *Economic and Political Weekly* 21, no. 25 (1986)

Bernstein, Thomas P., "The State and Collective Farming in the Soviet Union and China," in *Food, Politics, and Agricultural Development: Case Studies in the Public Policy of Rural Modernization,* edited by Raymond F. Hopkins, Donald Puchala, and Ross Talbot, Boulder, Colo.: Westview Press, 1979

Brown, Dorris D., *Agricultural Development in India's Districts,* Cambridge, Mass.: Harvard University Press, 1971

Discusses agricultural policy and Green Revolution strategy.

Chalmers, Douglas, "Corporatism and Comparative Politics," in *New Directions in Comparative Politics,* edited by Howard J. Wiarda, Boulder, Colo.: Westview Press, 1985

Dantwala, M. L., Ranjit Gupta, and Keith Souze, editors, *Asian Seminar on Rural Development: The Indian Experience,* New Delhi: Oxford University Press and IBH, 1986

Desai, D. K., "Lessons of IADP," in *Economic and Political Weekly* 4, no. 26 (1969)

Dogra, Bharat, "Can Seventh Plan Targets Be Achieved," in *Economic and Political Weekly* 21, no. 16 (1986)

Frankel, Francine R., *India's Political Economy, 1947–1977: The Gradual Revolution,* Princeton, N.J.: Princeton University Press, 1978

A very detailed analysis of India's political economy and the accommodative politics of the Indian state in reconciling the goals of social equity with a gradual nonconflictual mode of change.

Gaikwad, V. R., *Development of Intensive Agriculture: Lessons from IADP,* Ahemdabad: Centre for Management in Agriculture, Indian Institute of Management, 1977

The results of the author's evaluation studies of intensive and selective farming in the 1970s.

Hanumantha Rao, C. H., "Attack on Poverty and Deprivation: Role of Structural Change and Structural Adjustment," *Mainstream* 30, no. 10 (1995)

Khusro, Ali, *The Economics of Land Reform and Farm Size in India,* Madras: Macmillan India, 1966

A strong advocate of joint cooperative farming, Khurso explores the trade-offs between efficient size of landholding and equity.

Khusro, Ali, and A. N. Agarwal, *The Problem of Cooperative Farming in India,* New York: Asia Publishing House, 1961

One of the very few books written on joint cooperative farming. The authors very ably explore the concept of cooperative farming and its social and economic implications.

Kurian, N. J., "Anti-Poverty Programme: A Reappraisal," in *Economic and Political Weekly* 24, no. 12 (1989)

Ladejinsky, Wolf, *Agrarian Reform as Unfinished Business: The Selected Papers of Wolf Ladejinsky,* edited by Louis J. Walinsky, New York: World Bank and Oxford University Press, 1977

An excellent collection of essays by Wolf Ladejinsky concerning his experiences and conceptual ideas on various types of agrarian reforms.

Mathur, Kuldeep, "Politics and Implementation of Integrated Rural Development Programme," in *Economic and Political Weekly* 30, no. 41 (1995)

Minhas, B. S., I. R. Jain, and S. D. Tendulkar, "Declining Incidence of Poverty in the 1980s; Evidence Versus Artifacts," in *Economic and Political Weekly* 26, no. 27 (1991)

Rao, V. M., and R. S. Deshpande, "Agricultural Growth in India: A Review of Experiences and Prospects," in *Economic and Political Weekly* 21, no. 38 (1986)

Rath, Nilakantha, "Garibi Hatao: Can IRDP Do It?" in *Economic and Political Weekly* 20, no. 6 (1985)

Vashist, Purshotam D., *Planning Indian 8th Plan,* Jalandar, Punjab: ABS, 1991

One of the very few studies of the Eighth Five-Year Plan.

Vyas, V. S., and Pradeep Bhargava, "Public Intervention for Poverty Alleviation; An Overview," in *Economic and Political Weekly* (1995)

Reeta Chowdhari Tremblay is associate professor of comparative public policy and South Asian politics at Concordia University in Montreal, Canada, where she is also director of the graduate program in public policy and public administration. Her current areas of research are the secessionist movement in Kashmir, identity based politics and citizenship, and popular culture. She has published extensively in the areas of secessionist movements and public policy.

Chapter Seven

Federalism, Local Government, and Economic Policy

George Mathew

The Indian Constitution has created a federal structure, but nowhere in the Constitution is the term *federal* used. The Constitution says that India is a union of states: today, India has 25 states and seven union territories.

What do we mean by saying India is a union of states? B. R. Ambedkar, the architect of the Indian Constitution, said that the use of the word *union* was deliberate. The drafting committee wanted to make it clear that although India was to be a federation, it was not the result of an agreement by the states. The federation is a union because it is indestructible. India, federal in structure and functions during normal times, can be transformed into a unitary state during emergencies. So far, the Indian Constitution has functioned on a federal-unitary continuum. It has also been stated that India is a federation without federalism, a polity covered by some form of federalism, variously called cooperative, executive, emergent, responsible, parliamentary, populist, legislative, competitive, fiscal, restructured, reluctant, quasi, and so forth (Kashyap C. Subhash, "Building the Federal Union," Hamdard University, New Delhi, 1997, unpublished paper).

The Indian Constitution, adopted by the Constituent Assembly on 26 November 1949, envisaged a strong central government. At that time, the predominant concern of the founding fathers of the Constitution was preservation of the unity and integrity of India. In the last 47 years, however, several provisions of the Constitution have worked against the federal principle. The Constitution clearly demarcated the duties and the division of power between the union government and governments of the states. There are union lists consisting of 97 items; state and concurrent lists consist of 66 and 47 items, respectively. Over the years, the lists have been subjected to constitutional amendments in favor of the union. Five items have been omitted from the state list, five added to the concurrent list, and three added to the union list. Furthermore, the central government can usurp powers and jurisdiction of the state by parliamentary legislation. Under Article 249, if the Rajya Sabha declares, by a resolution supported by two-thirds of the members present and voting and for the sake of expediency and national interest, that parliament should make laws with respect to any matter enumerated in the state list, it could do so. Such a resolution remains valid for a year, and it can be extended by another year by a subsequent resolution. Similarly, under Article 250, parliament is empowered to make laws on any item included in the state list, for the whole or any part of India, while proclamation of an emergency is in operation.

Another instance of the central government's violation of the federal principle has been the arbitrary use of Article 352 and 356. Article 352 provides for a proclamation of emergency when "the President is satisfied that a grave emergency exists whereby the security of India or of any part of the territory thereof is threatened, whether by war or external aggression or armed rebellion." Article 356 contains provision for the assumption of legislative and executive powers when "a situation has arisen in which the government of the State cannot be carried on in accordance with the provisions of this Constitution." Using or misusing Article 356 over 100 times, "President's rule" (union government directly taking over the government of a state) has been imposed over states since 1950. It has given rise to severe criticism for violating the federal character of India. The Supreme Court of India ruled in 1994 that the power of the central government under article 356 to remove a state government from office was not an absolute but a conditional power (*S. R. Bommai and Others v. Union of India,* 1994). An emergency under article 352 was declared in October 1962 (Sino-Indian conflict). It was revoked in 1968. It was proclaimed again in December 1971 (war with Pakistan). A fresh proclamation of political emergency was made in June 1975 (on the grounds of internal disturbances) suspending enforcement of fundamental rights. Both these proclamations–1971 and 1975–were revoked in March 1977.

In financial matters, it may be said that the general tendency has been in favor of centralization. The union is financially more powerful than the states. Discussing the developments in the economic administration of the last 40 years, experts came to the conclusion that "economic institutions and trends have tended to encourage systems of administration which have not entirely been in tune with true federal concepts. The politico-economic structure that emerged has almost become feudal in spirit, if not in form" (Krishnaswamy, in Mukarji and Arora, 1992: 187).

It is widely recognized that the Indian system of governance rests on two basic principles: democracy and federalism (Mukarji and Arora, 1992: 267). In the last five decades, India has come a long way on both these counts. Only in a federal polity could the unique socio-cultural diversities of the country as a whole and the states in particular be held together as a nation. Rasheeduddin Khan, who has done extensive work on the subject, calls India an evolving "Federal Nation."

> A Federal Nation is a mosaic of people in which unified political identity is reconciled with socio-cultural diversities. Its hallmark is unity of polity and plurality of society. It is a conglomerate of segments whose diverse identities based on ethnicity, language, religion, region, etc., are nevertheless united politically into territorial sovereignty. (Khan, 1992: 29–30)

Given the manifold dimensions of India's pluralistic society, the federal principle offers the only viable basis for the maintenance of a strong and united Indian state. However, the extension of participatory democracy to the popular base, which reflects the social realities of a federal polity, was, unfortunately, lacking in the past due to centralizing tendencies. Therefore, the search for institutional arrangements for improving the federal system was at the top of the agenda of concerned intellectuals, the judiciary, thinkers, and political parties. By the late 1980s, it was realized that extension of the federal idea hinges on decentralization at the substate level. It may be recalled that a serious flaw of the Indian Constitution adopted by the Constituent Assembly was the absence of primacy for the local governments in the rural and urban areas. The local bodies for rural India are traditionally called *panchayats.* According to the 1991 census, 74% of India's population lives in rural areas, but the local bodies for this vast population had no constitutional status; they were mentioned only in the Directive Principles of the Constitution. They did not have any role–either developmental or governmental. According to Article 40 (Directive Principles), it was left to the states "to take steps to organize Village Panchayats and endow them with such powers and authority as may be necessary to enable them to function as units of self-government," but the state governments did not take this seriously.

The inseparable link between local government and democracy was in the spirit of the Constitution, the democratic struggles, and the ethos of the country in the 1940s and 1950s. In 1962, the Allahabad High Court heard a case concerning elections for municipalities in the state that had not been held for a long time. The government took the stand that it had the freedom to decide whether to hold elections or not. In its judgment, the Allahabad High Court said that without local self-government, the rest of the democratic structure would collapse.

> Local self-government was meant to go on progressively with greater participation of the people, whether in village or in urban areas. This continuity in bringing local self-government with people's participation at the grassroots level of democracy was to be protected and perfected and not to be forgotten. From down below, centripetally culminating in Federal Government, there was to be a process of representative government. Local self-government is the arch upon which parliamentary democracy in this nation rests, and upon it a dome with State Legislatures and the Federal Parliament. The structure will collapse without the arches. (All India Reports, 1992: 24–25)

Since the nation did not heed these words, the structure nearly collapsed. There were many signs of the structure cracking and collapsing: political emergency, suspension of democratic rights and their widespread violation even after the emergency, the rise of terrorism and violence, political assassinations, secessionist tendencies, and so on.

It took 43 years from the time the Constitution of India was adopted for the local bodies in the country to become statutory bodies. On 22 December 1992, the parliament passed the Constitution (73rd Amendment) Act 1992, and after several other formalities, on 24 April 1993, panchayats were brought under Part IX, "to enshrine in the Constitution certain basic and essential features of Panchayati Raj Institutions to impart certainty, continuity and strength." (G. Venkataswamy, "Statement of Objects and Reasons: The Constitution (Seventy Second Amendment Bill)," New Delhi: Government of India, Ministry of Rural Development, 1991). The Constitution (74th Amendment) Act was also passed at the same time to cover the urban local bodies.

In fact, the realization occurred by the mid-1980s that with concentration of power, Indian polity could crumble. In the late 1970s and early 1980s, West Bengal, Karnataka, and Kerala had given new status and power to panchayats, and they became political institutions in these states. Their success, as well as a general demand for decentralization of power with the slogan, "power to the people," also accelerated the pace for the Constitutional Amendment.

For the first time, on 15 May 1989, the 64th Constitution Amendment was introduced in parliament to give constitutional status to the panchayats by the erstwhile Prime Minister Rajiv Gandhi. The opposition in the parliament attributed political motives to this move by the prime minister and defeated the bill. Nonetheless, it could be seen as a response to the ungovernable conditions emerging in the country as a result of concentration of power at the center and the desperation articulated at the lower level, as nothing was working for the welfare of the people. The National Front government, which came to power in December 1989 following the Congress Party's defeat in the elections, introduced the 74th Amendment Bill (a combined bill on panchayats and municipalities) on 7 September 1990 during its short tenure in office, but the bill could not be taken up for discussion. Finally, it was the Congress Government, under Prime Minister Narasimha Rao, that made the constitutional amendments.

Significance of the Constitutional Amendments

So far, democracy in India has been confined to a parliamentary exercise, electing about 5,000 members to the parliament and the state assemblies once every five years. It was elitist in nature, with muscle power, money power, and caste as strong influences in the elections. Democracy had not taken deep roots for want of suitable support structures. In the last 50 years, elections to the parliament and the state assemblies meant very little in terms of meaningful democratic functioning or the creation of democratic culture at the grassroots level.

Now that the local bodies have constitutional sanction, elections to the three tiers of the panchayats and municipalities should be held every five years. Panchayats in states with a population above 2 million have three tiers: the village panchayats at the lowest level, the block panchayats at the intermediate level, and the district panchayats at the top. There should also be three types of municipalities in urban areas (depending on the population of towns and cities): the municipal corporations, the municipal councils, and the Nagar panchayats. India now has more than 500 district panchayats, about 5,100 block or taluka panchayats, and about 225,000 village panchayats. There are 3,397 municipal bodies in urban areas of the country. The panchayats, along with their urban counterparts, the municipalities, will be electing about 3 million representatives. This has certainly broadened the base of Indian democracy.

The change in the federal polity of the country that has taken place as a result of the local bodies becoming constitutional entities has far reaching consequences. It could be revolutionary. According to the constitutional provisions, the panchayats and municipalities have become the de facto "third tier of governance" in the country. The Eleventh Schedule and the Twelfth Schedule added to the Constitution (along with the amendments) have suggested transfer of 29 subjects to panchayats and 18 subjects to municipalities. The State Conformity Acts have more or less incorporated them, and when they are made fully operational, one can say that the local bodies at the district level and below will have become the third tier of governance in the Indian federal structure, giving the Indian Federation a new meaning. Of course, these local bodies have no legislative powers; nor do they have law and order (police) powers or judicial powers. The creation of local courts (*nyaya panchayats*) is mentioned in the Conformity Acts of some states

Articles 243 G and 243 P (e) of the Constitution define panchayats and municipalities as "Institutions of Self-Government," but nowhere is the scope of these institutions defined. Only three state conformity legislations (those of West Bengal, Tripura, and Bihar) unequivocally mention that the objective of their panchayat legislations is to endow panchayats with powers so as to enable them to function as vibrant institutions of self-government. The other extreme is that of the state of Haryana. The Haryana Act categorically says that the panchayat system is meant for administering the rural areas better. That is another way of saying that the state still believes in the "DC (Deputy Commissioner) Raj" and "SHO (Station House Officer) Raj" for administration at the district level and below.

The term *institutions of self-government* has been interpreted in two ways. The first is that when the Constitution says that panchayats are institutions of self-government, it implies that they must have autonomy and the power to govern in an exclusive area of jurisdiction. Governance by elected representatives of the people, according to the Constitutional provisions, is therefore its essential element. The 73rd Constitution Amendment gives panchayats this distinct status. Therefore, it is the de facto third tier of governance.

The second interpretation is that it is only strengthening "administrative federalism" and nothing more. The provisions of the 73rd Amendment strengthen administrative federalism in order to facilitate and encourage delegation of administrative and financial powers from the states to local bodies. Their administrative powers and responsibilities and the financial resources to exercise these powers and to discharge the responsibilities are entirely derived from legislation that will have to be passed by the state (S. Guhan, "Federalism and the 73rd

Amendment: Some Reflections," Institute of Social Sciences, New Delhi, 1995, unpublished manuscript). They have no legislative or judicial power and, according to this argument, just having constitutional status or regular elections does not confer on them the status of the third tier of governance.

In countries where local bodies exist, they are given powers of delegated legislations, such as budget, bylaws, and regulations. They also enjoy considerable powers of regulation attached to their functional responsibilities. Police and judicial powers can also rest with the panchayats.

> Two goals that have somehow remained outside the ambit of most thinking about local governance are decentralizing the system of justice and, paralleling that, decentralizing police functioning. There is clearly a linkage between the two. . . . The Constitutional Amendments that gave birth to the new panchayats and municipalities were, however, completely silent on the subject. . . . Very likely this was inadvertently done. Whatever the reason, decentralized justice now needs to be vigorously reviewed, for without a local "judiciary" local governance will remain glaringly incomplete. (Nirmal Mukarji, preface to S. N. Mathur, *Nayaya Panchayats as Instruments of Justice,* New Delhi: Institute of Social Sciences and Concept Publishing, 1997: 8)

It is evident that we cannot have a decentralized system of justice (*nyaya panchayats*) without a decentralized system of policing. Therefore, in the not-too-distant future, India will have village, block, and district police as well as city, town, and state police. Today, police is a state subject.

The moves to make the local bodies really the third tier of governance in India have been incremental, and there has not been a quantum leap. The passage of the Constitutional Amendments, however, could be considered a watershed because it has paved the way for the creation of district governments in the country. The fact that district planning responsibilities are also given to the local bodies through these amendments is a major step in that direction. The chairperson and members of the district panchayats elected by the voters of the district hold the key to a new polity. So far, districts were only administrative units. Now they are governance units. Until now, everything was centered around a district administrator, called a Collector. In the new dispensation, "Collector Raj" should give way to "Panchayati Raj"; that is, India will have about 500 more chief ministers in the country for each and every district in addition to the present 27 chief ministers (25 states, Delhi and Union Territory of Pondicherry have chief ministers).

After about half a century, doors are being opened for a reversal of India's political system. Until today, it was top heavy, everything flowing from top to bottom, and it depended on the goodwill of those managing the system. The people at the bottom did not matter. In the place of a few thousand representatives of the people who managed the affairs of the country, now a few million representatives will voice the concerns of people at different levels. Thus, the new amendments and local bodies are a leap forward. The edifice of parliamentary democracy is, for the first time, seen from bottom up. No more can the people at the lowest level–the villages and municipal wards–be taken for granted.

It is significant that today, "we the people of India" have given constitutional sanction to the arches of local self-government on which the dome of parliamentary democracy rests. It has opened up the possibility of looking at the federal structure from below.

Elections to the Local Bodies

The elections to the panchayats in India has been an excellent barometer of a functioning democracy. The polling percentage is quite high. In the panchayat elections held in January 1997, in Orissa, in some panchayats the voting was as high as 90%. The study of the panchayat election process and election issues in Karnataka in 1995 (Subha, 1997) and Tamil Nadu in 1996 (Institute of

Social Sciences, 1996) reveals some interesting facts about the democratic process at the grassroots level (Subha, 1997).

The most important aspect of these elections is that caste and religion, which were playing a prominent role in deciding the outcome of state and parliament elections in the last 50 years, have shown signs of declining in importance in the panchayat elections in some states. In Tamil Nadu, for example, during the 1996 panchayat elections, growing democratic consciousness of the people was clearly evident. Nearly 81% of the respondents of a survey said that the religious or caste leaders did not influence the people to exercise their voting right in one way or another. To the direct question of whether caste affinity was important for voting a particular candidate to power, an overwhelming majority of 73.6% said that the caste of the candidate was not an issue at all. In Karnataka, for 63.2% of the respondents, the caste of the candidate was not an issue at all, and 44.9% of the respondents said they voted for a particular candidate because he or she was a good person. These election studies show that the local body elections lessen the intensity of casteism and parochialism. This is mainly because people in general give priority to the welfare and development activities in the villages or blocks, transcending caste or party politics. For the general voter in the villages, the track record of the candidate on issues of corruption or ability to do some development work was more important. The May 1993 elections to the West Bengal panchayats also highlighted these positive aspects at the grassroots level.

Women and Weaker Sections in Local Governance

The new panchayats and municipalities provide opportunities for weaker sections–the Scheduled Castes (SCs) and the Scheduled Tribes (STs), who form 25% of India's population–to actively participate in them and get elected as members as well as chairpersons. The membership is decided by the proportion of their population in an area.

The terms *Scheduled Castes* and *Scheduled Tribes* refer to such castes/tribes, parts of, or groups within such castes/tribes as are deemed to be Scheduled Castes and Scheduled Tribes under Articles 341 and 342, respectively, of the Constitution of India. These categories have been created in the context of classification of castes and tribes in India in order to allocate reservations for government jobs, education in university colleges, etc., to economically underprivileged persons. Scheduled Castes are sometimes referred to as *Harijans* (people of God), as Gandhi called those who were considered "untouchables" by high-caste Hindus. Now these castes call themselves *Dalits* (oppressed). The use of the term *untouchables* is prohibited by Indian law.

The fifth and sixth schedules in the Constitution give special status and privileges to tribal areas. In reality, in tribal majority areas, nontribals have been controlling the affairs, dominating the scene, and destroying the tribal tradition. Tribal land has been appropriated by nontribals. The Union government in July 1994 set up a high-level committee under the chairmanship of D. S. Bhuria, member of parliament, to suggest proposals to extend the 73rd Constitution Amendment to the scheduled areas. The committee recommended (1) to constitute a village assembly in all tribal villages, because the community should be the basic unit of self-governance in tribal areas, (2) to reserve a majority of seats in all levels of the elected bodies for members of the Scheduled Tribes, and (3) only a tribal could be elected as a *sarpanch* (president).

Both houses of parliament adopted the Panchayats (Extension to the Scheduled Areas) Bill, 1996, framed in accordance with the recommendations of the Bhuria committee in December 1996. The act will benefit the scheduled areas referred to in Clause (1) of Article 244 of the Constitution, which include areas in Andhra Pradesh, Bihar, Gujarat, Maharashtra, Madhya Pradesh, and Rajasthan.

A unique feature of the new phase in panchayats and municipalities in India is that it has ensured one-third representation for women in the local bodies and one-third of the offices of chairpersons at all levels in rural and urban bodies. This has created the possibility for about 1 million women to get elected to the panchayats and municipalities. This is no mean achievement. It is important to note that in several states, more than the mandatory 33.3% women have been elected. The latest panchayat elections show that women responded in full measure to the opportunity provided to them. In West Bengal, on an average, nearly three candidates contested every seat reserved for women at the village panchayat and panchayat samiti levels, with the number going up to four at the district panchayat level. Out of 24,855 seats reserved for women in three tiers, only 561, or little over 2% went uncontested (Kumar and Ghosh, 1996). The last three years are replete with stories of how women panchayat members and women chairpersons asserted their rights, and things changed for the better. Today, even if a small percentage of the elected women representatives work successfully, it shows great promise.

Problems and Prospects

The reluctance of state-level politicians to recognize the importance of the lower levels of governance–their autonomy, their powers, and their areas of functioning–is creating serious problems for the healthy development of local government systems in India. The ministers, the members of the legislative assemblies of the states (MLAs), and senior political leaders are worried that the power they enjoyed so far will diminish if panchayats and municipalities become powerful. The MLAs feel threatened by this.

An MLA from Chattisgarh (Madhya Pradesh) said recently, "MLAs' role had been minimised in the new order and it had weakened them in their respective constituencies. A system of 50 years can't be changed like this and MLAs dumped." A woman MLA from Jabua, again from Madhya Pradesh, said, "Panchayati Raj, though good in many respects, had created unrest among MLAs" (*Hindustan Times* [New Delhi], 20 March 1997). The chief ministers and the ruling parties at the state level feel frightened at the possibility of people asserting their rights and punishing them. Therefore, they have tried time and again to postpone the panchayat elections. They do not want active and functioning local bodies to be "nurseries" of leadership, and they put hurdles in the way of the smooth functioning of panchayats to prevent them from blossoming into full-fledged local governments. They are afraid that the leadership that may come up from the lower levels will pose challenges to them in due course. In Orissa, when the new government came to power in early 1995, it decided to dissolve the duly elected panchayats and municipalities. The real reason for this action was that the MLAs were impatient to wrest full control of large sums of money coming to the panchayats through central government schemes for rural development. The case of the recent drought relief measures was no different. If panchayats function properly with a large number of elected representatives and under the critical eye of the opposition at the local level, people will become aware of their rights through regular participation in the panchayat programs and activities, resulting in the decline of the powerful position MLAs enjoy today.

Government officials and government employees prefer to work with a distant control mechanism–that is, the state capital. They do not want to be closely supervised under the panchayati raj. Therefore, their noncooperative attitude toward elected panchayat members is a major issue. Even in a state such as West Bengal, with a long history of panchayati raj, whenever staff is placed under panchayats, court injunctions come against such actions. The minister for panchayats in West Bengal, Dr. Surjyakanta Misra, said recently that hundreds of such court injunctions are pending against the department orders (Surjyakanta Misra, "A Note on Achievements on the Panchayats and Rural Development Fronts in West Bengal," Department of Panchayati Raj, Government of West Bengal, 1996, mimeographed). Another related issue is that officials working at the district level and below are reluctant to take orders from the elected panchayat executives, such as the district panchayat president, block panchayat president, or village panchayat president. Here, India needs a new culture of democracy to make the local governments work.

A low level of political consciousness in many parts of the country is another factor that will pull the new panchayati raj backward. The states of Bihar, Madhya Pradesh, Rajasthan, Uttar Pradesh, and Orissa, with a population of about 370 million (1991 census), have a low panchayati raj performance rating. The main reason is the low level of political awareness and the prevalence of feudal authority and feudal values. Madhya Pradesh was the first state to hold elections to the panchayats after the 73rd Constitution Amendment, and elected local bodies came into existence, but reports began to appear in newspapers that all was not well with their functioning. A chain of events was reported from different parts of the state: a woman president was stripped naked, another woman president was gang raped, a lower-caste vice president was tortured, and a Scheduled Caste panchayat member was beaten up. A sociological investigation of these incidents showed that "a Panchayat is a microcosm of the society of which the village forms a part. The noble ideals of 'institutions of self-government' as expounded by the 73rd Constitution Amendment, cannot be translated into reality in the present inequitable society" (George Mathew and Ramesh C. Nayak, "Panchayats at Work: What It Means for the Oppressed," in *Economic and Political Weekly* 31, no. 27 [1996]). All the case studies illustrate that there exists a social system that violates the dignity of the individual, a social value system that does not accord any respect to the human person or the office he or she holds.

In many places, the panchayats themselves are working as oppressive instruments. Absence of land reforms, low levels of literacy, especially among women, patriarchal systems, etc., go against weaker sections in the villages. All these factors seriously affect women's functioning in the elected bodies. The majority of people suffering from the effects of traditional oppressive power structures are unable to utilize effectively the new opportunities provided by the panchayats. After the new panchayats have come about, serious conflicts have taken place during elections and afterward in their functioning in the villages. The panchayat elections in Orissa had widespread violence, resulting in the loss of life of 12 persons.

The central government itself creates situations that are not conducive for the growth of panchayats. The provision in the Constitution providing MLAs and members of parliament (MPs) to participate in the meetings of block and district panchayats is contrary to the independence of these bodies. Moreover, any program, scheme, or organization created parallel to the functioning of panchayats will undermine the local government system. An example of a serious offense is the disbursal of Rs. 8,000 million out of the Consolidated Fund of India at the rate of Rs. 10 million per member of parliament, popularly known as MPs Constituency Development Scheme (M. A. Oommen and Mahi Pal, "Local Area Development Scheme: Dangerous Portent," in *Economic and Political Weekly* 29, no. 5 [1994]: 223–25). This not only interferes in the constitutional provisions but also in the federal spirit with which the new local bodies will be operating. The former chief justice of India, E. S. Venkataramiah, has said that "It has the effect of interfering not merely with the federal scheme, but also with the healthy constitutional principle of separation of powers" (E. S. Venkataramiah, "MPs' Constituency Development Scheme: Assaulting the Constitution," in *Indian Express* [New Delhi], 13 February 1997).

In spite of the preceding account, after the Constitution Amendments, several developments point to a situation of hope and optimism.

The nongovernmental organizations (NGOs), community initiatives, people's organizations, and the civil society in general are playing an important role in strengthening the panchayats and municipalities. After the 73rd and 74th Constitution Amendments, a large number of NGOs in India are helping to create enabling conditions for the success of the panchayats through awareness-building programs, training elected members (especially women) and ensuring their active participation in elections, and assisting panchayats in planning and implementing social development strategies and programs. The local bodies in India, with their constitutional legitimacy and interaction with citizens' groups, and voluntary organizations present an ideal meeting point between the state and the civil society.

After the establishment of the new panchayats, new programs with the participation of villagers are being taken up in a meaningful way. The participatory and sustainable Panchayat Level Development Planning (PLDP), with the slogan "planning by the people for the people," that is taking place in Kerala panchayats is a case in point. People's Campaign for the Ninth Plan was a landmark program. The approach paper stated that it was an attempt to empower the panchayat bodies by ensuring that the panchayati raj/municipal bodies prepare and prioritize a shelf of integrated schemes in a scientific manner (Government of Kerala, 1996: 1). By all accounts, this program has assumed the dimensions of a movement and has attracted nationwide attention. In the 1997–98 budget, the Kerala government has allotted 36% of plan funds for panchayati raj and Nagar Palika institutions with guidelines that it should be spent on productive sectors, social service sectors, and on infrastructure in a 40:30:30 ratio in rural areas.

Even the states that had a disappointing history of panchayati raj are coming forward to give panchayats adequate powers and finance. Madhya Pradesh and Tamil Nadu are examples. Immediately after the October 1996 panchayat elections, the Tamil Nadu government, "as a measure of its anxiety to endow the newly elected panchayat with constructive responsibilities in regard to development programmes" (Government of Tamil Nadu, 1997: 9–10), asked the State Planning Commission to constitute a special group to advise the state government on which of the powers and functions (which are now being carried out by the state departments) may be entrusted to the local bodies in an orderly scheme of governance of development.

On the basis of the recommendations of this special group, the government has given orders to the effect that the local bodies will make decisions, and the officials will have only the responsibility of implementing these decisions. The government declared that the panchayat president is the executive authority. In this year's Tamil Nadu budget, the allocation to the local bodies has gone up to 82% more than last year's budget from the revenue account.

The Constitution Amendment provides necessary conditions for bringing into existence vibrant local governments. Political will is an equally important factor for panchayats to take roots, but what will happen if there is no political will? In the post–73rd Constitution Amendment scenario, hitherto impossible developments to strengthen panchayats are taking place. In Orissa, the state government and MLAs tried all that they could to withhold elections. But the constitutional authority–the Election Commission, the press, and the people–wanted the elections and the judiciary lent a helping hand. Against considerable odds, the Orissa government has completed elections in January this year, and the state has about 100,000 people's representatives in rural areas (George Mathew, "Orissa's Tryst with Panchayati Raj," in *The Hindu* [Madras], 21 February 1997). All this is happening because of the strength of the Constitution Amendment and the new climate created since 1993.

An important contribution the panchayati raj can perhaps make to politics is reducing corruption, which has become a bane of national life in India. There is a viewpoint that along with political and economic decentralization, corruption will also be decentralized. One can see that accountability to the people at the lower levels is higher in front of the watchful eyes of the people.

> At the decentralised level of governance, transparency is forced on the functioning of the locally elected representatives and bureaucrats and it is easy for the local people to identify and expose corruption. This has been proved in some states where the proportion of intended benefits reaching the target groups has increased substantially ever since they have been channelled through decentralised local self-governments. (G. Thimmaiah, "Decentralisation: Is It a Danger or Virtue?" Hamdard University, New Delhi, 1977, mimeographed: 11–12)

In many cases, *gram sabhas* (village assemblies) and village panchayats are becoming forums of social audit, making the elected representatives alert.

The new panchayati raj is opening up possibilities for a better flow of information. Information is power, and the dominant class in the country for a long time kept the ordinary people in the dark. Transparency in public dealing was not there because everything was officially secret and confidential. Panchayats will break this centralized information system when 1 million elected members ask for information on a variety of matters that affect people's lives.

New Economic Policy and Panchayat Finance

Since 1991, India has launched a New Economic Policy (NEP) in the wake of a severe balance of payment crisis. At that time, inflation was running high, the external solvency was at the lowest ebb, investors' confidence was negative, and fiscal deficits were going out of control. The main features of the NEP were (1) devaluation of the rupee by 21%, (2) a new industrial policy allowing more foreign investments, (3) opening up more areas for private investment and referring sick public sector units to the Bureau of Industrial and Financial Reconstruction, (4) sale of part of government equity in profitable public sector enterprises to the private sector, (5) closing sick public sector enterprises, (6) an exit policy for the private sector, (7) reform of the financial sector by allowing foreign banks to conduct business, (8) decentralization of all imports, (9) indiscriminate export promotion, and (10) market-friendliness and less government intervention.

The new economic policy has created a better liquidity situation in the country due to capital inflows. Today, the foreign exchange reserves have touched US$26.49 billion, excluding gold (Reserve Bank of India, *Weekly Statistical Supplement,* 21 April 1997). India's creditworthiness has improved.

It was widely felt in India that the NEP affected the poor and vulnerable sections and their access to basic services. This was consequent to retrenchment in the organized sectors and fewer opportunities for employment in the unorganized sectors of industry as well as inflation and reductions in overall public expenditure. As a result, it is widely felt that the economic policy for the poor must include the Public Distribution System, poor-oriented employment and income generation programs, primary education, primary health care, drinking water and sanitation in rural areas and urban slums, child nutrition, and maternal and child health services (Guhan, in Arora and Verney, 1995). It is evident that the implementation of these tasks depends on the states, and the state's capacity depends on the extent to which the central government is able to increase its overall revenue transfers to the states.

In India, it is now recognized that the poor-oriented programs (or in a larger sense the common-people-oriented programs) are best implemented only through decentralized bodies. Here, the local bodies–the panchayati raj institutions and municipalities–with their constitutional status and with more powers vested by the State Conformity Acts, assume great significance. The decentralization through local bodies, according to Thimmaiah, is not only intended to achieve allocatable efficiency in the provision of public goods but also to involve the local people in setting local development priorities and enabling them to provide basic local public services: "Since these basic needs are not likely to be provided adequately by the private sector at affordable price, they become local public goods. Decentralisation is an effective means of involving local people in improving the local delivery system of local public goods" (G. Thimmaiah, "Decentralisation: Is it a Danger or Virtue?" Hamdard University, New Delhi, 1977, mimeographed: 11–12). With the NEP, there has been a clamor for privatization of all service delivery systems. Its logical end is that all basic services go beyond the capacity of the poor. Only the local bodies can protect the interests of the disadvantaged through their democratic participation and decision making. In this context, the financial strength of these institutions of self-government at the local level assumes importance.

The crux of the autonomy of panchayats and municipalities in India is related to finance (M. A. Oommen, "Panchayati Raj: A New Challenge to Federal Finance," in *The Hindu* [Madras], 19 November 1995). The state finance commissions (SFCs) have been established in all the states according to the provisions of the Constitution Amendment. These commissions are expected to look into the tasks relating to substate-level fiscal devolution that are critical for a federal polity. Based on articles 243G, 243H 243I, and 243ZD of the Constitution, these could be summarized as follows:

1. To formulate clear ideas regarding the divisible pool after taking into account the functional domain of the state on the one hand and of the panchayati raj institutions (PRIs) and urban local bodies on the other
2. To broadly evaluate the vertical gap at the PRI and urban local bodies levels, taking into account their revenue assignments and potentials on the one hand and expenditure responsibilities on the other
3. To suggest measures to reduce the vertical gap that includes tax sharing and grants
4. To design methods for determining the inter se share of PRIs as well as urban local bodies on an equitable and efficient basis (M. A. Oommen, "Panchayati Raj: A New Challenge to Federal Finance," in *The Hindu* [Madras], 19 November 1995).

Most of the State Finance Commissions have submitted their reports along these lines.

The Tenth Finance Commission of the central government, which submitted its report in 1995, has recommended an ad hoc per capita grant of Rs. 100 for rural populations (based on 1971 Census figures) for the period 1996–2000. This grant-in-aid works out to Rs. 438,093 million (Rs. 109,523 million per year). These allocations are meant to be only an "additional amount over and above the amounts flowing to the local bodies from state governments" (Government of India, *Report of the Tenth Finance Commission 1995–2001,* New Delhi: Government of India, Ministry of Finance, 1994: 48). In addition, the Panchayat Ministers' Committee has recommended that all state governments make 5% of their tax and nontax collection available to the panchayats.

The panchayats and municipalities will have the following sources of funds: (1) allocations by the central government through the Central Finance Commission, (2) allocations by the state government, (3) central government rural development schemes, and (4) taxes and other revenues that the local bodies themselves raise. It has been found in states such as Karnataka that when people realize that local bodies function responsibly and democratically, their capacity to raise resources also increases substantially.

District planning (Article 243ZD) is required to maintain an organic link with state and national planning. It also envisages a bottom-up planning exercise, building blocks comprising the plans of village panchayats, block/Taluka-level panchayats, district panchayats, and municipal bodies. (West Bengal already has a District Planning Committee Act, which envisages a district council chaired by the leader of the opposition.) However, the interlinkages between the Union Finance Commission (UFC), the Planning Commission, the state finance commissions (SFCs), the state planning boards, and the district planning committees is not yet clear.

To conclude, federalism is central to Indian democratic polity, particularly in order to accommodate India's unique diversities. Undoubtedly, panchayati raj is the logical extension of federalization to the grassroots level. Only the local governments can be effective liaisons between the ordinary people and the central and state governments. Elected local representatives with the closest proximity to the people are best placed to articulate their constituents' grievances and respond to problems. Interestingly, in the context of the new economic policy, the local government system assumes a critical role. Its responsibility for social sectors has increased manifold. Furthermore, in the emerging federal finance structure, the local bodies are acquiring powers and unprecedented financial strength. All this augurs well for strengthening the grassroots level democracy and rapid rural and urban development in India.

Further Reading

All India Reports, *Karnataka High Court Judgement,* New Delhi: Institute of Social Sciences and National Dairy Development Board, 1992

A document that gives the text of the court's judgment, delivered on 10 April 1992, on the petition to delay holding the elections to the district panchayats in the state. The High Court found that the action of the state government in delaying the election was contrary to the Constitution of India.

Arora, B., and D. V. Verney, editors, *Multiple Identities in a Single State: Indian Federalism in Comparative Perspective,* New Delhi: Centre for Policy Research, Centre for the Advanced Study of India (Philadelphia), and Konark Publishers, 1995

This book contains the papers presented at a seminar held at the Centre for Policy Research, New Delhi, on 5 and 6 January 1993 and at the University of Pennsylvania on 21–23 October 1993. It is divided into three parts: (1) "Political Structures, Parties, and Governance," (2) "Economic and Financial Issues," and (3) "Towards a Research Agenda." It contains 11 papers, including S. Guhan, "Federalism and the New Political Economy in India."

Government of Kerala, *People's Campaign for 9th Plan: An Approach Paper,* Trivandrum: Government of Kerala, State Planning Board, 1996

A booklet introducing the concept of planning from the village level and of people's participation in the planning process. The people's campaign for the ninth plan began in October 1996, in Thiruvananthapuram, capital of the state of Kerala.

Government of Tamil Nadu, "Entrustment of Powers to Rural Local Bodies State Planning Commission (First Report)," Madras: Government of Tamil Nadu, State Planning Commission, 1997

A report submitted to the Government of Tamil Nadu by the State Planning Commission in January 1997. The report deals with how the powers may be devolved to the district, block, and village panchayats in the state of Tamil Nadu.

Khan, Rasheeduddin, *Federal India: A Design for Change,* New Delhi: Vikas, 1992

This book places the problem of federalism in the global context and focuses attention on the problems of federal polity in India. It is a policy-oriented study covering themes such as the sociocultural dimensions of India's federal polity, political federalism, and the pattern of federal nation building in India. It is published under the auspices of the Indian Institute of Federal Studies, New Delhi.

Kumar, Girish, and Buddhadeb Ghosh, *West Bengal Panchayat Elections 1993: A Study in Participation,* New Delhi: Institute of Social Sciences and Concept Publishing Company, 1996

This book, based on field surveys in West Bengal during and after the May 1993 panchayat elections, offers a critical analysis of the extent of popular participation in the panchayat election process and the working of the panchayat institutions. The authors base their analysis primarily on the state's successful and long-standing experience with local self governance since 1978.

Mathew, George, "Restructuring the Polity: The Panchayati Raj," New Delhi: Indian Renaissance Institute, 1997

Mathur, S. N., *Nyaya Panchayats as Instruments of Justice,* New Delhi: Institute of Social Sciences and Concept Publishing Company, 1997

This book looks at the village courts in the emerging context as a system of decentralized justice in relation to the broad objective of local governance.

Misra, Surjyakanta, "A Note on Achievements on the Panchayats and Rural Development Fronts in West Bengal," Department of Panchayati Raj, Government of West Bengal, 1996, mimeographed

Mukarji, Nirmal, and Balveer Arora, editors, *Federalism in India: Origins and Development,* New Delhi: Centre for Policy Research and Vikas, 1992

This book contains the papers presented at a conference held in New Delhi on 11 and 12 December 1987 at the Centre for Policy Research in collaboration with the Centre for Ethnic Studies, Sri Lanka. The papers analyze and interpret the dynamics of federal processes in different spheres, especially in the post-Nehruvian Phase. It provides an interdisciplinary perspective on the mainsprings of India's federal polity. The book contains ten papers and an epilogue.

Subha, K., *Karnataka Panchayat Elections 1995: Process, Issues, and Membership Profile,* New Delhi: Published for the Institute of Social Sciences and Concept, 1997

This book critically examines the important issues that influenced the voting pattern in the 1993 and 1995 panchayat elections in the state of Karnataka and the level of popular participation in the *gram* (village), *taluk* (block), and *zilla* (district) panchayats. It is based on extensive field investigations in all the districts of the state.

George Mathew received his Ph.D. in sociology from Jawaharlal Nehru University, New Delhi. He has been a visiting fellow at the University of Chicago South Asian Studies Center (1981–82) and a visiting professor at the University of Padua (1988). He was awarded a Fulbright Fellowship in summer 1991 to work at the University of Chicago, and he has participated and presented papers in international conferences on religion and society, political process and democracy, and human rights. His studies and articles on state and society have appeared in national dailies, journals, and books. He is currently Founder Director of the Institute of Social Sciences, New Delhi.

Chapter Eight

Changing Indian Agriculture: Agrarian Society, Economic Planning, and Development since 1947

Allen Kornmesser

Agriculture and the Village

One out of every ten people on the face of the earth lives in the Indian countryside and depends, directly or indirectly, on agriculture to survive. Seventy percent of India's work force is directly employed in agriculture, either as owner-cultivators or agricultural laborers, but there are many more villagers who work in various other occupations related to agriculture, and there are many, particularly women and children, who labor on their own and others' fields but are not recognized in official statistics as "employed" in agriculture. Add to these a vast population of urban residents employed in agriculture-related occupations–traders in farm produce, producers and distributors of seeds, fertilizers, and other agricultural technologies, bankers, bureaucrats, and agents of government departments administering development plans and programs–and one begins to get a sense of the importance of agriculture to India and its economy, and indeed to the world.

Although agriculture employs more than two-thirds of the Indian work force, the growth of industry and the nonagricultural sector has reduced its share to barely one-third of the nation's gross domestic product (GDP). Even so, because of its centrality in the lives of the majority of the population, and its interdependence with the rest of the economy, agriculture remains a kind of keystone of the Indian economy. When agricultural production is down, the nation's economy suffers. It is only in the last decade that industry has become sufficiently independent of agriculture that a failure of the monsoon season did not necessarily mean a drop in industrial production and productivity.

Agriculture is also in many ways a keystone of Indian politics. Political power rises and falls in part on the strength of the economy, which in large part depends on agriculture, and there may be some natural basis for observing that political fortunes in India seem to change with the weather. But politics in democratic India also depends on the electoral support politicians can rally. With the majority of the voting population living in rural areas, peasants, farmers, and the rural poor have become important constituencies to be courted with promises of resources for agricultural development and village assistance. These plans and programs are well enough appreciated by Indian politicians, for whom they are "wishing cows" (*kam dhenu,* in idiomatic Hindi), giving forth benefits, jobs, and favors to distribute as patronage, reward, or payment for services rendered. Critics, however, have expressed growing doubt as to their effectiveness in accomplishing their stated goals, especially the goals of benefiting those at the bottom of the rural hierarchy of wealth and power.

In India, agriculture is based in more than half a million villages that dot the countryside, some with population densities equivalent to urban densities in the west. There are pockets where large-scale farming with urban or corporate ownership is occurring as well, often alongside more traditional, village-based agriculture, but land ceilings and regulations have kept agriculture a relatively small-scale operation compared with modern farming in North America. Although most rural Indians live in villages, what that means varies significantly from one village to the next. Villages are thought of as complex networks of social, personal, economic, and familial ties–groups of people bound together by blood and tradition, to the soil and each other. In India, this characterizes many villages, though others have become less cohesive habitations of people no longer so closely bound by customary ties.

Villages in some regions consist of several small hamlets, often inhabited by a few families of similar caste or occupational status, while elsewhere they may each consist of one large settlement with thousands of inhabitants. Most villages have a few hundred people at most. Agricultural land in the village may be owned primarily by independent small cultivators, using mostly their own family labor, or it may be monopolized by one or two families, with subordinate castes employed either as tenants or as landless agricultural workers. In some regions, members of a single dominant caste will own all or most of the land over vast tracts, creating culture zones built on their political and economic power and ritual centrality in the affairs of the villages within them.

The villagers survive on agriculture, and their primary occupation is cultivation of the soil, either as owners or laborers, or both. Most villages historically have included some number of artisans who make tools, ropes, pots, and such for farmers. Due to the growing availability of modern tools and plastic buckets, artisans have been replaced by vendors of goods in many villages and have been forced to change jobs or migrate to the cities. Many villages will have a priest or two, a barber, and nowadays, often a school teacher or perhaps a postman. Many villages have members whose occupations force them to live and work outside the village; these members provide income to cash-starved subsistence farm families and serve as a conduit for other villagers seeking urban employment. Many of these villagers return regularly to their villages for harvests and important ritual events.

Another notable feature of Indian agriculture, compared with much of the West, is the small size of most landholdings. There were once the enormous "estates" of the old zamindars, inheritors of the Mughal system of tax farming, which the British tried to mold into a landed gentry. But zamindari was abolished shortly after independence, and the primary tenants were given ownership rights to their sometimes considerable holdings, while their innumerable erstwhile subtenants, whose tenancy was upgraded, could mostly lay claim to only very small plots of land, many too small to feed a whole family.

The abolition of zamindari after independence and the dismantling of the multitude of princely states that had survived by some compromise with colonial power, though symbolically revolutionary, could hardly compare with the social revolutions that transformed France or China. Indeed, there has been much speculation about the lack of a serious, broad-based revolutionary tradition among Indian peasantry in spite of sometimes inhumanely harsh, exploitative, and unequal treatment of the village poor, who are usually of low caste status. There have been local "revolutions" in such insurrection-prone regions as Telanga, in Andhra Pradesh, or Naxalbari, in West Bengal (the latter giving us the term *Naxalite,* which is applied rather loosely to violent militants in India). But even when inspired by Marxist/Leninist ideology, such militancy often has an ethnic tinge, such as organizing tribal plantation workers against nontribal owners, or aligning mostly Hindu tenants against their mostly Muslim landlords. Perhaps the best explanation for a lack of revolutionary tradition is the strength of caste identity, or rather that of *jati* (subcaste), followed by local/regional or religious identity, all of which tend to eclipse any very strong feelings of class consciousness that might otherwise extend beyond these narrower bounds and threaten class warfare or revolution in India (Brass, 1994: 320–31).

Any generalizations about Indian villages and agricultural practices must be qualified by acknowledging the overwhelming diversity of forms they take. Villages represent local adaptations both to the cultural and historical backgrounds of their members and to the particular needs and conditions of their environment. The low-tech productive practices of village agriculture have made them models of "sustainability," their purpose being to provide the bulk of their own needs without dependence on outside markets and inputs. Villagers have traditionally managed their own genetic resources by saving their own seeds and preserving the local varieties of crops that best suit the specific needs and conditions of their regions. The diversity and relative independence of village economies and the sheer number of owners of very small plots no doubt contribute to the slowness with which change and development have occurred in Indian agriculture. Whether this is a good thing or a bad thing is a matter of some speculation.

Agriculture and the life of the village also hold a central place in the national consciousness. It is a persistent theme in popular culture, from folk tales to classical literature and modern urban cinema. For Gandhi and his ideological heirs, the village was the moral center of Indian civilization and the hope for a future different from that which soulless capitalism had bequeathed on Europe and America. The values of the village were simple living, modest virtue, and respect for family and tradition (though Gandhi rejected what he considered to be the "excesses" of caste tradition). The village embodied the essence of independence as a moral virtue, which Gandhi called *swadeshi*. Much more than merely a strategy against the British of buying Indian and boycotting imported goods and even weaving one's own cloth, *swadeshi* implied consuming local products rather than what was brought from far away, even within India. In self-sufficiency there was freedom, and the goal for Gandhi was always greater than mere national sovereignty. It was sovereignty of self and the corporate local sovereignty of the village from dependence on outside forces, foreign or Indian (Fox, 1989: 45, 56–59; Ashe, 1968: 179).

Though the institution of the village has survived, in India, its relation to the city has grown ever closer. Villagers still express their pride in, and preference for, their local produce and the relative independence from the outside world that comes from providing one's own basic necessities. But more and more one also sees how much the luxuries of urban life and the technologies of the developed world are becoming necessities in the minds of many villagers. Whatever independence the fabled "village republics" of India may have had has been diluted by the growing interdependence between rural and urban economies, populations, and cultures. Wealthier farmers have become significant consumers of commercial products. In addition to agricultural technologies, they buy watches, radios, motor scooters, televisions, refrigerators, etc. They travel more to cities for trade, to visit family, and for education and recreation. Meanwhile, poorer farmers and landless rural families send members to work in the cities, while others travel through urban centers en route to labor in commercial farm belts, often hundreds of miles from home.

The village is romanticized as well in the imaginations of urban Indians. Even those born and raised in the city are often conscious of their rural roots. Many can tell you the name of "their village," whether or not they have actually been there. Some urbanites travel periodically to ancestral villages for ceremonial occasions, to maintain family ties, or to survey inherited properties. The urban working poor, servants and casual laborers in the so-called unorganized sector of the economy, include many who have family ties to the villages or who will return to their villages as needed, when ill, or when they become too old to survive in the city. The villages function in this way to provide a kind of social security, lessening the economic burden on the state.

After Independence: Land Reform and Community Development

At the time of India's independence, the villages occupied a somewhat ambiguous position. In a political movement led by mostly urban, even Westernized, elites, the support of the rural masses had been nonetheless necessary to a democratic nationalist movement. However, the villages were an alien world for many in the Indian National Congress, Jawaharlal Nehru among them. Leaders of the Congress felt awkward among the unrefined and illiterate masses, from whom they or their fathers had perhaps struggled to separate themselves through education, occupation, and the adoption of a more Western style of life in the cities. From a village perspective, there was little innate difference. Both were outsiders, whether urban Indian elites or agents of the Raj. Both were regarded with suspicion. But Gandhi, with his rustic style of dress and his saintly visage, had an ability to communicate with Indian villagers in terms they understood. He was just the kind of charismatic leader the nationalist movement needed in order for the Congress to extend its base of support to the rural masses. But Gandhi's vision for India's future gave priority to the villages, even above the nation, which put him sharply at odds with Nehru and others in the Congress who saw sovereignty of state and nation as their fundamental priority (Chatterjee, 1986: 148–53).

Whatever their differences, both sides could agree that conditions in the villages were far from ideal. Agricultural production had declined over the years to a dismal low. Nationalist historians such as Dadabhai Naoroji and R. C. Dutt had argued that colonialism drained off most of the economic surplus that might have been invested in agriculture. Famines, which had struck several times in the late 1800s, demonstrated to them that India had been enfeebled to the point that it could no longer feed itself. Gandhi saw the decline as dating from Mughal times, indeed earlier, whenever foreign rule had weakened the integrity of India's ancient institutions. Colonialism had been the worst of successive waves, if also the one to unleash the spirit of independence on India's people. The import of cheap machine-made cloth and other goods had severely undermined village crafts, leaving the countryside, indeed the country, weakened and dependent. The villages, which had been bastions of abundance and independence under the mythic rule of Ram–*ram rajya*–had, under colonialism, fallen into poverty and dependence. Such was the state of the majority of the rural population at the time when independence came in 1947 (Tomlinson, 1993: 13–14; Dutt, 1906; Morris, 1963: 607–8).

It was Nehru, however, who took the helm of the new nation, though initially his party's commitment to certain Gandhian principles constrained his own instincts. For Gandhi, the nation had to be reformed from the ground up, starting with the villages and the rural poor, but Nehru loved science and technology and, as much as he was obligated to respect the villages as part of India's heritage, he saw them as antithetical to the spirit of modernity that inspired him. Nehru wanted to see India follow a rapid course of industrialization. He was impressed by the swift industrial development of the Soviet Union and hoped to use the mechanism of the Indian state similarly to organize a very disorganized economy for rapid development, albeit without the political repression of Stalinism. The creation of a Planning Commission, which gave the government primary responsibility for coordinating and planning the country's economic development, was justified at the time as a means to implement the lofty goals of the Constitution's Directive Principles, especially the alleviation of rural poverty and the reduction of inequalities between rich and poor, and between regions within India (Tomlinson, 1993: 175).

Among the aims of the First Five-Year Plan (1951–56) was the goal of "more intensive and diversified agricultural production and of a more diversified occupational structure in rural areas . . . [in order to] increase the volume of rural employment and bring increasing opportunities to agricultural workers." The primary vehicle for this effort was to be the Community Development Programme, launched in 1952. Community Development sought to extend the benefits of development, modestly conceived, to every village in India. It established a parallel structure to the old district administration, with power exercised at the level of development block, each comprised of about 100 villages. With the reestablishment of the village panchayats, some hoped that Gandhi's vision of development planning would be realized, and eventually of government itself, moving "upward" from the villages to the level of implementation. Village workers organized at the block level were authorized to set goals, coordinate, and oversee programs for providing water, education, and meeting the productive needs of farmers–acquiring seeds, fertilizer, water, etc. (Rao, 1996: A54; Brass, 1994: 138–39; Jain, 1985: 17–59).

When zamindari abolition created 20 million new owner-cultivators out of statutory tenants, the Congress paid off its debt to the rural elite and intermediate caste groups that had supported the nationalist movement rather than siding with their landlords, the pro-British zamindars, princes, and maharajas. These largest landholders in the complex rural hierarchy were seen at the time as the most likely to adopt modern agricultural methods. Though they constituted only a fraction of the rural population, they controlled more than 40% of the land available for cultivation. As for the disenfranchised landlords, as much as they resisted zamindari abolition, it paved the way for them to become more efficient capitalist farmers as well. Most were able to hold on to their best land through various technicalities and subterfuges, while being freed by the state from their traditional obligations and expenses under the zamindari system (Dantwala, 1986: 73–74; Whitcombe, 1980: 156–80; Rao, 1996: A54).

Nehru might have preferred a more aggressive role in directing the economy during the first years after independence, but the First Five-Year Plan tended rather to formalize policies that Congress had already approved or committed itself to. The government did seek more decisive land reform, urging the states to enact land ceilings and redistribute land among the poor, but state Congress leaders, many of whom had just gained ownership of land under zamindari abolition, were loathe to inflict such punishment on themselves and their constituents, and so the effort mostly floundered. The center was constrained both by constitutional limits on its powers and by a budget stretched by the exigencies of attaining political independence and going to war with Pakistan, which until then had been part of British India. There were also the costs of restoring order where partition had displaced vast populations. Some of the best and most productive agricultural land in South Asia, the fertile Punjab, had been a battle zone, which resulted in food shortages that threatened not only human survival but seemed to threaten the survival and stability of the newly emerging nation. As a result, plan outlays in the beginning favored agriculture over industry as a matter of necessity and reflected the expressed goals of the Congress to serve the masses, especially the rural poor (Kohli, 1987: 63; Frankel, 1978: 18; Brass, 1994: 278–79).

Agriculture and Industrialization: The Nehru-Mahalanobis Strategy

The failures of the early years after independence were at least as numerous as the successes. The lack of authority for the center to impose land ceiling laws on the states, or to require the states to enforce them, made a farce of further land reform beyond zamindari abolition. The Community Development Programme had failed to show much success in getting the agricultural economy growing. There were too few resources, which when spread evenly over all the villages of India left too little to make much difference in any of them. Nehru and his primary economic adviser, P. C. Mahalanobis, were persuaded that the return from investment in industry would be more immediate. Beginning with the Second Five-Year Plan (1956–61) and continuing into the third (1961–66), the Planning Commission began a strategy of fast-paced industrialization, led by the so-called public sector, which would dominate for the next 30 years. The weakness of the Indian "bourgeoisie" assured even the support of the business community for a government-directed economy. But the new strategy of Nehru and Mahalanobis reversed the early commitment to balanced growth, agricultural development, and employment for the rural poor. Instead, by concentrating its resources on industrial development, outlays to agriculture declined by nearly half as a percentage of the budget. The beneficiaries of this strategy were mostly in the cities, with industrialization having little impact on unemployment beyond them (Kohli, 1987: 56; Brass, 1994: 275).

The Mahalanobis strategy was successful in helping India build up a heavy industrial base, which has remained quite strong, but the early burst of economic growth it unleashed, spurred by large investments of foreign aid from the United States and the Soviet Union, began to weaken in the 1960s. The British Raj having been banished, the era of the planned economy ushered in what critics and deregulators have called a "permit-license-quota raj" of government regulation and red tape. The heavily bureaucratized state-run industries of the 1950s were already showing declining efficiency by the 1960s. Meanwhile, growth in the private sector was constrained by government overregulation, which in turn created an atmosphere that fostered corruption.

The lack of investment in agricultural development contributed to a decline in that sector that was aggravated by several monsoon failures. Between 1964 and 1966, there was an overall decline of 27% in foodgrain production. With the population growing steadily and the government focusing its developmental efforts on industry, India, in an effort to keep food supplies high and prices low, began importing huge quantities of foodgrains, much of it from the United States. The US Agricultural Trade Development and Assistance Act of 1954, also known as the PL480 plan, funneled US foodgrains into India for a decade beginning in 1957 (Dandekar, 1994: 219; Rao, 1996: A54).

Growing debt and the spiraling inflation that resulted caused the Indian government–on the advice of the World Bank–to devalue the rupee, leading to an economic recession and growing public doubt about the government's ability to control the economy. Popular support of the government was held together in part by patriotism in the face of two more wars, one with China over India's borders in 1962, and another with Pakistan over Kashmir in 1965. The latter conflict led the United States to cut off funds to India, exposing to a critical public the vulnerability that India's foreign aid dependence had created. Amidst these various crises, the planning process was stressed to the breaking point. A "plan holiday" was announced, from 1966 until 1969, when the government submitted its plans on a year-by-year basis until, it was hoped, some collective "vision" would be regained to lead the country out of this morass. Unfortunately, during this time, the gods continued to frown. Monsoons failed, producing severe food shortages and near famine conditions in some areas. It was clear that more needed to be done to assure that India could feed its own people, and this meant something had to be done for agriculture. But the government had committed itself to creating an industrial sector that employed millions and could not easily be abandoned, even if its momentum was continuing to decline. Besides, the growing and powerful urban middle classes were demanding the benefits that more advanced economies enjoyed and were not satisfied with the second-class status to which colonialism and an agrarian economy had relegated them. The government needed a solution that would enable it to increase agricultural production without significantly reducing its contribution to industry.

India's Green Revolution

A possible answer to this quandary appeared to emerge in the early 1960s from test fields in Mexico where Norman Borlaug, sponsored by the Ford Foundation, was conducting experiments on some new hybrids and Japanese dwarf varieties of wheat. These "high yielding varieties" (HYVs), given appropriate conditions, could consume higher levels of fertilizers and produce more food, expending less energy on leaf and stem growth than standard varieties. The appropriate conditions they required included, at minimum, a timely supply of irrigation water, well-drained soil, and chemical fertilizers and pesticides. Farmers also needed to be connected to the relevant infrastructure for access to markets, electrical power, fuel, etc. Besides that, farms had to be large enough to be considered "economical," and it helped if farmers were literate in order to read and follow instructions on using the new technological components. For these reasons, it was clear that HYVs were unlikely to benefit the rural poor and illiterate masses in remote and underdeveloped areas and were not suited to unirrigated farms or the almost wholly unirrigated regions that together constitute the greater majority of rural India. Only in the sense that India benefited "as a whole" for having more food to go around did this approach adhere to the Directive Principles of the Constitution by which the government justified its involvement in planning and leading the economy (Brass, 1994: 303–4; Frankel, 1978: 276–77; Kohli, 1987: 74–75).

With candidates for the new technology limited in number and few resources to go around, it was decided to reverse the "extensive" approach of the Community Development Programme, which spread resources thinly but evenly, and adopt an "intensive" approach, which sought greater returns and a possible breakthrough in productivity that would encourage other farmers to give it a try. If extensive efforts at community development had failed to reduce poverty, production might still be raised enough in some intensive regions to avoid the food shortages that had periodically threatened the country. The new seeds were provided only to those best suited to maximize their return, those in areas that were already the most agriculturally advanced. If the new strategy's intensive approach worked, then the Congress government's prayers would be answered, since it would mean that a "Green Revolution" could be sparked in Indian agriculture without any significant divergence of funds from established programs for industrial development.

The intensive approach had been quietly set up as early as 1960 in 28 districts of 12 states with some guidance and funding from the Ford Foundation. The Intensive Agricultural District Programme (IADP) was created to expand on the efforts of the Community Development Programme by seeking more aggressive ways to increase food supplies, specifically rice and wheat. By the time of the Fourth Five-Year Plan (1969–74), the Community Development Programme had been largely phased out and its name changed to reflect a new emphasis on "rural" over "community" development. The IADP lived on, however, and the new agricultural strategy of introducing HYVs of wheat and later rice was grafted onto it. The IADP districts were already identified as best situated to benefit from the new seeds, with their access to the whole package of accompanying technologies and "progressive" farmers desiring change. It was top-down development, both in catering to districts and to farmers who were already better off and in abandoning any pretense of control or planning from below, as had been the goal of the Community Development Programme (Frankel, 1978: 179–82; Jain, 1985: 44).

The New Strategy for Agriculture, which launched the so-called Green Revolution, was never intended to directly satisfy the needs of the rural poor or to lessen inequalities between individuals, groups, or regions. It was a technocratic effort to solve a growing problem of food shortages, the threat of which was keenly felt by political leaders when public confidence eroded. Furthermore, it did not require any major land reform or agrarian restructuring. The introduction and use of HYVs in India did in fact help to boost foodgrain production from about 95 million tons in the 1967–68 season to 130 million tons by 1980–81. Wheat showed the most dramatic increase, from 10 million tons in 1965–66 to 26 million tons by 1971–72, an increase of 160%. Rice production grew more slowly, from 31 million tons in 1965–66 to 42 million tons in 1971–72, or about 35% (Rao, 1996: A55)

Although the overall annual growth rate for Indian agriculture (2.7% from 1950–85) barely exceeded population growth (2.1% over the same period), the increase in foodgrains production allowed the government to reduce net imports from a high of more than 6 million tons of foodgrains a year between 1966 and 1970 to virtually none from 1976 to 1980. Opinions naturally are divided over whether such modest and limited growth should be characterized as a "revolution." At least food production did not fall behind population growth during this period, which would have required greater imports of foodgrains and created greater indebtedness and dependence for the country. Nonetheless, the per capita availability of foodgrains in India, which from 1956 to 1960 stood at about 161 kilograms per year, was unchanged from 1976 to 1980 (Brass, 1994: 310; Rao, 1996: A55–A56).

Perhaps the most damaging critique of the Green Revolution strategy in Indian agriculture is that it not only failed to address the lofty aspirations of the Directive Principles, it may have actually heightened inequalities–both between rich and poor within certain regions and between regions–that the government's role in economic planning was intended to reduce. Though it has been argued that there is nothing inherent in use of HYV technology to prevent it from being used by and benefiting poor farmers, the reality is that the larger, wealthier landowners have had the means to adopt the new technologies first, and expansion of HYV use has benefited them first. For the majority of rural landholders, whose properties are often minuscule, fragmented, unirrigated, or in some other way less than ideal, or who are wholly or primarily subsistence oriented, or whose villages are too remote to have access to seeds and accompanying supplies (fertilizers, pesticides), the Green Revolution has not been a blessing. To the extent that they have been left behind as larger landowners and other regions become wealthier and more "developed," it may seem to them a curse (Dhanagare, 1987).

Due in part to the fact that the first HYVs were mostly wheat, the benefits of the new strategy largely went to the wheat-growing regions of the northwest, especially the Punjab, Haryana, and western Uttar Pradesh. Irrigation being a necessity even for traditional wheat production, it was already well established in these areas. Several colonial canal systems irrigated the hinterland around Delhi, and there was a well developed transportation and communication infrastructure. Commercial agriculture and wheat production being already well established there, the introduction of HYVs of wheat meant little change for farmers. Returns, at their best, could be impressive, and wheat production in India grew rapidly, tripling in less than a decade after the introduction of HYVs while expanding into areas where previously rice and other crops had dominated. As an example, in one part of northeastern Uttar Pradesh, the area under wheat production grew from about 27% of the season's crop in 1950–51 to over 83% by 1982–83, most of it HYVs (Allen Kornmesser, "The Magical Rational Peasant: Discourses of Power and Agriculture Productivity in Village India," Ph.D. diss., University of Washington, 1994: 851–52, 869–71; Brass, 1994: 310–12).

High yield varieties of rice were introduced, but with less success than wheat for a number of reasons. The primary reason is that rice is a rainy-season, *kharif* crop and is usually not irrigated by other means. Most rice lands lack the irrigation and control of drainage needed to achieve the increased yields of the new seeds. Innumerable local rice varieties are adapted to a wide range of local conditions. Some are sprouted in nurseries and transplanted in standing water held back by low mud walls, other varieties are broadcast on upland tracks to sprout where they land when it rains. Many rice growers are also subsistence farmers, producing for their own consumption on tiny plots, never generating the capital from agriculture that would justify investing in HYVs. Rice remains the single most important food crop in India, and the staple food for a large majority of the population, yet most experts agree that the "revolution" in rice production has barely begun. Surprisingly, for crops other than wheat, including rice and the so-called coarse grains essential in the diet of the rural poor, the growth rate over the height of the Green Revolution, 1966 to 1985, was actually less than it was in the pre–Green Revolution years, 1950 to 1965.

Irrigation is in many cases the critical factor. If adequate irrigation is provided, either through canal systems or individual tube wells, it usually means the infrastructure for marketing and communication is also already in place, increasing the opportunities for a successful adaptation to the new technology. In some areas where HYVs of rice have flourished, they have been converted from a rainy season (*kharif*) crop into a winter dry season (*rabi*) crop so that irrigation can be almost entirely controlled. High yield varieties of other foodgrain crops have been similarly successful on irrigated land in parts of western and southern India. The problem remains that without another strategy to extend the technology for increasing agricultural productivity in India, the Green Revolution will remain a revolution at best only for those farms and those regions that have adequate irrigation, capital, and infrastructure to make use of the new crop varieties.

This is not to say that India has been wholly lacking in alternative approaches to agricultural development. Although "community development" fell by the wayside, there remain critics, Gandhians, and some farmers who continue to believe the government has sacrificed the broad-based economic development of the countryside in order to pursue a strategy of industrial development that primarily benefits urban middle-class and business interests, powerful forces within the Congress party. The peasant leader Charan Singh, who rode a wave of opposition to the heavy-handed rule of Indira Gandhi's emergency (June 1975 to December 1977) to become prime minister briefly in 1979, was the only one of India's leaders who challenged the status of agriculture relative to industry in the allocation of resources for development.

Charan Singh argued that pursuing industrial development before securing a solid foundation in agricultural development was backward and ill-suited to a poor country such as India. With its huge population, agriculture, which is labor-intensive, is a more appropriate focus for development than industrialization, which is capital-intensive and may even displace labor (Singh, 1959). He was able to incorporate some of his ideas into the writing of the Sixth Five-Year Plan (1980–85), despite the conflicts within the fractious Janata coalition that prevented any very directed change of course. Still, allocations for agriculture, irrigation, rural electrification, and development of village and small-scale industries were increased only temporarily. This change of direction was reversed again when the Janata disintegrated and Indira Gandhi's Congress Party was voted back into office. The priority of industrial development was restored formally in the seventh plan, introduced under Rajiv Gandhi's leadership after the assassination of his mother in 1984, but it was clear that the government's and the public sector's role in industry would have to be curtailed. "Liberalization" became the new slogan of the day: reduce government regulation and the economy would flourish.

Present Trends and the Future of Indian Agriculture

Throughout the period of rapid economic growth in the 1980s and despite its outward call for liberalization and deregulation of the economy, the Indian government continued to finance a top-heavy public sector through excessive borrowing. This included borrowing from the International Monetary Fund (IMF) and foreign commercial banks at high interest rates and forcing Indian financial institutions to buy government loans. By the end of the decade, the fiscal deficit had grown so large that it threatened domestic interest rates, prices, and the balance of payments. The 1990s began in economic crisis, with an inflation rate of 17% being reported in August of 1991 and a balance of payments that had reduced foreign exchange reserves to dangerous levels–at one point only enough to cover about three weeks of imports (Dehejia, 1993: 77; Khatkhate, 1992: 49–50).

All this provided the government with an extra incentive to pursue liberalization in earnest. It was not until P. V. Narasimha Rao was handed the reins of the Congress after the assassination of Rajiv Gandhi during the 1991 election that the government was set on a firmer course of economic deregulation. To this end, Rao selected as finance minister Manmohan Singh, a respected academic and experienced adviser to several international agencies, charging him with rapid reform of the regulatory regime that appeared to be stifling the Indian economy. The reforms he implemented seemed to succeed in getting the Indian economy back on the right track: inflation declined to 7% and economic growth increased to 4% in 1992–93, up from 1.2% the year before (Dehejia, 1993: 75).

In agriculture, although performance had been good, the economic crisis of 1991 increased public criticism of the high levels of input subsidies going to agriculture. Subsidizing agricultural inputs means that prices charged to users do not fully cover the costs of providing them, which, though politically popular with certain beneficiaries, has been damaging in other ways. Heavy subsidies of irrigation works, while maintenance and operating costs rose rapidly, resulted in farmers paying barely 10% of the actual cost of irrigation. This has led to excessive and inefficient use of irrigation water, with facilities falling into disrepair due to lack of funds. Subsidies of fertilizers have similarly encouraged their excessive use, sometimes with dangerous or damaging consequences. The government, answering its critics, initially announced that it would eliminate all subsides on fertilizers, but pressure from some farm groups and opposition political leaders led it to eliminate subsides only on phosphate and potassium based fertilizers (nonetheless, a significant decrease). Despite all the calls for limiting government subsidies, 40% of agricultural spending in 1994–95 went to input subsidies, a figure nonetheless well below the high of 60% in 1981–82 (Dev and Mungekar, 1996: A41–42; Dehejia, 1993: 77, 81).

In the 1990s, even with the higher agricultural growth of the 1980s, India's agricultural productivity lags well behind many other countries in Asia. To the extent that expanding the use of irrigation-dependent HYVs remains the primary means to achieve greater productivity, growth will require significant expansion of irrigation. The obstacles to further expansion have, however, been growing along with the perceived need for irrigation. In the past, the government has tended to favor "big projects"–dams, canal systems, etc.–because of their expected scale of return, because they have kept administration centralized, and because they have provided political leaders with large numbers of jobs, contracts, and other resources to distribute as patronage. Even where it has been relatively successful, canal irrigation frequently bypasses the neediest farmers and their tiny, often fragmented, plots of land. The logistics of providing such farmers with well irrigation or of extending well irrigation to India's vast dryland tracts are daunting in other ways. Some large-scale projects that were planned and begun in the 1970s and 1980s–including dam-building efforts, such as the Sardar Sarovar project on the Narmada river–have faced mounting costs and growing local, even international, resistance in the 1990s. Donors and foreign lending agencies have needed to respond to critics and activists in various ways, setting clearer standards for resettlement of displaced villagers and even reexamining the viability of projects likely to arouse serious opposition, all in an environment of increasing international skepticism about the involvement of governments in economic control and development planning (Pant, 1982; "Sardar Sarovar Project," 1993).

There is little doubt that the policy of economic liberalization, which has been embraced in principle by most major political parties, will continue into the future. The Congress, which led the country toward socialism under Nehru's leadership in the 1950s, began a turnaround under his grandson, Rajiv Gandhi, in the 1980s. The economic crisis that greeted Narasimha Rao's new Congress government in 1991 forced it to pursue liberalization more seriously. Even the Hindu nationalist Bharatiya Janata Party (BJP) has accepted liberalization as a means to restore India's wealth and power in the global economy, despite reported differences between so-called Brahman and Thakur factions over its extent. The Communist Party of India (Marxist), ensconced in power in West Bengal since 1971, and the communist parties important in Kerala, in south India, may reject liberalization out of ideological necessity, but in practice they have welcomed international capitalism with growing enthusiasm and have recognized the economic value of deregulation for attracting business. But while deregulation of business in the cities will likely move steadily ahead, there are greater limits to deregulating agriculture, especially relating to land ownership and farm size.

Despite wide support for liberalization in the nonagricultural economy, support has been mixed in the countryside and among India's agricultural communities. Certain groups operating on behalf of farmers, such as Sharad Joshi's Shetkari Sanghatana, based in Maharashtra state, have been vocal in their support for liberalization of the agricultural sector of the economy. They see government regulation as consistently favoring industry over agriculture, the city and urban classes over the rural areas and their populations. For Joshi, the market would be at least a more impartial master than the state. To a large extent, support for liberalization among farmers has been based on the belief that it would lead to higher prices being paid for agricultural produce. Indeed, Joshi's organization has from the beginning emphasized a "one-point programme," that point being not liberalization but "remunerative prices" for agricultural work and produce. It is fair to wonder, in light of this, just how strong farmers' commitment to liberalization will be when it results in lower prices for certain agricultural commodities.

Deregulation has attracted foreign corporate investors. Some, such as Enron and Kentucky Fried Chicken, have faced serious opposition to their doing business in India, while others have found eager markets and little organized resistance. In general, the agricultural sector is attractive to foreign investors because of the enormous market it represents, but it may be more prone to political resistance than urban and high-tech markets. This is due in part to the strong cultural value of agrarian self-sufficiency in India, which people equate with freedom and sovereignty and which is a potent symbol for mobilizing people politically. While some farmers' groups have embraced economic liberalization, others–such as followers of Karnataka's Nanjundaswamy–have been at the forefront of efforts to keep out foreign businesses. They argue that it can only harm farmers to have to compete for survival with multinational corporations and agricultural producers in other countries.

It was only 50 years ago that British rule was driven from India, based in part on a popular perception that "right rulership" produces self-sufficiency, hence sovereignty, whereas colonialism had produced just the opposite. Whether expressed in Nehruvian terms, as national sovereignty against foreign imperialism and dependence, or as Gandhian *swaraj,* epitomized by the self-sufficiency of village agriculture, this is a theme with roots as ancient as the Arthashastra and the Ramayana. This "sovereignty issue" has reemerged again of late in the debate over foreign involvement in India's food production through the ownership of genetic resources (crop varieties) patented as "intellectual property" by foreign and multinational companies. This is part of an effort, according to activist Vandana Shiva, "to replace the small peasant- and farmer-based agricultural economy of India with agribusiness-controlled industrial agriculture." According to this argument, big foreign and multinational corporations do not create new crops to suit the local needs and interests of farmers but rather to meet the needs of industrial processing and of corporations to sell chemical fertilizers, pesticides, herbicides, and other components that their "improved" crop varieties require to achieve full productivity. Some farm leaders advocate organic farming as a means to liberate growers from dependence on chemical-industrial Green Revolution technologies and to foster biodiversity and protect local ecosystems, and there is a growing market for organic produce domestically and for export (Shiva, 1996: 1621).

There are limits to agrarian liberalization due to widespread rural poverty in India. Government plans and agricultural development programs often double as antipoverty efforts. These are hard to eliminate partly for political reasons. The rural poor vote and their electoral support is courted assiduously. But the disparity between nonagricultural and agricultural incomes, as measured in per capita GDP, has grown over the years from a ratio of about two to one in 1950–51 to about four to one in 1990–91. It has been suggested that nonagricultural employment is an "organized sector" to which entry is restricted such that it "does not take in any more people than it can remunerate at a relatively high level. All the rest must stay behind in agriculture," which as a result has become, according to V. M. Dandekar, "a parking lot for the poor" (Dandekar, 1994: 14; Rao, 1996). Even those most optimistic about the recent indicators of agricultural growth and expanding commercialization hesitate to call for a completely open market, noting that the majority of small and marginal landholders would be especially vulnerable to the risks of unmediated exposure to market forces (Nadkarni, 1996: A72). Besides, political leaders fear creating inhospitable conditions in the countryside that could lead to mass migrations of the rural poor to the already overcrowded cities, something that the urban upper and middle classes have nightmares about.

What is likely to emerge is a kind of "compromise liberalization" in the countryside, with exemptions to certain regulations considered on a case-by-case basis. For example, although regulations in most places still make it impossible for corporations to buy enough agricultural land to set up large-scale operations, some state governments–for example, Madhya Pradesh and Maharashtra–have begun creating schemes to bypass land ceilings for large corporate ventures in agro-business (Mahendra Dev and Mungekar, 1996: A46). Of course, such adjustments will be an invitation to old-fashioned bureaucracy, middlemen, and corruption, but even so, it may prove more politically palatable to rural constituents than a wide open market.

More than corporate large-scale farming operations, ready opportunities for growth in the agricultural sector are available from private investment in the technological and infrastructural means to efficiently carry, warehouse, and process India's huge agricultural harvests. While India is the largest producer of fruits and the second largest producer of vegetables in the world, it accounts for only 1% of world exports of fruits and vegetables. In large part, this is due to huge postharvest losses in handling, transport, and storage of produce. Foodgrains face similar losses from insecure storage facilities. It is estimated that 10% to 14% of India's foodgrain harvest is lost each year to rats, which well-built modern grain silos would prevent. Refrigerated storage and transport facilities for fruits and vegetables, as well as for flowers and horticultural commodities, would greatly increase marketability and exportability of farm produce. There are currently only about 50 large cold storage plants in all of India, and almost no refrigerated transport to get produce to market in peak condition.

Many see the best bet for the future of agriculture and India's rural population in supplementing farming with small-scale production units for processing food. There is significant room for growth, especially in processing fruits and vegetables (canning, freezing, drying, etc.), both for urban and export markets. Only 1% of these products are currently processed in India, compared to 70% in countries such as Brazil. The advantage of processing in small production units is that they can be dispersed throughout rural areas, facilitating easy access both to farm produce and to farm labor and providing work for unemployed or semiemployed rural dwellers in their own home areas.

Agriculture in the 1990s continues to be the primary employer of Indians, while it contributes one-third of the nation's GDP. It has shown slow constant growth over the years, increasing recently to about 5% per year (1992–93). Agriculture in India can become much more prosperous, but limited deregulation that allows the growth of commercial opportunities without destroying the social, cultural, and economic strengths of the villages is more likely to occur than a sudden liberalization of the rural economy. Regulations and all, agriculture remains one of the most vibrant sectors of the Indian economy and one in which India is plotting its own course. Despite alleged underproductivity and inefficiency, India has been able to maintain a largely village-based agrarian system that has the potential to increase productivity and profitability while remaining more "sustainable" and more environmentally sound than the high-tech industrialized agriculture that feeds much of the rest of the world.

Further Reading

Ashe, Geoffrey, *Gandhi,* London: Heinemann, and New York: Stein and Day, 1968

This biography of the political and spiritual leader of the independence movement remains a classic, in large part for revealing the complexity and humanity of a man who became a myth and a symbol of modern India.

Brass, Paul R., *The Politics of India since Independence,* 2nd edition, Cambridge and New York: Cambridge University Press, 1994

An indispensable analysis of Indian politics over nearly 50 years, Brass includes an examination of agriculture's significance to Indian politics, changes in government policy on agriculture, and the political consequences of agrarian change.

Chatterjee, Partha, *Nationalist Thought and the Colonial World: A Derivative Discourse?* London: Zed Books for the United Nations University, and Minneapolis: University of Minnesota Press, 1986

One of the most challenging contemporary analysts of colonialism and Indian nationalism shows the contrasts between Gandhi and Nehru that shaped postindependence policies.

Dandekar, V. M., *The Indian Economy 1947–92,* volume 1, *Agriculture,* New Delhi: Sage, 1994

The first volume of this important new survey by an influential Indian economist is dedicated entirely to agriculture's place in India's economy since independence.

Dantwala, M. L., editor, *Indian Agriculture Development since Independence,* New Delhi: Oxford and IBH, 1986

Articles by leading Indian scholars examining various aspects of agricultural development in postindependence India.

Dehejia, Jay, "Economic Reforms: Birth of an Asian Tiger," in *India Briefing, 1993,* edited by Philip Oldenburg, Boulder, Colo.: Westview Press, 1993

A brief and optimistic look at India's economic potential assuming continued liberalization and deregulation of the economy.

Desai, Meghnad, Susanne Hoeber Rudolph, and Ashok Rudra, editors, *Agrarian Power and Agricultural Productivity in South Asia,* Delhi and New York: Oxford University Press, and Berkeley: University of California Press, 1984

Articles by eight leading scholars on the relationship between political economic power and productivity in Indian agriculture. Most see local power as inhibiting needed productive growth and development.

Dev, S. Mahendra, and B. L. Mungekar, "Maharashtra's Agricultural Development: A Blueprint," in *Economic and Political Weekly* 31, no. 13 (1996)

Considers policies that have increased agricultural production and decreased government subsidies in a single state.

Dhanagare, D. N., "Green Revolution and Social Inequalities in Rural India," in *Economic and Political Weekly* 22, no. 19–21 (1987)

Surveys the literature on the Green Revolution and concludes that it has failed to reduce socioeconomic inequalities in rural India.

Fox, Richard G., *Gandhian Utopia: Experiments with Culture,* Boston: Beacon Press, 1989

Fox's anthropological analysis of Gandhi and Gandhian philosophy takes a postmodern turn, examining them in terms of the cultural conditions that gave them context and meaning.

Frankel, Francine R., *India's Political Economy, 1947–1977: The Gradual Revolution,* Princeton, N. J.: Princeton University Press, 1978

A meticulous examination of government policy and the planning of the Indian economy after independence, if at the cost of overlooking other factors.

Jain, L. C., *Grass without Roots: Rural Development under Government Auspices,* New Delhi and Beverly Hills, Calif.: Sage, 1985

Jain argues that Gandhian village-directed development was eclipsed by a state-directed bureaucratic regime that became more an obstacle than an aid to rural development.

Khatkhate, Deena, "India on an Economic Reform Trajectory," in *India Briefing,* edited by Leonard A. Gordon and Philip Oldenburg, Boulder, Colo.: Westview Press, 1992

The author argues that the Indian economy improved in the 1980s, poverty went down significantly, and the climate favoring economic liberalization promised accelerating economic growth through the early 1990s.

Kohli, Atul, *The State and Poverty in India: The Politics of Reform,* Cambridge: Cambridge University Press, 1987

Seeing resistance to reform in the classes that dominate Indian political power, Kohli presents a critical analysis of the Indian government's failure to significantly reduce poverty after 40 years of independence.

Morris, Morris David, "Toward a Reinterpretation of Nineteenth Century Indian Economic History," in *Journal of Economic History* 23 (1963)

Dated but nonetheless articulate argument against the Indian "nationalist" position that colonialism caused the decline of the villages and rural artisanship.

Nadkarni, M. V., "Accelerating Commercialisation of Agriculture: Dynamic Agriculture and Stagnating Peasants?" in *Economic and Political Weekly* 31, no. 26 (1996)

This article examines whether liberalization can bring economic growth without exposing vulnerable rural classes to undue hardships with political as well as social costs.

Pant, Niranjan, "Major and Medium Irrigation Projects: Analysis of Cost Escalation and Delay in Completion," in *Economic and Political Weekly* 17, no. 26 (1982)

Pant finds large development projects (dams, etc.) to be often less efficient and more expensive than medium-scale alternatives

Rao, V. M., "Agricultural Development with a Human Face: Experiences and Prospects," in *Economic and Political Weekly* 31, no. 26 (1996)

Rao examines whether and to what extent the poor have benefited from technological change in Indian agriculture.

Rudolph, Lloyd I., and Susanne Hoeber Rudolph, *In Pursuit of Lakshmi: The Political Economy of the Indian State,* Chicago: University of Chicago Press, 1987

An ambitious analysis of the Indian state in relation to economic classes and demand groups, including agricultural producers.

"Sardar Sarovar Project: Review of Resettlement and Rehabilitation in Maharashtra," in *Economic and Political Weekly* 28, no. 34 (1993)

Collectively written by a group of Indian social scientists, this article finds government efforts insufficient to address problems of resettlement of villages in areas affected by the building of the Sardar Sarovar dam.

Shiva, Vandana, *Monocultures of the Mind: Perspectives on Biodiversity and Biotechnology,* Dehra Dun: Natraj, and London: Zed, 1993

Shiva, an outspoken feminist environmental activist, depicts property rights on genetic material (by Western and capitalist agrobusiness) as enabling a postcolonial assault on India's biodiversity and native environmentalism.

Tomlinson, B. R., *The Economy of Modern India: 1860–1970,* Cambridge and New York: Cambridge University Press, 1993

An excellent survey of the evolution of the Indian economy from colonial times through the first decades of independence. Tomlinson emerges as a moderate between positions established by British colonialist historians and by Indian nationalists.

Varshney, Ashutosh, *Democracy, Development, and the Countryside: Urban-Rural Struggles in India,* Cambridge and New York: Cambridge University Press, 1995

Uses "rational choice" to analyze the slow emergence of agriculture as politically significant in India and of agriculturists as a political force.

Whitcombe, Elizabeth, "Whatever Happened to the Zamindars?" in *Peasants in History: Essays in Honor of Daniel Thorner,* edited by Hobsbawn et al., Bombay: Oxford University Press, 1980

Demonstrates that many traditional "feudal" landlords were able to benefit from the elimination of their official rights (and obligations) under policies of zamindari abolition.

Allen Kornmesser received his Ph.D. in political science from the University of Washington, where he has taught in the political science department and the Jackson School of International Studies. He has recently written an article on "Ithiel de Sola Pool's Unfinished India Research: A Critical Review" (1997) to accompany Pool's papers in the manuscript archives of the Massachusetts Institute of Technology.

Chapter Nine

India's Increasing Integration in the World Economy: The Tensions of Nationalism and Globalism since 1990

Alan Heston

The Context

Following the lead of many other low- and middle-income countries, India embarked on major programs of economic liberalization in the 1990s that signified a departure from previous experience with economic planning. By the 1980s, policies that were adopted following independence appeared to many Indian leaders to be inadequate to cope with the changing world economy. The breakup of the Soviet Union and the widespread reforms in Eastern Europe, China, and Latin America also hastened the call for reforms in India. A very significant part of the reforms involved India's foreign exchange, trade, and investment policies, which will be the focus of this chapter.

In his survey of Indian reform efforts, Jagdish Bhagwati asks "How Did We Get into This Mess?" (1993: chapter 2). Why look at the past, one might ask. Because in order to understand reactions to reform policies, it is necessary to know how India got to where it was in 1991. China presents an interesting parallel in its development of special economic zones (SEZs) on the Southeast coast, seemingly an obvious policy for development. However, in the Chinese context, the SEZs were formerly all treaty ports that were forcefully created in the nineteenth century and that guaranteed extraterritoriality to foreigners, so any discussion of special treatment of these areas had to overcome some painful memories. In the Indian context, many leaders have to retreat from previous policy positions if they are to support present reforms. Other officials must lose power or access to resources in the process of reform, and any politician can appeal to nationalist sentiments rooted in the colonial experience to generate opposition to international involvements. Consequently, understanding the present requires some understanding of the past.

After independence, India attempted to guide its economic development within the framework of five-year development plans that were implemented by specific policies aimed at controlling the direction of industrial expansion. A legacy of the British was the attitude of Indian business that it was useful to have access to government to obtain the economic scarcity rents created by government licensing procedures. Indian businessmen resented what they considered the favored role of British business in India (Kidron, 1965). British firms often had access to contracts through the social network of their clubs, or to government created monopolies such as mail transport. Indian businesses often felt that obstacles were put in their path if their interests conflicted with those of British business; for example, Jamshed Tata's first attempt at building a steel mill in India met opposition, but after Belgian steel imports into India outdistanced those from England, the British supported the venture.

One of the major economic statements following independence was the Industrial Policy Resolution of 1948, which set aside sectors for only the public enterprise firms, sectors for either public or private firms, and sectors principally for private sector firms, including producers of most consumer goods. The Industries Development and Regulation Act (IDRA) established the framework for implementing the Industrial Policy Resolution. The act suggested priorities for industrial expansion and indicated scheduled industries in which the private sector could expand capacity with a license. It called for registration of undertakings and approval of all increases in capacity by a Licensing Committee set up in 1952. Apart from the licensing, imported capital goods had to be cleared through the Capital Goods Licensing Committee, and where government finance was required, additional approvals were necessary. Once these hurdles were cleared, it was again necessary to obtain a license for foreign exchange from the Reserve Bank of India. The system of controls also extended to exports and to imports through quotas, tariffs, or both.

A pivotal agency in the licensing procedure was the Directorate General of Technical Development (DGTD). The lower-level civil servants in DGTD, as in many organizations, wielded power by saying no and risked review by saying yes. The DGTD had many checkpoints in the licensing procedure. These checks included the technical soundness of the project, the environmental impact of effluents, the need for additional capacity given existing production and export possibilities, the relation of the capacity sought and the capital stock of the enterprise, and the degree to which the additional capacity would make use of indigenous inputs. Based on government reports and interviews, Bhagwati and Desai (1970: 249–73) produced a sharply critical appraisal of the whole licensing process, including the practice of treating applications sequentially, without regard to alternative proposals in the pipeline; the DGTD was singled out for failure to establish criteria for their decisions, poor handling of information, and failure to review implementation of approved projects.

This set of regulations heavily affected the foreign sector, especially on the import side. First, import substitution was pushed so that quotas were often put on imports and domestic production could be used in their place. Second, many items were proscribed from import, including many luxury goods. Foreign investment was also limited first to those sectors where the private sector was permitted to produce and then to minority ownership. India, like many other countries, emerged from the colonial experience feeling that its exports of primary products, such as jute, raw cotton, and tea, had not been to their advantage but rather to the advantage of the British. Further, there was a pessimistic attitude toward the growth of export demand that in turn reinforced the idea that it would be necessary for India to become as self-sufficient in agricultural and industrial production as possible.

There were many proposals for reform in the late 1960s, but in fact, the early populist years of the Indira Gandhi era added important additional hurdles to foreign investment in India. In 1973, a Secretariat for Industrial Approvals (SIA) was established that was to coordinate and expedite applications. Under the Monopolies and Restrictive Trade Practices Act (MRTP) and the Foreign Exchange Regulation Act (FERA), foreign or large firms faced additional committees for approval. While there were time limits set for turning around applications at each stage, the impact of the powers granted to various ministries and committees was to require an application to be reviewed by up to 35 persons in up to 12 different departments (Kochanek, 1985). Rarely would an application be approved in less than six months, and delays of a year were frequent. The tradition that one department would not overrule the decision of another department also added to the complications of obtaining approval. Of the many committees established by the Lok Sabha in the last 20 years to review the licensing procedures, all have called for extensive liberalization and reform.

Strong vested interests continued to make reform difficult in India; their lobbying efforts could be easily financed by the rents from the licenses and quotas they had obtained. Other vested interests included the bureaucrats who had strong powers that bred bribery and corruption. However, the confrontational character of some business-government relations of the 1970s had given way to a situation of accommodation by the 1980s. Under Rajiv Gandhi, several attempts at sidestepping the bureaucratic apparatus were introduced, and with some successes. These included broader interpretations of capacity, so that a firm with a license to expand bus production, for example, may be allowed to expand production of any vehicles, where previously that was not allowed. New firms of small size or firms seeking small increases in capacity were approved more quickly, and whole areas of production were exempted from a number of the licensing hurdles. However, because these reforms often substituted political exemptions for regulatory reform, charges of high-level corruption undermined Rajiv's reform efforts and his subsequent election bid.

Economic Reforms since 1990

For India, the Gulf War meant losses of the foreign exchange earnings of migrant workers in the Gulf. India had attracted accounts from non-resident Indians (NRI), accounts that paid attractive rates but that could be withdrawn freely; these accounts were drawn down heavily in 1991 as confidence in the Indian economy dwindled. The subsequent foreign exchange crisis that India faced in spring 1991 clearly required an immediate response. What appeared different in the policies of Prime Minister Narashima Rao and Finance Minister Manmohan Singh was their sense of commitment to reforms, including key administrative appointments, and their very explicit policy changes, many in response to the foreign exchange crisis. The present United Front (UF) government, led by Prime Minister Deve Gowda, has thus far not departed substantially from the directions of Rao and Singh, so in this essay, Indian reforms since 1991 encompass both governments.

Worldwide patterns of economic reform have many common elements, such as convertibility of currency on current (and capital) account, removal of quantitative and other trade restrictions, elimination of subsidies, privatization or restructuring of state-owned enterprises, encouragement of private foreign investment, and achievement of macro stability, typically calling for reductions of budget deficits. However, these common elements of reform play themselves out very differently around the world, depending on the political structures of the country, the initial conditions, and the extent to which reforms are a response to crisis. Vested interests can often blunt reform efforts or transfer government monopolies into private monopolies with little net benefit to consumers. Many of the reforms bear significant political and human costs, and this often influences the speed with which countries undertake their reforms. In countries that operate within a democratic framework, such as India, opposition parties can easily capitalize on policies that appear to hurt certain groups, such as workers laid off from nonprofitable public enterprises or beneficiaries of subsidies that have been eliminated in efforts to balance government budgets. And the bureaucracy itself is hard to dismantle; it is often said that India has gotten rid of hundreds of regulations but not a single regulator. The foreign investment issues raise a number of questions of internal policy including privatization, labor legislation, and capital market reforms. Similarly, trade policies can directly affect employment in previously protected industries.

Currency Convertibility

One of the major goals of India in the 1990s was to achieve convertibility on current account, and this was achieved by 1994. The exchange rate of the rupee was not allowed to move freely, but the rupee was devalued by an amount thought to be more than adequate to achieve balance in short-term transactions. Exporters were free to convert a part and later all of their earnings to foreign currencies if they chose, and importers faced few restrictions on what could be brought into the country, though tariffs on many items remained. The rupee underwent a major devaluation of nearly two to three in 1991, and many import restrictions and cash subsidies for exports were removed. These policies, as well as the freedom of exporters and importers to retain or obtain foreign exchange, have greatly facilitated export growth and have generally made it much more attractive to carry out business in India. Firms can now more easily borrow short-term from abroad to finance production and exports or to finance imports.

Capital account convertibility has been another matter, and the gradual approach of India has been reinforced by the problems faced by the Mexican peso. India has seen its foreign exchange reserves go from the dangerously low level of US$1.1 billion in June 1991 to over US$20 billion in 1994 and to settle at about US$17 billion recently. While US$20 billion is a good cushion against current account fluctuations, it would do little against a speculative attack on the rupee of the type that Mexico experienced against the peso. Further, the NRI accounts remain a potential item for immediate withdrawal at any time, and having been burnt by the NRIs in 1991, India still feels that NRIs leave their reserve position vulnerable. With total foreign debt of about US$100 billion today, India is not overly indebted, but its economic policy makers have continued to take a gradual approach to capital convertibility.

What this has meant is that Indian nationals are not fully free to move capital to other countries, and foreign nationals are not free to acquire any type of asset in India. India has encouraged some types of foreign financial investment in terms of mutual funds owning stock in Indian firms and has allowed marketing of such funds outside of India. However, India has not been willing to allow the rupee to be freely floating but has chosen a more managed system that has maintained the rupee at about 31.5 to the dollar from 1991 to 1995 (from its previous rate of about 19.5 to the dollar). Since 1995, the rate has moved up to 36 rupees to the dollar and remains at a level that leaves Indian exports attractive.

One common failure of exchange rate management is to allow inflows of foreign long-term capital to be used for imports and to appreciate the local currency. India has consciously avoided this pitfall during the 1990s. While conservative in its policies, India has certainly maintained an external stability during the last six years that is attractive to foreign investors.

Private Foreign Investment

The major obstacles to direct foreign investment, such as percentage of ownership, repatriation of profits, and approvals required, have been greatly simplified since 1991. The result has been apparent in the heightened interest of multinationals and a significant flow of investment funds and technology and management practices from abroad. The results have been very striking in a number of sectors that often make use of India's highly educated labor forces, not simply unskilled workers. However, with the inherited distrust of foreign firms, there remain a number of issues, among which are the communications sector, the power sector and the importance of state politics as illustrated by the Enron case, and the case of middle-class consumption items, including franchise foods, and related questions of perceived cultural and economic imperialism.

Telecommunications

India has suffered the fate of most countries with central telephone authorities: poor service, slow response to demand, and long delays in obtaining installation. The Department of Telecommunications (DOT) does not, it is claimed, have access to capital to finance the estimated US$42 billion that would be needed to expand India's telephone usage from 8 million in 1994 to the probable 40 million by the turn of the century. Although call completion rates have notably improved under DOT in the past ten years, the rate of incomplete calls and subscriber breakdowns per year (two per year, which is ten times international standards) remain high.

In 1994, reforms were instituted that would permit private firms, including foreign firms (up to 49% equity) to supply local service in competition with DOT. The DOT was to retain a long distance monopoly for five years, at which time the situation would be reviewed. Given the large number of arrangements to reverse long distance calls initiated in India or other countries, this is not a major restriction on consumers obtaining service. With the recent World Trade Organization agreement on long distance rates, the pace of change and reduction of rates in India may be quickened. This can only serve to improve opportunities for foreign involvement in the Indian economy.

The bidding in the various circles in which phone service is provided in India began in 1995 and has been the subject of much controversy. Bidding for both basic phone service and cellular service in the 20 circles were to be subject to a technical review and a financial review. In the cellular competition, two suppliers were to be chosen in each circle. The difficulties have included the following.

Choosing successful bidders has not been straightforward, and there are legal actions underway against officials in the Ministry of Communications. It has sometimes been claimed that Indian companies would successfully win bids on the basis of foreign participation for half the capital when in fact the Indian company had not secured collaboration and could not expect to raise its half of the capital until foreign collaboration was secured. In this type of case, the restrictions on equity ownership by foreigners act as an opportunity for promoters in India who have no particular expertise in the field to make windfall rents by obtaining contracts.

The overall impression in the telecommunications field is that this is an attractive area for foreign investment. There is probably less political capital to be made of opposing entry by foreign firms as compared to the power industry, for example. However, while there has been improvement in service, thus far it has not been as rapid as it might have been, and the principal obstacle seems to have been domestic politics. In fact, the problem seems to be that obstacles to majority foreign ownership have served to make transparency in awarding contracts more difficult to achieve.

More generally, the past regime of controls continues to cast a long shadow over attempts at reforms. In particular, the practice of "fast-tracking" some approvals is a two-edged sword. It is a way of bypassing a cumbersome approval apparatus, which is probably good. However, fast-tracking often neglects sound technical reviews of past procedures and passes on decisions to politicians who come under pressure or temptation to award contracts on some other basis than delivery of a given output at the least cost to India. Some aspects of the power sector have the same problems.

Political Decentralization and the Power Industry

Economic reform is usually associated with political decentralization. India has a highly centralized administration, and for the reforms to spread widely from Delhi, state and local administrations must become involved. However, the incentives often do not work that way. When the central government reduces its budget deficit in India, often it is at the expense of transferring responsibilities to the states. As in the United States, if unemployment results from plant closures, the impact falls within local jurisdictions, so central policies that produce closings by lowering import restrictions, for example, impose a burden the center may only partially bear. Also, many of the central licensing controls that were eliminated had their counterparts in the states of India. Fortunately, the competition of states to attract new domestic and foreign investment is likely to insure that these controls are also eliminated in the states. The same local initiatives can also turn to protectionism as has occurred in both China and India, where imports from other provinces or states may face taxes from which local production is exempted. In order for state governments to generate the kind of business confidence necessary to locate businesses in their jurisdictions, their political leaders will need the type of political vision that has fueled reform policies at the center, coupled with a minimum of protectionist restrictions.

Foreign investment can often act as a focus of political confrontation between the state and center, as has been the case with the Enron project in Maharashtra. All parties agree that India needed to quickly expand electric power generation, though many projections appeared to be overestimates. The public, which had suffered from intermittent supplies and power for only a few hours a day, tended to favor policies that would rapidly expand electricity supply. However, electricity in India has been underpriced, and there is somewhat less agreement on how quickly India should raise the price of electric power to reflect its real cost. Most State Electric Boards (SEBs) in India, which are linked to a national grid through five regional power systems, have operated at a loss for many reasons, including excessive staff and political interference. One important cause is that much electricity is taken from the system by nonapproved connections (open theft) and approved connections that are not billed or shut off for lack of payment (less open theft). The subsidized rates for electric power have typically been justified on some type of equity grounds, though the consumer beneficiaries are typically in the middle- or upper-income groups. Electric power for manufacturing firms at subsidized rates may keep firms in business and provide employment, which is also a frequent justification.

With the arrival of foreign firms, such as Enron, who offer to build and operate plants to generate needed power, more complications arise. To begin with, any power project needs clearance from the Central Electric Authority (CEA) and the central ministries of power, finance, environment and forests, petroleum and natural gas or coal, and possibly railways, shipping, and surface transport. An Investment Promotion Cell (IPC) within the Ministry of Power was to be a single window at which these clearances could be obtained, but so far that has not worked out. After central clearance, the SEB and state government must also approve the project. The 1991 reforms permitted private firms, domestic or foreign, to operate power generation projects, to earn a return of 16% in foreign exchange (for a plant load factor of 68.5%), to freely repatriate earnings in the case of foreign firms, and to sell power to public or private users. A private firm might well sell power to the grid at rates that will cover their costs, and these costs may often be above current rates to users because of existing subsidies; often, the central government has agreed to make up the difference to the SEB. It remains easy to accuse a private firm, whether it be foreign, like Enron, or domestic, of exploitation, since the power authority in the end may need to raise electricity rates to users.

The Enron scenario illustrates the importance of past policies and of state-center interactions. After the Enron venture (which was a fast-track agreement) had been approved at the center and by the state of Maharashtra, Enron began work on the Dabhol project. The BJP-Shiv Sena alliance that came to power in Maharashtra in 1995 campaigned on a platform of canceling the contract, which they did on taking office. Under the agreement, after cancellation, renegotiation was an option within six months. Also, since Enron had begun work, they asked for US$100 million in cancellation charges through arbitration and curiously were not asked to curtail construction during the subsequent two years.

An immediate reaction to the cancellation was to treat this as a setback for foreign investment in India, but the Enron lessons are not so simple. One early suggestion that it was a sweetheart contract was that General Electric indicated it could bid much lower for the Dabhol project. This was a break in the cartel of fast-track foreign power firms seeking contracts in India and elsewhere and led to renegotiations in several of the eight fast-track power projects in India and some unilateral cost estimate reductions by foreign firms in other states.

During 1995, neither Enron nor the Government of Maharashtra indicated they were unwilling to negotiate further. At the end of 1995, the Enron arbitration was temporarily suspended in London, a suit was pending in the Bombay government against Enron and some Maharashtra SEB officials for fraud, the government of Maharashtra was quelling local protests against continuing construction of the Dabhol project, and the renegotiation of the project took final form. The main changes are that a different fuel source is to be used and the SEB will share equity with Enron (both previously offered by Enron), the cost estimates have been reduced by US$375 million (about 15%), and the power purchase agreement (PPA) rate is to be reduced by 20%. Many critics remain unhappy because they believe environmental issues have not been addressed and returns remain too high.

It is easy to see how the overlapping approval system in India combined with the present unprofitable operation of publicly owned power firms creates an environment in which the private firm, even more so if foreign, can be cast as the bad guy. Further, a private firm bidding in an environment that is quite uncertain is likely to seek a contract that provides them a large margin for potential cost increases or revenue shortfalls–in short, a contract that can easily be criticized as a give away by government with the implications of kickbacks and the like. All of these factors seem to have entered into the Enron case. However, the long-run effect of a high profile case like this would appear to have made the other states in India much better negotiators in dealing with private power projects, whether domestic or foreign.

Consumer Goods for the Middle and Upper Classes

The ideology in India in the planning period had been to resist foreign collaboration in many consumer products and services. For example, provision of hotel services was not an area to encourage foreign capital. Some domestic chains, such as the Oberois or the Government Tourist Organization, could handle such matters. Let foreign capital go to areas where their technology was not available in India. Symbols such as Coca Cola were presented by some politicians as precisely the type of products that India either did not need or could make just as well themselves. Where consumer items such as automobiles were made in India with foreign collaboration, they typically fixed on a model and retained it while trying to produce as many components as possible within India.

The most strikingly visible change in India in the 1990s is the reversal of these restrictions. Not only has foreign investment been welcomed in service areas such as restaurants and hotel chains, but multinational corporations (MNCs) producing many of the popular brands of products from around the world have found the possibility of investing and producing in India increasingly attractive.

If one compares the auto industry in India with that of Korea, Brazil, and China, ones finds that India has made arrangements to produce a wider variety of cars (Suzuki, Peugeot, General Motors, Ford, Fiat, Daewoo, Mercedes, Rover, and Volkswagen) than any of the other four countries. This is a remarkable turnaround for India, which previously had offered a very limited choice to consumers. The list could be expanded to many other consumer goods, much to the dismay of some critics.

The Tensions of Foreign Involvement and Domestic Politics

One of the tensions that has existed in political discourse in India from independence onward has been between economic growth and equity. The issue is not that these two goals are valued so differently by the different political parties but rather how to attain them. The Gandhian and socialist traditions have been interventionist and have sought to raise the lot of the poor by distribution schemes and allocation of productive resources in an egalitarian manner. They have felt that leaving matters to the market will not provide improvements for the poor.

Critics have claimed that most of the interventionist policies have resulted in subsidies that rarely get to those they are intended to benefit or in laws, like land reforms, that were written and undermined by the large land holders and result in less secure tenure and production than had existed before. Advocates of economic growth point out that all rapidly growing states in India have had declining numbers in poverty compared to slower growing states. Many advocates of growth also push for more widespread education, basic health services, and employment schemes for the rural landless. What they do not support are policies that attempt to do good for the poor by holding back those earning high incomes.

These same issues arise in the international trade and investment spheres. For over a century, a theme in India has been that exports are to benefit the rich at the expense of the poor in rural areas. It is of course true that if land is allocated to export crops at the expense of other crops, the price of the latter are likely to rise, which in some circumstances may raise costs to poor households (that may or may not be offset by more imports at lower prices). It does not follow, however, that increased exports should be discouraged; a preferred solution would be to find more jobs for the poorer households.

Similar issues arise with respect to permitting imports at lower duties and forcing public and private firms to meet the competition of the international market. In many sectors, these policies force larger subsidies or losses for sick industries that would close if market forces were to operate. India has taken a gradual approach to these issues by lowering tariffs most on the sectors that can compete internationally and maintaining protection for the less efficient industries for at least five years, hopefully long enough to reduce by attrition the labor and capital committed to these sectors.

With respect to foreign collaborations, liberalization is often thought to hurt the poor because the types of goods produced are typically for middle- and upper-income families. It is felt that the employment gains are small and that the type of open economy that is engendered by foreign investments and collaborations serves to widen the gap between the rich and the poor. When Deng was criticized for his reforms in China on these grounds, he said that if all Chinese are going to get rich, someone has to get rich first. Clearly it would be easy in present day India or China to find those who have gotten rich first. What is less clear is whether the conditions of the poor have deteriorated. As in most trade agreements or innovations, the benefits are diffused as small gains to large numbers of people while the losers are few, but the few are rightfully highly vocal about their losses. The tension that India faces is that there are groups, such as workers in factories who face job losses, who object to India entering the global economy, even at the relatively slow pace it has done so compared to other Asian countries. The political-economic contents of these discussions are very similar to those in other democracies, such as the North American Free Trade Agreement (NAFTA) debates in the United States. However, the fear of foreign economic dominance remains very strong in the domestic politics of India, and the road to economic liberalization will have its share of rough spots, especially when it comes to foreign investment.

Further Reading

Alagh, Yoginder K., *Indian Development Planning and Policy,* New Delhi: Vikas, 1991

This study reviews the planning process in India from the perspective of an academic economist and practitioner of policy.

Bhagwati, Jagdish N., *India in Transition: Freeing the Economy,* Oxford and New York: Oxford University Press, 1993

These lectures at Oxford reflect the long-held and critical views of the author.

Bhagwati, Jagdish N., and Padma Desai, *India, Planning for Industrialization: Industrialization and Trade Policies since 1951,* London and New York: Oxford University Press, 1970

This seminal study was one of the first to look at the licensing policies for industries in India and to make clear that much of the justification for long delays and seemingly questionable procedures was not well based.

Economic and Political Weekly

One of the most useful publications on the Indian economy covering a variety of subjects. While having an editorial point of view to the left, a wide variety of views are expressed in the articles included.

Financial Express and *Financial Times*

Two of the leading daily financials in India. The in-depth articles and contributed op-ed pieces represent many points of view.

Government of India, *Economic Survey 1995–96,* New Delhi: Government of India, Ministry of Finance, 1996

This annual publication by the GOI is very informative of government policy and while partisan, ranks with the *Economic Report of the President* for the United States as a valuable and influential source on the Indian economy.

Hanson, James A., and Samuel S. Lieberman, *India: Poverty, Employment, and Social Services,* Washington, D.C.: The World Bank, 1990

An in-depth study of social policies in India in the spheres of education, health, and poverty alleviation.

Heston, Alan, "Poverty," in *India Briefing,* edited by Philip Oldenburg, New York: Asia Society, 1990

This article describes the important policies that India has undertaken to combat poverty and the influence that these policies have had on thinking on these issues elsewhere.

Kidron, Michael, *Foreign Investments in India,* London and New York: Oxford University Press, 1965

A study of foreign investments in India prior to and immediately after independence.

Kochanek, Stanley A., "The Politics of Regulation: Rajiv's New Mantras," in *Journal of Commonwealth and Comparative Politics* 23 (1985)

A thorough summary of the licensing labyrinth as it existed in India in the early 1980s from the perspective of a political scientist.

Srinivasan, T. N., editor, *Agriculture and Trade in China and India: Policies and Performance since 1950,* San Francisco: International Center for Economic Growth, ICS Press, 1994

This study, with contributions by Justin Yifu Lin and Yun-Wing Sung, on agriculture and foreign trade in China is an excellent comparative study. It treats the agriculture and foreign trade sectors within the context of Indian planning and recent reforms.

Srinivasan, T. N., and Pranab K. Bardhan, editors, *Rural Poverty in South Asia,* New York: Columbia University Press, and Delhi: Oxford University Press, 1988

This volume contains contributions by many of the leading authorities on poverty and includes some aspects of how international trade policies may impinge on rural areas.

World Bank, *Economic Developments in India: Achievements and Challenges,* Washington, D.C.: The World Bank, 1995

This report is particularly strong on legal and administrative changes at the center and state level as they affect labor, the power, transport, and communications sectors, and education. The report also provides a review of recent macro developments in the Indian economy.

Alan Heston is professor of economics and South Asia regional studies at the University of Pennsylvania. He has recently published comparisons of postindependence development in India and Pakistan and an account of recent scholarship on economics and economic history in India.

Social and Cultural Aspects of Economic Growth

Chapter Ten

Population, Health, and Development: Policy Debates, Directions, and Dilemmas

Barbara D. Miller

Studies of the relationships between population, health, and development offer multiple and often contradictory views and implications for policy. Most scholars would agree that the relationships are complex and contextually variable, but agreement stops there. The widely varying models range from Malthusian or neo-Malthusian approaches on one side to Marxist or neo-Marxist approaches on the other, with other possibilities in between (see Ester Boserup, "Development Theory: An Analytical Framework and Selected Applications," in Population and Development Review 22 [1996]: 505–15).

The Malthusian and neo-Malthusian approaches take population growth as a key constraint to economic growth and development. Marxist or neo-Marxist approaches reverse the direction of causality, emphasizing the determinant role of the economy in affecting patterns of population change and quality. A third view could be called the Boserup model. It posits that population growth has prompted economic growth. Ester Boserup's influential writings over several decades have showed how population growth spurred innovation and change throughout history (*The Conditions of Agricultural Growth: The Economics of Agrarian Change under Population Press*, Chicago: Aldine, 1965, and *Population and Technological Change: A Study of Long-Term Trends*, Chicago: University of Chicago Press, 1981). This view is reflected in some national policies today; the Malaysian government, for example, currently supports population growth as a means for development ("India vs. Malaysia: Two Population Policies," in *Futurist* 30 [1996]: 44).

Such determinative models exist alongside those that emphasize mutual effects between several variables, such as poverty, population, and environment (Partha Dasgupta, "Population, Poverty, and the Local Environment," in *Scientific American* 27 [1995]: 40–45). Cross-national studies analyzing the potential impact of population growth on economic measures have produced mixed results. One analysis of the possible effect of population growth on poverty levels found little direct evidence for a population growth impact (Dennis A. Ahlburg, "Population Growth and Poverty," in *The Impact of Population Growth on Well-Being in Developing Countries*, edited by Dennis A. Ahlburg, Allen C. Kelley, and Karen O. Mason, New York: Springer Verlag, 1996). Indirect evidence suggests that for some countries, reduced poverty seems to be the result of a growing population, while for others, population growth appears to be a contributing factor to difficulty in reducing poverty.

Increasingly, microlevel research indicates that the determinants of population change in different contexts are diverse and that it is difficult to point to one general theory that applies universally. Nevertheless, in varying degrees, key variables, including economic entitlements, education, and access to and availability of health services, continue to show an important relationship to population growth, depending on context. Social class and gender roles cross-cut all of these as important variables everywhere. Policy emphases also vary between taking a Malthusian-guided path of focusing on reducing population growth through noneconomic means (such as provision of family planning programs) or a more Marxist-inspired approach of pursuing economic reforms (such as entitlement adjustments) as the best way to improve population and health indicators. From the Malthusian perspective, population and health are clearly seen as constraints to economic development, and from the Marxist perspective, the economic structure is seen as a constraint to widespread improvements in human welfare. Behind all the model building and theory testing of both Malthusianists and Marxists, however, international consensus generally exists among scholars and policy makers that a smaller population is better than a larger one for economic development (Alaka Malwade Basu, "The 'Politicization' of Fertility to Achieve Non-Demographic Objectives," in *Population Studies* 51 [1997]: 5–8); the key question is how to achieve that goal.

Population and Health in South Asia

South Asia presents a diverse picture of population and health dynamics. It encompasses some of the world's largest and most densely populated countries (India, Bangladesh, and Pakistan, which are among the world's ten most populous countries) and also one of the smallest and most sparsely populated (Bhutan). India is the second-largest nation in the world in terms of population, with its 1990–95 estimated population over 944 million people. It is likely that India's population will surpass China's in the near future. Pakistan has the seventh largest population internationally, with nearly 142 million people. Bangladesh is ninth, with over 117 million people. The other three South Asian nations have smaller total populations: Nepal is 46th, with almost 21 million people, Sri Lanka 51st, with 18 million people, and Bhutan 141st, with over 1.5 million people. Three of the world's largest cities–Mumbai (Bombai), Calcutta, and Delhi–are in India. The South Asian nations also rank high in global comparison in terms of population density, with the exception of Bhutan. At one extreme, Bangladesh has 2,384 people per square mile, and at the other, Bhutan has 48 people per square mile (in between, India's population density is 827, Sri Lanka's is 737, Pakistan's is 449, and Nepal's is 440).

According to the United Nation's Human Development Index (HDI), a composite measure based on life expectancy, educational attainment, and income, Sri Lanka stands out as leading in human development measures in South Asia (United Nations Development Programme, *Human Development Report 1994,* New York: Oxford University Press, 1994). Sri Lanka ranks in the category of "medium" human development (number 89) out of 173 developing countries. The other five South Asian nations are in the "low" human development category: India (133), Pakistan (134), Bangladesh (143), Nepal (151) and Bhutan (159). According to the 1993 data used by the United Nations in constructing its HDI, life expectancy at birth in the South Asian nations is 72 years in Sri Lanka, 61.8 years in Pakistan, 60.7 years in India, 55.9 years in Bangladesh, 53.8 years in Nepal, and 51 years in Bhutan. The average life expectancy at birth in all developing countries is 61.5 years. Each South Asian nation, except for Sri Lanka, is either close to or substantially below the global average. Sri Lanka far exceeds the rest of South Asia on all other human development measures, including literacy for both males and females (see Table 10.1).

Focus on India

India has a dominant position in South Asia in terms of political and economic power, location, area, and population. Along with China, it is the second "Asian giant." With scarcely 2.4% of the world's land mass, India has over 16% of the world's population. For this reason alone, Western and other experts have long worried, in a Malthusian way, that India's population will "inevitably" and "inexorably" (such words often appear in the press and in scholarly literature on India) outstrip its natural resource base and continue to exceed and overburden any economic progress India manages to achieve. These sentiments appeared, for example, in the Ministry of Health and Family Welfare's 1992–93 yearbook (Government of India, *Family Welfare Programme in India, Yearbook 1992–93,* New Delhi: Government of India, Ministry of Health and Family Welfare, 1994: 17).

In the ten years between 1981 and 1991, India's population increased by 163 million people, almost the same number of people that were added during the 30-year period between 1931 and 1961. Now, about 17 million people are added each year, exceeding China's 14 million annual increase and nearly equaling the total population of Australia. This high level of increase creates demand for jobs, food, health care, education, and housing.

An Indian Population Bomb?

For several decades, so-called population alarmists, such as Paul Erlich, with his popular concept of the "population bomb," have broadcast disastrous images of population growth, especially for India and Bangladesh. Crowded streets, pregnant and poor women, and skinny kids figure prominently in representations of India, tightly linking First-World images of India with "population problems" that demand solutions, usually from outside experts. The 1950s and 1960s brought to South Asia substantial US involvement in the form of promoting Western methods of family planning. These programs were based on the largely mistaken assumptions that Western family planning methods would meet with widespread demand and that delivering family planning technology would reduce birth rates and defuse the "population bomb." Village-level studies conducted by anthropologists revealed instead that Indian families have their own ideas about family composition goals and strategies and are not "empty vessels" into which Western values can be easily poured (Ketayun H. Gould, "The Twain Never Met: Sherupur, India, and the Family Planning Program," in *Culture, Natality, and Family Planning,* edited by John F. Marshall and Steven Polgar, Chapel Hill: Carolina Population Center, University of North Carolina, 1976). Other village-level studies showed that population growth rates were declining at the same rate in villages where family planning services had been provided as in the "control" villages where they had not (Mahmood Mamdani, *The Myth of Population Control: Family, Caste, and Class in an Indian Village,* New York: Monthly Review Press, 1972).

Recent trends suggest that no "population bomb" lurks within India. Data from the 1991 Census of India show that the average exponential growth rate slowed from 2.22% to 2.14% between 1981 and 1991. Furthermore, the mean age at marriage for both males and females continues to rise. In 1971, it was 22.4 years for males and 17.2 years for females; in 1981, it was 23.3 years for males and 18.3 years for females. In addition, the infant mortality rate of 104 in 1984 declined to 79 in 1992.

An analysis of longitudinal data on population, poverty rates, and other measures in India since the time of independence offers a closer look at how well Malthus's model works in India (Nigel Crook, "Population and Poverty in Classical Theory: Testing a Structural Model for India," in *Population Studies* 50 [1996]: 173–85). Specifically, the following three propositions are examined.

1. Population growth will outstrip food supplies and result in high mortality in the absence of a regime of late marriage that will reduce population growth.
2. Poverty in households is a consequence of excessive fertility caused by "improvidence."
3. Poverty in society reflects household poverty, with high fertility translating into high rates of natural increase (see Mamdani, 1972: 174).

Between 1951 and 1991, India's natural population increase has been sustained at about 2% per year, lower than Pakistan's annual 3% increase. This pattern is largely the result of declining mortality, first, and then a more modest decline in fertility (with regional variations). Since independence, the growth of food production has kept up with, or even exceeded by a small margin, population growth. As Nigel Crook comments, the classical Malthusian argument has thus failed in the case of India, a rather large failure. The neo-Malthusian argument substitutes growth in the economy for food production alone. Again, the data indicate that India's gross national product (GNP) has grown at a rate exceeding that of population growth, even during the agricultural slump of the 1960s and the industrial slump of the 1970s. In addition, there is no evidence to indicate that the proportion of the poor has increased, as Malthus would have predicted. State-level analysis for India shows that poverty affects population growth very weakly, and there is no evidence that population growth affects poverty.

The big picture of Indian population change and development provides no evidence of a Malthusian disaster or an impending "population bomb." Nevertheless, microlevel studies of social dynamics often reveal serious and sustained conflict over resources and human welfare deprivation that might make Malthus feel that his model still has a chance. The following discussion moves progressively to more localized levels of analysis, beginning with data on India's major states and then village data.

Regional Variation

Analyzing regional variation in population and health dynamics in India at the state level is not optimal because many of the states are huge and all of the larger states considered here comprise substantial regional variation within them. District-level data are preferable (in 1991, India had 466 districts), but this discussion is forced to rely mainly on state-level data, mainly because recent data on key population and health variables have not been examined at the district level. Nevertheless, even at the state level, important patterns emerge.

A long-standing contrast exists between northern states (especially those in the plains region stretching from Gujarat in the west through Rajasthan, Punjab, Haryana, Uttar Pradesh, Madhya Pradesh, and Bihar) and the southern states (including Karnataka, Andhra Pradesh, Tamil Nadu, and Kerala). In general, northern states have the highest fertility rates and highest infant mortality rates (see Table 10.2). In almost every measure, northern plains states present the strongest challenges to achieving broad-based human welfare and development. The northern states are joined in this generalization by the eastern state of Orissa, which is marked by having the highest infant mortality rate (IMR) in the nation. In the southern states, led by Kerala, fertility rates and IMRs are low. Kerala's IMR of 23.8 is lower than that of many other developing countries in the "high" human development category, and it is similar to some industrialized nations as well (according to other sources, Kerala's 1991 IMR was 17 for boys and 16 for girls; see Amartya Sen, "Radical Needs and Moderate Reforms," in *Indian Development: Selected Regional Perspectives*, edited by Jean Drèze and Amartya Sen, Delhi: Oxford University Press, 1997).

Striking similarities and differences in population and health variables by state have attracted research attention in order to discover explanations for these patterns. One such study, using National Family Health Survey data on fertility and contraceptive use, compares the unfavorable situation in the northern state of Uttar Pradesh with the more favorable conditions in two southern states, Andhra Pradesh and Tamil Nadu (Robert D. Retherford and B. M. Ramesh, "Fertility and Contraceptive Use in Tamil Nadu, Andhra Pradesh, and Uttar Pradesh," in *NFHS Bulletin* 3 [1996]). The total fertility rate in Uttar Pradesh is 4.8. In Tamil Nadu, it is 2.5, and in Andhra Pradesh, 2.6, fairly close to the government's desired replacement fertility level of 2.0. Strong parallels exist with the CPU, or contraceptive prevalence use rates (the percentage of currently married women aged 13 to 49 who are currently using any method of contraception). The CPU is about 50% in Tamil Nadu and Andhra Pradesh but only 20% in Uttar Pradesh. In addition, female sterilization accounts for a greater percentage of all forms of contraception in the two southern states: 95% in Tamil Nadu and Andhra Pradesh and 66% in Uttar Pradesh.

What accounts for these differences? First, infant mortality rates are lower in the southern states, and a lower IMR is generally an antecedent of lower fertility, as families gain confidence that their children will survive. Age at marriage of females is often considered a crucial factor, though in this comparison it appears to be less of a factor, since female age at marriage is high only in Tamil Nadu (18.1 years), while in both Andhra Pradesh and Uttar Pradesh, it is the same (15.1). Women in Andhra Pradesh, however, start and end their reproductive phase early; they have the youngest average age at sterilization of women in any Indian state (24.5 years). Female labor participation (FLP) is a key underlying factor associated with women's status and fertility rates. In this comparison, a strong parallel exists between high FLP and low fertility. In the two southern states, the percentage of married women working is 47% in Tamil Nadu and 53% in Andhra Pradesh, while it is only 13% in Uttar Pradesh (women in this northern state have high "work" loads, but their work is more likely to be unpaid household labor than wage-earning work). Female literacy and school attendance, like age at marriage of females, shows a mixed pattern in these three states. Again, Andhra Pradesh resembles the northern pattern, with its 62% of females illiterate compared to Uttar Pradesh's 69% and in contrast to Tamil Nadu's 44%. Clear north-south differences also appear in the higher rate of access to electricity and mass media in the southern states. The simple exercise of comparing states that differ markedly and states that share some features but not others forces us to relinquish a tight hold on a single model or policy for all of India and to consider the utility of locally specific models and policies.

Comparisons of variables and outcomes between villages yield important findings for scholars and policy makers. One South Indian study highlights the role of women's employment and independent income in relation to other variables such as women's education and autonomy and the use of family planning (A. Dharmalingam and S. Philip Morgan, "Women's Work, Autonomy, and Birth Control: Evidence from Two South Indian Villages," in *Population Studies* 50 [1996]: 187–201). The two villages contrast markedly in women's employment opportunities. In Village I, nearly all women are employed making *beedis* (hand-rolled leaf cigarettes). The women pick up the raw materials in the morning and deliver the finished products in the evening. They work in groups of their own choosing. This work provides both an independent source of income and opportunities for spatial movement in the village and substantial interaction with other women. In Village II, no such work opportunities are available. Women are primarily involved in domestic production; only a few do agricultural labor for wages. According to widely accepted models, the higher level of economic and social autonomy of women in Village I should result in higher rates of family planning. Results were, however, surprising. Tubectomy is the only birth control method reported in both villages, and its prevalence is nearly the same in both. In addition, fertility levels vary little between the two villages. What accounts for this similarity, given the underlying differences in women's autonomy in these two villages?

The answer is a sharp reminder that a similar outcome can be produced by different causes. Girls' schooling emerges as an important factor, but not because it increases female autonomy. Higher rates of girls' schooling in Village II are in fact correlated with lower female autonomy. Girls' schooling increases the costs of children and thereby stimulates fertility reduction. Reduced fertility in Village I occurred in a context of high rates of female employment and autonomy and low rates of schooling. In Village II, it occurred in the context of low female employment rates and high rates of schooling that increase child costs. This seemingly counterintuitive finding about education demonstrates how the social context defines women's status and thereby filters and restructures the impact of particular interventions. Schooling increases child costs directly through such things as fees, books, sometimes uniforms and transportation, and opportunity costs. In addition, schooling is associated in many contexts in India with other features of upward social mobility such as spatial restrictions on females, their removal from wage labor, and their reduced social and economic autonomy. These relationships have a strong bearing on contemporary attempts to reduce child labor in India and increase primary school attendance.

Policy and Program Priorities

Throughout South Asia, governments view their countries as having population growth rates that are too high and that require intervention (Table 10.3). The one exception is Bhutan, with a growth rate considered too low but no intention of policy interventions at this time. In the other countries, a variety of policies and programs attempt to reduce population growth mainly through reducing fertility. Governments rely to varying degrees on internal and external funding and have emphasized or de-emphasized different strategies over time. For example, the predominantly Muslim nations of Pakistan and Bangladesh (and the heavily Muslim state of Kashmir in India) have been more restrictive of the use of abortion than the South Asian countries more heavily populated by Hindus (*Encyclopedia of Bioethics,* s.v. "Population Ethics: Religious Traditions in Hinduism," by Barbara Miller).

The Indian government's Family Welfare Programme, launched in 1951, is focusing on reducing infant, child, and maternal mortality, and increasing attention to low-birth weight babies, prenatal care, the number of trained birth attendants, family planning for couples, child immunization, and attention to particular health problems such as tuberculosis, leprosy, and blindness (Government of India, *Family Welfare Programme in India, Yearbook 1992–93,* New Delhi: Government of India, Ministry of Health and Family Welfare, 1994: 17). It provides services through a network of subcenters, primary health centers (PHCs), and community health centers (CHCs) in rural areas, and hospitals and dispensaries in urban areas. An expanding number of postpartum centers are being established at district and subdistrict levels. At the subcenters, the auxiliary nurse and midwife (ANM) is the key paramedical worker providing family welfare services in the community. She is supported by a male multipurpose worker. Located at the PHC is the lady health visitor (LHV) who supervises all ANMs in her area. Currently, the government hopes to increase the number of subcenters, primary centers, and community centers for every 5,000 people (3,000 in "tribal" and more remote areas) by the year 2000. Some of the Family Welfare Programme's major initiatives are outlined here.

1. In 1992–93, an integrated program called the Child Survival and Safe Motherhood Programme was launched with assistance from the World Bank and UNICEF. This program includes an array of activities, including strengthening immunization coverage throughout the nation, augmenting oral rehydration therapy (ORT) use, controlling anemia in pregnant women, preventing blindness in children, initiating a program for controlling children's respiratory infections, and increasing the number of trained birth attendants. Some of these initiatives are already being pursued in all districts, while others are in the early stages of promotion and were implemented in 1992–93 in only a few dozen districts.
2. Area Development Projects, started in 1973, continue to seek to improve the delivery of health and family welfare services through improved infrastructure in selected districts and states and

by improving the skills of medical and paramedical personnel. In 1992–93, 11 projects were underway. Financial assistance is provided by the World Bank, United Nations Fund for Population Activities (UNFPA), the Overseas Development Administration (ODA) of the United Kingdom, and the Danish International Development Agency (DANIDA).

3. A project called Innovations in Family Planning Services in Uttar Pradesh, funded by the United States Agency for International Development for a ten-year period, seeks to extend family planning services in the public sector and through nongovernmental organizations (NGOs), and promote social marketing of contraceptives, expand the choice of contraceptive methods, improve the technical competence of personnel, and increase public understanding of the benefits of family planning.
4. Funded by the World Bank, the five-year Social Safety Net Scheme provides financial assistance to upgrade PHCs in 90 identified districts with the specific goal of reducing maternal mortality.
5. The All-India Hospitals Post-Partum Programme, initiated in 1966, seeks to provide a package of maternal and child health (MCH) services, including attempts to motivate couples to support a small family norm. The program includes training for personnel and rural outreach. As of 1992–93, over 1,500 centers were functioning. (Government of India, *Family Welfare Programme in India, Yearbook 1992–93,* New Delhi: Government of India, Ministry of Health and Family Welfare, 1994: 19–22)

Several other schemes include expanding the availability of facilities for Pap smear testing, reserving a number of beds in government hospitals for tubectomies, improving outreach services in urban slum areas, strengthening aid to states for medical termination of pregnancy (MTP) services, establishing laparoscopic training centers for medical personnel, supporting a network of 18 population research centers, improving plans for NGO involvement, and increasing attention to targeted states and districts for education and communication among the rural population.

In addition to this set of national programs, thousands of local population and health programs and projects are ongoing, many of them sponsored by the mushrooming number of NGOs. It is difficult, if not impossible, to envision how one would evaluate the effects of this vast pool of activity for India as a whole. Most attempts look at a few selected measures, such as percentage of couples practicing contraception, as indicators of wider patterns. Localized case studies of the impact of a particular project help reveal in-depth evaluation information.

From reviewing the above policies and programs, three problems are clear. First, as currently structured in India, public population and health programs do not themselves address the underlying causes of poverty and deprivation. Looking at population and health programs in isolation from policies and activities that seek to address entitlement failures thus offers a very partial view of the complex whole. Second, while a certain amount of regional and social targeting seems to be occurring, more locally contextualized policies and programs are needed. Third, these policies and programs fail to address specifically India's continuing, and perhaps increasing, scarcity of girls compared to boys in the population. All three of these problems are linked. The following section highlights son preference and skewed sex ratios in relation to development goals.

Economic Development as a Constraint on Human Welfare

Since the 1970s, many anthropologists and other social scientists have led a critique of top-down economic development in terms of the negative impacts it has on the poor, women, children, and indigenous peoples. Two decades later, such impacts are more widely recognized, but structural adjustment and privatization policies have proceeded largely unimpeded. The voices of protest and the amount of data have grown, but so has the power of top-down development institutions and interests. One should not, however, overlook the fact that many grassroots efforts have made progress, especially in microcredit schemes for workers in the informal sector (see, for example, Sidney Ruth Schuler, Syed Mesbahuddin Hashemi, and Ann P. Riley, "The Influence of Women's Changing Roles and Status in Bangladesh's Fertility Transition: Evidence from a Study of Credit Programs and Contraceptive Use," in *World Development* 25 [1997]: 563–75). One area where growth-driven, top-down economic change appears strongly implicated is in the unbalanced juvenile sex ratios in India (especially in the north).

The issue of the gender gap among children in India has several dimensions, including long-standing differences between the levels of boys' and girls' health and survival and in investments such as schooling and clothing. The newest trend is the increasing ratio of boys to girls who are born, a result of increasing resort to female-selective abortion (parts of the following draw on Miller, 1997: 197–217).

Prior to the mid-1980s, when sex-selective abortion began to be used in India to a noticeable degree, unbalanced sex ratios among children were due solely to excess child mortality. The dominant pattern of unbalanced sex ratios among children is one of male bias in survival, especially in the densely populated northern plains region. In this area, juvenile sex ratios (the number of boys per girl under the age of ten years) have reached levels as high as 118–20 boys per 100 girls. This figure means that of every five or six girls born, one suffers from excess female child mortality (EFCM). This excess mortality is mainly the result of higher levels of illness and malnutrition among girls that goes uncared for or untreated at medical centers. In some cases (how many will always be unknown), direct female infanticide may occur.

Much development and demographic theory of the past and present has blandly (and blindly) assumed that economic and demographic "modernization" will automatically bring with it a decrease in female disadvantage in health and survival. Increases in child survival in general have been assumed to spell increases in survival for all. Over the past 20 years, research on girls' health and survival in India indicates that often the opposite has occurred: the gender gap in health and survival has either failed to decrease or has, instead, increased, often among groups benefiting from economic development but not in the so-called backward and undeveloped areas or among the poorest of the poor.

Continued High Levels of Excess Female Child Mortality (EFCM)

Overall, for girls and boys combined, mortality rates have declined in India since 1980. As noted above, IMRs vary by state (with the lowest in Kerala). Generally, regions with the highest IMRs also have the highest juvenile sex ratios (JSRs). This association is logical because excess female infant (and child) deaths contribute to a mortality rate that is higher than would otherwise be the case. A juvenile sex ratio of 105 boys per 100 girls is evidence that cultural manipulation of the sex ratio in favor of boys is occurring. According to the 1961 census data, more than 30% of all rural districts had juvenile sex ratios above 105. In both the 1971 and 1981 district data, JSRs above 105 increased to 40% of all districts (as of this writing, no one, to my knowledge, has analyzed district-level data on juvenile sex ratios using 1991 census data). Whether or not one accepts these district patterns as evidence that lethal discrimination against daughters is increasing in India, they seem nevertheless to suggest strongly that such discrimination is not decreasing. In strongly patriarchal cultures that value and support the lives and welfare of males more than females, benefits of economic development such as education, health infrastructure, and improved communications are skewed toward enhanced survival of males.

Regional Spread

India's rural juvenile sex ratios have long been unbalanced toward boys in a clear regional cluster of districts in the northwestern plains. Analysis of change in the regional pattern using 1961 and 1971 census data shows that the area of daughter scarcity expanded outward from the northwestern core in all directions: north into a few of the Himalayan districts, east into Bihar, south into Madhya Pradesh and Maharashtra, and west into more districts in Rajasthan and Gujarat. The 1981 rural data indicate that the pattern spread further into Bihar, Maharashtra (where the number of districts with JSRs over 105 increased to six), and Tamil Nadu (where two districts appeared in the over-105 category and another two were close, with JSRs of 104).

Explaining the diffusion of the northern pattern is difficult due to lack of district-level longitudinal data on causally related factors. One possible explanation that merits examination is the potential impact of the Green Revolution. The Green Revolution affected landholding patterns, female labor force participation, income distribution, and consumption. Exactly how such changes translate into more extreme son preference is not clearly understood. Localized analyses will no doubt reveal other economic factors at work as well. For example, Green Revolution agriculture has been little promoted in the Himalayas, yet some Himalayan districts' juvenile sex ratios masculinized at the same rate as in the plains. A major research task is to attempt to construct a longitudinal dataset that could be used to test such hypotheses; microlevel studies of key variables are also needed.

Another theory involves what anthropologists have labeled "Sanskritization," a process of social change in which lower classes and groups attempt to raise their status by adopting lifestyles of the upper strata, including northern-style dowry marriage, purdah, removal of women from wage labor, and prevention of widow remarriage. These practices would promote increased son preference and daughter aversion.

Upper–caste/class marriage practices in India's northern kinship system have long been more expensive for the bride's side among propertied groups. Since the 1960s, costs to the bride's side have risen dramatically, especially in the north, but elsewhere as well. The destination of the money and goods has shifted away from the newly married couple to the family of the groom, and what was formerly called dowry is now more accurately termed *groomprice* (Michael Billig, "The Marriage Squeeze on High-Caste Rajasthani Women," in *Journal of Asian Studies* 50 [1991]: 341–60). The groom's family may use the incoming money and goods for the marriage of their own daughter if they have one. Cross–sectional district-level data on marriage costs do not exist, but Kishor used district-level data on marriage distance as a proxy measure of groomprice marriage (Sunita Kishor, "'May God Give Sons to Us All': Gender and Child Mortality in India," in *American Sociological Review* 58 [1983]: 247–65). These data revealed a regional pattern that closely matches the pattern of greatest masculinization of JSRs between 1961 and 1971. It is likely that as economic development has occurred, some people have increased both their wealth and their social aspirations and have adopted northern-style marriages. This change would then logically be accompanied by increasing son preference and daughter disfavor.

Spreading Class Distribution

In the 1931 census data on sex ratios of selected castes in the northwest and the south, high JSRs most characterized the propertied castes of the northwest, while propertied castes in the north had balanced ratios. Unpropertied castes in both the north and the south had moderately high JSRs. Microstudies also indicated that gender disparities were most characteristic of northern propertied groups. This class model of discrimination against daughters has been highly contested: a persistent and tenacious belief that son preference and daughter neglect are caused by poverty continues to pervade much popular and scholarly thinking.

District-level 1971 census data on sex ratios for three rough "class" groupings–Nonscheduled Castes (propertied), Scheduled Castes (unpropertied), and Scheduled Tribes (unpropertied)–have several problems, the most important of which are that age data are not provided for under-ten-year-olds but rather for under-fifteen-year-olds (labeled youth sex ratios or YSRs as compared to JSRs); within-group economic variation exists, especially in the first two categories, since caste labels do not always correlate well with wealth; and the listing of castes as Scheduled or Nonscheduled varies from state to state. However, from a tentative analysis, two informative generalizations emerge. First, in the northwest, YSRs of the Nonscheduled and Scheduled Castes are similar, with both groups having very high sex ratios and the latter sometimes exceeding the former (there are no significant numbers of Scheduled Tribes reported in these northwest districts). Second, among the Scheduled Tribes, YSRs are either balanced or biased toward females. Agnihoti reports on similar findings from his even more recent analysis of 1981 data (Satish B. Agnihotri, "Missing Females: A Disaggregated Analysis," in *Economic and Political Weekly* 30 [1995]: 2074–84).

Meta-analysis of local case studies provides complementary information on class differences in EFCM. Fourteen microstudies conducted in the 1980s that contained relevant, gender-differentiated data revealed greater disparity in survival and nutritional status between boys and girls in propertied groups than among unpropertied and "tribal" groups, for whom there was little evidence of disparity (Barbara D. Miller, "Social Class, Gender, and Intrahousehold Food Allocations to Children in South Asia," in *Social Science and Medicine* 44 [1997]: 1685–95). In central and southern areas, some evidence exists of nutritional discrimination against daughters during periods of seasonal food scarcity. Higher education levels were also positively associated with more preference for sons and discrimination against daughters in food allocation and nutritional outcomes.

Female-Selective Abortion

Given the strong preference for sons among many families in India, it is not surprising that prenatal sex selection has found a massive market there. The question of how extensively such techniques are used in India is important but, like the others, difficult to answer conclusively. One indication is from estimates of the national sex ratio at birth in India, which place it "as high as 112" (Sidney B. Westley, "Evidence Mounts for Sex-Selective Abortion in Asia," in *Asia-Pacific Population and Policy* 34 [1995]: 1). Beneath this national average, substantial regional and class variation are likely to exist, following patterns in the degree of son preference. A local study of sex ratios at birth was conducted from 1983 to 1988 in several hospitals in the city of Ludhiana, Punjab (R. K. Sachar et al., "Sex-Selective Fertility Control: An Outrage," in *The Journal of Family Welfare* 36 [1990]). This study found that sex ratios at birth increased substantially from 105 in 1983 to 119 in 1988.

Another study conducted in Ludhiana involved interviews with mothers of 596 infants selected at random from those born or admitted to Brown Memorial Hospital between 1990 and 1991 (Beverly E. Booth, Manorama M. Verma, and R. Singh Beri, "Fetal Sex Determination in Infants in Punjab, India: Correlations and Implications," in *British Medical Journal* 309 [1994]: 1259). Reported use of fetal sex determination was common, involving 13.6% of the mothers of boys. Only 2% (five cases) of the mothers of daughters in the sample reported using fetal sex determination; of these, the pregnancies either involved a male twin or had been misdiagnosed as male. Female-selective abortion (FSA) in this sample increased with the level of household income. In this study, education was positively correlated with use of FSA. None of the mothers with no formal education had undergone prenatal testing, while among those with formal education, the frequency of FSA was constant.

Advancements in medical technology support the demand for sons. In the early 1980s, FSA was available in some cities in India, mainly in the north but probably also in Bombay. By the 1990s, ultrasound had replaced amniocentesis as the predominant means for determining the sex of the fetus in India. The profits from manufacturing and distributing ultrasound machines are huge. In India, ultrasound equipment constituted 20% of the total market in medical technology in 1993, and the market is growing by 20% a year. A new General Electric model offers most of the conventional functions in a 20-pound unit that can fit in the back seat of a car and can thus reach even relatively remote villages.

Son Preference, Unbalanced Juvenile Sex Ratios, and Development

At the same time that economic growth and development are occurring in many regions of India and leading to substantial income growth for many, son preference is strong and increasing, as is daughter scarcity. This situation is driven in part by economic growth and development (and enabled by medical technology). Two other relationships are also crucial for our consideration. First, without the effects of EFCM or FAS, son preference boosts fertility. One estimate is that India's national fertility would be reduced by 8% in the absence of son preference (R. Mutharayappa, Minja Kim Choe, Fred Arnold, and T. K. Roy, "Is Son Preference Slowing Down India's Transition to Low Fertility?" in *NFHS Bulletin* 4 [1997]). The demand for sons, therefore, has two effects: it causes higher fertility than would otherwise be the case, and it causes EFCM and FSA. Second, development goals have long included attempts to reduce fertility. They should now be designed to target EFCM and FSA. Among other concomitants of a gravely unbalanced sex ratio in the population is a heightened level of sociopolitical violence, as demonstrated by a strong correlation between incidents of violence and skewed juvenile sex ratios in districts of Uttar Pradesh (Philip Oldenburg, "Sex Ratios, Son Preference, and Violence in India: A Research Note," in *Economic and Political Weekly* 27 [1992]: 2657–62).

Solutions are situated in highly sensitive domains. Governments must take a strong stand in promoting policies and programs that give top priority to distributional questions instead of sheer growth, especially in working toward gender equality in economic entitlements. In regions and groups of South Asia where female economic entitlements are most secure, sex ratios tend more toward equality.

Further Reading

Basu, Alaka Malwade, *Culture, the Status of Women, and Demographic Behaviour: Illustrated with the Case of India,* Oxford and New York: Oxford University Press, 1992

Survey-based study of low-income slum-dwellers in Delhi of both north and south Indian origins. Considers aspects of women's status, maternal characteristics, child health, and fertility. In terms of fertility, differences between the two cultural groups emerge in the initiation, pace, and termination of child-bearing.

Das Gupta, Monica, Lincoln C. Chen, and T. N. Krishnan, editors, *Women's Health in India: Risk and Vulnerability,* Delhi: Oxford University Press, 1995

Eleven chapters on women's health in India in addition to an overview chapter and a concluding chapter on policy implications. Topics include child mortality, reproductive health, maternal mortality, risks of STD and HIV transmission to women, widowhood, and health status of older women. Includes case studies of the state of Manipur and Tamil Nadu.

Drèze, Jean, and Amartya Sen, *India: Economic Development and Social Opportunity,* Delhi: Oxford University Press, 1995

Analysis of links between poverty, population, health, and education in India using mainly state-level data. The authors focus on the issue of social inequality and gender inequality as key challenges facing policy makers. Comparisons are offered with China.

Drèze, Jean, and Amartya Sen, *Indian Development: Selected Regional Perspectives,* Delhi and New York: Oxford University Press, 1997

Introduction discusses key issues facing India as privatization proceeds. In-depth studies of three states provide regional contrasts in development: Kerala (success), Uttar Pradesh (failure), and West Bengal (mixed). Final chapter offers a district-level analysis of many variables related to development and concludes that female literacy and female labor participation rates have significant impacts on fertility.

Fricke, Thomas E., *Himalayan Households: Tamang Demography and Domestic Processes,* Ames: Iowa State University Press, 1986

An example of demographic anthropology, this local study of population patterns and change in one region of Nepal includes chapters on the subsistence economy, fertility and mortality, the life course, household dynamics, and recent changes.

Hartmann, Betsy, and James Boyce, *Needless Hunger: Voices from a Bangladesh Village,* San Francisco: Institute for Food and Development Policy, 1979

Based on fieldwork in rural Bangladesh, the authors argue that poverty and hunger in Bangladesh are primarily caused by severe class inequalities in economic entitlements. Includes a critique of the role of foreign aid in perpetuating these inequalities and suggestions for change.

Jeffery, Roger, and Alaka M. Basu, editors, *Girls' Schooling, Women's Autonomy, and Fertility Change in South Asia,* New Delhi: Sage, 1996

Chapters address women's education, autonomy, and fertility in Bangladesh, India, Sri Lanka, and comparative issues in South Asia. The editors find evidence that no universal causal relationship exists between rising levels of female schooling and declining levels of fertility or that female schooling necessarily enhances female autonomy. The goal of reducing fertility should be served by increased provision of "female-friendly" family planning methods and services.

Miller, Barbara D., *The Endangered Sex: Neglect of Female Children in North India,* 2nd edition, Delhi: Oxford University Press, 1997

Examines regional and class differences behind rural India's unbalanced juvenile sex ratios. Considers history of direct female infanticide in India, contemporary patterns of intrahousehold allocations of food and medical care, marriage practices and dowry, and women's labor force participation. A new postscript chapter describes the continuing gender gap among children in health and survival and the increased use of female-selective abortion, and proposes that balanced sex ratios be considered a public good.

Wyon, John B., and John E. Gordon, *The Khanna Study: Population Problems in the Rural Punjab,* Cambridge: Center for Population Studies and Harvard University Press, 1967

Classic report on the design, implementation, and outcome of Harvard's multiyear, multivillage attempt to promote the use of cosmopolitan family planning methods in India's Punjab state.

Barbara Miller is associate professor of anthropology and international affairs at George Washington University, where she is also director of the women's studies program. Her publications include *The Endangered Sex: Neglect of Female Children in Rural North India* (2nd edition, Oxford University Press, 1997), an edited volume, *Sex and Gender Hierarchies* (1993), and a coedited volume, *Talking Hair: Asian Studies* (forthcoming, 1998).

Table 10.1: Selected Statistics on Health and Population in South Asia

*Country**	*Population mid-1996 (millions)*	*Births per 1,000 population*	*Natural increase (annual %)*	*"Doubling time" in years at current rate*	*Infant mortality rate*	*Total fertility rate*	*Percent of married women using contraception*	*Secondary school enrollment (%)*	
								Male	*Female*
Pakistan	133.5	39	2.9	24	91	5.6	12	28	13
India	949.6	29	1.9	37	79	3.4	41	59	38
Nepal	23.2	39	2.6	26	98	5.2	23	46	23
Bhutan	0.8	39	2.3	30	121	5.4	-	8	2
Bangladesh	119.8	31	2.0	35	88	3.7	45	25	13
Sri Lanka	18.4	20	1.5	47	18.4	2..3	66	71	78

*Counties are ordered in a general regional pattern starting in the northwest
Source: Population Reference Bureau, *1996 World Population Date Sheet,* Washington, D.C.: Population Reference Bureau, 1996

Table 10.2: Population and Health in Selected States of India

*State**	*Population (1991 census in thousands)*	*Percent increase in population 1981–91 (1991 census)*	*Total fertility rate 1990–92*	*Infant mortality rate 1988–96*	*Current use of contraception (all methods)*	*Percent of females 20–24 married before age 18*	*Percent of females literate age 6+*	*Percent of children undernourished (weight for age) under age 4*
Himachal Pradesh	5,171	20.8	3.0	55.8	58.4	24.2	57.4	47.0
Punjab	20,282	20.8	2.9	53.7	58.7	14.9	52.0	45.9
Haryana	16,464	27.4	4.0	73.3	49.7	57.3	45.9	37.9
Rajasthan	44,006	28.4	3.6	72.6	31.8	69.5	25.4	41.6
Gujarat	41,310	21.2	3.0	68.7	49.3	33.4	51.3	50.1
Uttar Pradesh	139,112	25.5	4.8	99.9	19.8	63.9	31.5	59.0
Madhya Pradesh	66,182	26.8	3.9	85.2	36.5	73.3	34.3	57.4
Bihar	86,374	23.5	4.0	89.2	23.1	69.1	28.6	62.6
West Bengal	68,078	24.7	2.9	75.3	57.4	56.4	55.2	56.8
Orissa	31,660	20.1	2.9	112.1	36.3	45.5	41.4	53.3
Maharashtra	78,937	25.7	2.9	50.5	53.7	33.4	51.3	50.1
Karnataka	44,977	21.1	2.9	65.4	49.1	51.2	46.5	54.3
Andhra Pradesh	66,508	24.2	2.6	70.4	47.0	68.6	38.5	49.1
Tamil Nadu	55,859	15.4	2.5	67.7	49.8	36.1	56.1	48.2
Kerala	29,099	14.3	2.0	23.8	63.3	19.3	82.4	28.5

*The states are ordered regionally starting from the northwest. The smaller states of Goa and those in the far northeast are not included.
Source: Population Reference Bureau, "Health and Family Welfare" (wall chart), Washington, D.C.: Population Reference Bureau, 1996

Table 10.3: Summary of Population Policy in South Asian Countries

*Country**	*Population growth*	*Fertility level*	*Contraception use*	*Mortality level*	*Spatial distribution*	*Internal migration*	*Immigration*	*Emigration*
Pakistan	Too high; intervene to lower	Too high; intervene to lower	No major limits; direct suport	Acceptable	Satisfactory	To decelerate trend	Satisfactory; no intervention	Too low; intervene to raise
India	Too high; intervene to lower	Too high; intervene to lower	No major limits; direct support	Unacceptable	Major change desired	To decelerate trend	Satisfactory; intervene to maintain	Satisfactory; intervene to maintain
Nepal	Too high; intervene to lower	Too high; intervene to lower	No major limits; direct support	Unacceptable	Major change desired	To decelerate trend	Too high; intervene to lower	Satisfactory; no intervention
Bhutan	Too low; no intervention	Satisfactory; no intervention	No major limits; direct support	Unacceptable	Minor change desired	To accelerate trend	Too high; intervene to lower	Satisfactory; intervene to maintain
Bangladesh	Too high; intervene to lower	Too high; intervene to lower	No major limits; direct support	Unacceptable	Major change desired	No intervention	Satisfactory; no intervention	Satisfactory; no intervention
Sri Lanka	Too high; intervene to lower	Too high; intervene to lower	No major limits; direct support	Unacceptable	Minor change desired	No intervention	Satisfactory; intervene to maintain	Satisfactory; no intervention

*Countries are ordered in a general regional pattern starting in the northwest.
Source: United States Population Fund, *Inventory of Population Projects in Developing Countries around the World,* New York: United Nations, 1995

Chapter Eleven

Living Multiculturally in a Federal India

Reeta Chowdhari Tremblay

Independent India's federal project was as much a product of its colonial legacy as of its need to respond to nation-building requirements. The Government of India Act, 1935, introduced the concept of federation in British India. The founding fathers of the independent nation expected this imported institutional framework to address simultaneously the complex diversity of the country and the building of a new nation on the three fundamental principles of democracy, secularism, and socialism. Although the Indian Constitution has been characterized as quasi-federalist (federalist in name but unitary in practice), the federal process in India has exhibited both centripetal and centrifugal tendencies. Indeed, the 1947 partition event, which meant the large-scale cross-migration of Hindus and Muslims between the newly formed countries of Pakistan and India, widespread communal violence, and a large refugee population, weighed heavily on the minds of the members of the Constituent Assembly. State formation was to be accompanied by nation formation in which the Indian state would play a decisive role in the construction of a political community, a complex task given the embedded loyalties of individuals to religions, castes, languages, tribes, and regions. At the time of independence, India inherited 14 geographical linguistic groups, more than 3,000 caste groups, a large Muslim minority, and tribal loyalties in the northeast. The task of integrating more than 500 princely states into the new state made the process of nation building even more challenging.

The state's excessive reliance on the central government was a result of its nation-building requirements, a pervasive paternalistic ideology, a disinclination to break away from tradition, and the absence of a tradition of states rights. The Indian Constitution, promulgated in 1950, in its various provisions empowers the national parliament to intervene in the affairs of the regional government and assigns to the Union government supervisory powers over the legislative and executive authority of the states. Articles 2 and 3 empower parliament "to admit into the union or establish new states" by splitting or altering the boundaries of any existing state. Although this provision was intended to facilitate the alteration of the colonial boundaries of the regions that had been constituted "on considerations of military, political and administrative exigencies or the convenience of the moment" (Chanda, 1965: 44) and to fulfill the Congress Party's promise during the national movement of linguistic reorganization of the states, it also laid the foundations of "administrative federalism" where regional identity was denied constitutional protection. Articles 352, 356, 358, and 359 endow the central government with overriding powers in its relations to states by allowing the declaration of a national emergency on the grounds of national security, a failure of constitutional machinery in states, or a financial emergency. In the duration of the emergency, Article 250 empowers the parliament "to make laws for the whole or any part of the territory of India with respect to any of the matters enumerated in the state list." Powers granted to the central government in the suppression of regional governments and the suspension of the Constitution in the state, particularly with regard to the exercise of fundamental rights, are extraordinary. Invoking national security and national integration, the central government's excessive powers in the suppression of rights have been continually further enhanced by various pieces of legislation, such as the Armed Forces Special Powers Act, the Maintenance of Internal Security Act, the National Security Act, the Unlawful Activities Prevention Act, the Terrorist Affected Areas Ordinance, and the widely used Terrorist Activities and Preventive Detention Act. The effect of these constitutional and legal provisions has been further strengthened by political processes, such as the dominance of the Congress Party until the mid-1960s at both the regional and national levels, followed by the centralized personal leadership of Indira Gandhi and the role of the Planning Commission in directing the economic development of the country, all of which have effectively placed tremendous powers in the hands of the central government.

Nevertheless, the centralizing features and tendencies of Indian federalism coexist with pluralism, regionalism, and decentralization in India's contemporary politics. Regional and cultural pluralisms have crystallized and have been sharpened due to several factors: the linguistic reorganization of the states, the granting to minorities of constitutional rights to cultural and educational privileges, the exclusive jurisdiction of regional governments in the two crucial domains of education and agriculture (responsible, respectively, for the failure to transform Hindi into the sole official language and for defeating the national policy of land reforms), the extension of preferential policies in employment and education to local ethnic majorities, the tribal struggle for autonomy in the Northeast hills, and the secessionist Khalistan and Kashmir's Azadi (freedom) movements. Strong regional parties, such as the Telugu Desam in Andhra Pradesh, Dravida Munnetra Kazhagam in Tamil Nadu, Akali Dal in Punjab, and National Conference in Jammu and Kashmir, have been extremely successful in mobilizing voters on the basis of regional, linguistic, and religious identities. At both the national and the regional levels, political parties such as Janata Dal, Bhaujan Samaj, and Samajwadi have emerged as the representatives of Other Backward Castes (OBCs), Muslims, and the Scheduled Castes. Undoubtedly, complex identities of caste, language, religion, and region constantly challenge independent India's founding and homogenizing myths of democracy, socialism, and secularism.

While the discourse on federalism in India has been traditionally formulated within the framework of center-state relations, thus focusing on centralization and decentralization, this chapter makes a departure and views federation as both a territorial and nonterritorial project, the former addressing the fragile equilibrium to be maintained between indestructible union and indestructible units, the latter directed to the issues of cultural representation and identity within the concept of a multicultural society.

One of the tasks of a multicultural federalism is to provide cultural recognition and ensure that differences from the dominant regional norms do not result in the powerlessness and marginalization of the identities of the minority groups. Over the last four decades, Indian federalism has had to face the challenge of balancing territorial with nonterritorial requirements of the Indian nation. Furthermore, given its cultural diversity and social pluralism, it has constantly struggled to reconcile the claims of equal citizenship with group identities and interests. The problem has been how to maintain an egalitarian democratic society when the universal franchise is used by political parties and leaders toward the mobilization of primordial loyalties in order to secure and maintain power and to shift public policy in favor of their constituencies. The consequences of this double-edged federalism have been, on the one hand, the reinforcement of cultural and social pluralism, further sharpening group identities, and on the other, pushing the national leadership to a centralist ideology, making the state more hegemonic. This in turn generates resistance by regional and cultural groups, thus challenging the nation-state project. In his discussion on ethnicity and nationalism, Paul Brass quite aptly remarks that "the process of consolidating power (by the Central government) is inherently tenuous and that power begins to disintegrate immediately at the maximal point of concentration. At that point, regional political forces and decentralizing tendencies inevitably reassert themselves unless the national leadership chooses to attempt a more definitive consolidation by bringing into play the full range of unitary powers provided in the Constitution of India" (Brass, 1990: 117). Brass's contention is that these centralizing efforts are doomed to failure "in the long run" and regional and cultural pluralism will eventually reassert itself against any consolidating regime. Indeed, the Indian state has been unable to transcend the hegemonic concept of the nation-state and innovate structures for celebrating multiple cultural forms where recognition of difference is combined with the rights of a full citizenry. While the Indian state finds itself simultaneously autonomous and constrained, pursuing its goals of integration and development "relatively insulated from societal forces" and yet constantly blocked "by the representation of organized social (and cultural) forces" (Rudolph and Rudolph, 1987: 61), the ideological balance of power has always remained in India in favor of a nation that would transcend the "parochial" forces of caste, language, and religion. The following discussion provides a brief overview of the crisis of Indian federalism and reveals the complex reality of struggle, on the one hand, for political power and economic security by the Backward Castes, the upper-caste Hindu nationalist majority, and the regional linguistic majorities, and, on the other, for the preservation of religious/ cultural and tribal identities.

Construction of Political Community and Accommodative Politics

The Constitution of independent India, promulgated on 26 January 1950, ushered in a new political community in which the principles of a liberal state were carefully combined with the historical legacies of the British Raj and traditional India. Although it has been aptly asserted that the new political institutions, "to be acceptable by entrenched elites across diverse regions, had to be constituted in a form that presented little danger to the socio-cultural foundations of dominance" (Frankel and Rao, 1989–90: 501), an overriding concern of the founding fathers was, as Nehru stated, to "promote the unity of India and yet preserve the rich diversity of our inheritance" (Austin, 1993: 111). To ensure religious harmony, particularly after the violence accompanying partition, the Western concept of secularism was Indianized by referring to the traditional Hindu pluralism and syncretism as expressed in the idiom *Sarva Dharma Sambhava* (respect for all religions), or Gandhi's well-known prayer *Ishvar Allah Tere Nam, Sabh Ko Sanmati De Bhagvan* (Ishvar [Hindu] and Allah [Muslim] are two names for the same God who should provide wisdom to all). Similarly, national symbols, such as the Indian flag and the national anthem, were constructed to reflect both the modern liberal project and India's commitment to its traditional heritage of religious harmony and diversity. The tricolor flag, where deep saffron stands for Hinduism, green for Islam, and white for peace or neutrality (of the state), is the best articulation of the Indian version of secularism. The *chakra* (the wheel) in the center of the flag, which originally appeared on the abacus of Sarnath Lion, capital of Ashoka, signifies the ideology of peace since the days of the Maurya kingdom and the universal state of the Mauryan ruler, Ashoka. To celebrate both the unity and diversity of India, the national anthem recognizes the regional traditions of Punjab, Sind, Gujarat, Maratha, Dravid, Orissa, and Bengal. A small section of

Tagore's rendition of the national anthem reads, "thy name rouses the hearts of Punjab, Sind, Gujarat and Maratha, of the Dravida and Orissa and Bengal."

It is interesting to note that Western academic discourse has only recently begun to explore the issues of citizenship, cultural pluralism, and a multicultural political community. A major theoretical concern of leading North American scholars such as Charles Taylor, Amy Gutman, Nancy Fraser, and Carol Pateman has been to show how the politics of recognition of identity and group rights can be reconciled within a liberal democratic and cohesive political community. One of the major preoccupations of the Indian Constituent Assembly was precisely with this concern, which the present multicultural Western societies have recently begun to address. Consequently, the Indian Constitution makers were not only sensitive to group identities but were innovative in generating a difficult and a challenging nontraditional discourse on political community. Relying on indigenous Hindu traditions, emphasizing collective identities such as family, caste, and tribe, and borrowing from "imported" liberal theory revolving around the concept of individualism, the founding fathers of the Indian state constitutionally attempted to balance contradictory principles of equal citizenship with collective rights: secularism with religious community rights, fundamental equality for all citizens with preferential privileges for backward classes/castes, and an official language with the protection of minority linguistic rights, etc. Although these contradictions in the Constitution have had the adverse effect of fragmenting the society by creating narrow primordial loyalties and generating violence as the dominant mode for the resolution of community conflicts, these present-day realities of the Indian polity result as well from the failure of the overloaded and overpoliticized Indian state to respond effectively to the economic demands of the mobilized caste/community collectivities; from the nature of a political leadership whose interest in political power is fundamentally guided by the consideration of distribution of economic privileges to their narrowly based political constituencies; and from state repression to implement three overriding unifying principles. These principles are (1) the Indian union is perpetual and no demands for secession would be entertained, (2) no religious consideration would form a part of the state agenda, and (3) India's territorial integrity in relation to its border states and vis-à-vis its neighbors would at all costs be protected.

Article 29 of the Indian Constitution recognizes the rights of "any section of the citizens of India" who have a distinct language, script, or culture "to conserve the same." While Article 350 allows any linguistic minority to "submit representation for redress of grievance to any central or state authority," Article 350A makes it obligatory for all regional and local governments "to provide adequate facilities for instruction in mother-tongue at the primary stage of education to children belonging to linguistic minority groups" (Brass, 1990: 155). Along with the religious communities, the linguistic groups whose minority status has been recognized in the Eighth Schedule have been given the right, under Article 30, to establish and administer their educational institutions while barring the state from any discrimination against them when it grants financial support to private educational institutions. In conformity with the constitutional principles, two types of denominational schools have been organized in India by the four major religious communities: (1) religious institutions aided by a particular community (for example, Maktabs and Madrasas of Muslims, Gurukhulas, Pathshalas, or Sanskrit schools of Hindus, Gurmukhi schools of the Sikhs, and Mission schools of the Christians), and (2) religious schools aided by private or government agencies (for example, Islamic schools or colleges of the Muslims, Arya Samaj or Santan Dharam schools of the Hindus, Khalsa schools of the Sikhs, and Convent schools of the Christians). While the former category of schools impart religious education, the latter, in addition to a limited number of hours for religious instruction in the classroom and during the morning assemblies, follow the secular government-prescribed curriculum. In the spirit of accommodation of religious minorities, the secular Indian Constitution allows the religious communities to adhere to their personal laws in the governance of their communities in spheres such as marriage, divorce, and inheritance. The 1950 Constitution opted to continue the British practice to leave "personal, or family, law to adjudication by the various communities unless the individuals concerned opted to place themselves under British law" (Austin, 1993: 123). Although in the early 1950s, Nehru tried to secularize the Hindu law, he faced opposition from both within and outside the Congress Party. And although a series of parliamentary acts were passed, collectively known as the Hindu code, no such attempt was made to modernize the Muslim personal law, generating a lot of resentment by the Hindu majority against the Muslims.

The democratic egalitarian ideology of the Congress Party guided the constitutional decision to provide guarantees of political representation (Articles 330–34) in parliament and regional legislative assemblies and reservation in educational and administrative institutions (Article 335) for Scheduled Castes and Tribes. (By retaining the list complied by the British government in India in 1935, the Constitution placed a large number of "Untouchable" castes and tribes on a schedule or list. Thus, *Scheduled Castes* and *Scheduled Tribes* are technical/formal terms.) Of all central government jobs, 22.5% are reserved for this targeted group. Of the 543 Lok Sabha seats, 78 are reserved for the Scheduled Castes and 41 for the Scheduled Tribes. The legislative assemblies have reserved a similar proportion of seats for this group (Hardgrave and Kochanek, 1993: 190). Although these guarantees were to end in 1960, they have been renewed and were expanded to include the Backward Castes, officially designated as "Other Backward Castes" (OBCs), a largely rural group that accounts for approximately half of India's population. In response to their political power, initially at the regional level and now at the national level, in 1990, 27% of all central government jobs were reserved for the OBCs. In addition to the preferential treatment for the Scheduled Castes and Tribes, the Constitution has also created flexibility for the Indian parliament to protect the job interests of the local majorities. In defense of this provision, B. R. Ambedkar pointed out to the members of the Constituent Assembly, "You cannot allow people who are flying from one province to another, from one state to another as mere birds of passage without any roots . . . just to come, apply for posts, and so to say take the plums and walk away" (Weiner and Katzenstein, 1981: 24). Thus, in India, although the Constitution does not allow dual citizenship and clearly states that there should exist equality of opportunity for all citizens in matters of employment and that no discrimination should be allowed on the basis of religion, race, sex, caste, descent, place of birth, or residence, common citizenship has come to coexist with regional citizenship, which is effectively a prerequisite for employment.

Until the mid-1960s, conflicts arising out of contradictory constitutional proposals and policies emanating from them were generally avoided. This is attributable to the accommodative leadership style of Nehru, the dominance of the Congress Party at both the regional and the central levels, the Congress culture of centralist politics and of absorbing pressures from the margin, and the party's successful mobilization of the landed interests, the poor, and the Muslim minority. Frankel correctly notes, "the entire national political structure after independence was, in fact, built upon accommodation of these linguistic, religious, and caste sentiments and structures as the only way to accelerate national integration, enhance the legitimacy of the political system, and maximize the possibility of peaceful adjustments of social conflicts that arise during the development process" (1978: 20). However, by the end of the 1960s, the political arena was beginning to witness a shift toward the separation of politics between the two levels of government. The Congress Party began slowly to be replaced by regional parties at the state level. Half-intentioned land reforms, through the abolition of intermediaries, had given rise to an influential class of small- and middle-level farmers, the so-called bullock capitalists, translating their economic strength into political power. The rural and vernacular leadership first slowly took charge of the regional government and then began to assert themselves at the national level. The poor economic performance and limited distributive rewards activated political competition among various groups to influence public policy in their favor. While in the last three decades, the response of the national leadership to the turmoil has been a further consolidation of central power and increased recourse to state repression, the Indian political system has been subjected to power struggles, often resulting in violence, where "the subjects of contention have been positions of power and status, state-controlled patronage and access to educational institutions, and economic rewards to be derived from land and wages" (Kohli, 1990: 14).

Linguistic Federalism and Minorities

The Indian federation is divided into 25 states, each with a dominant linguistic group. Recognizing linguistic homogeneity as an important factor conducive to administrative efficiency, the State Reorganization Commission, in its 1955 report, recommended that regional boundaries be redrawn in accordance with the linguistic divisions of the country. Consequently, in 1956, 14 states and six territories were created, each state's boundary coinciding with that of the dominant regional linguistic group. In 1960, Bombay was divided into Gujarat and Maharashtra. In 1963, as a result of the Naga rebellion, the state of Nagaland was created by separating the tribal territory from the Assamese administration. In 1966, in response to Akali Dal's agitation for Punjabi Suba and a growing demand for a separate state for Hindi areas in Punjab, two new states were created, Haryana and Punjab. In 1971, Himachal Pradesh, the hill districts of Punjab, was given the status of full statehood. In 1972, tribal violence in the Northeast was responsible for the creation of Manipur, Tripura (by providing full statehood to these two Union Territories) and Meghalya (by separating this part from Assam). In 1984, to squelch political insurgency associated with the demand for independence, Mizoram, another Union Territory, was granted statehood. In 1987, Goa was added to the large list of states in India.

The impact of linguistic federalism has been an increased sense of regional identity of which the leaders of the dominant language groups have been able to take advantage. Recognition of a dominant language within a particular geographical boundary has provided politicians with open opportunities to capture and control power locally. By favoring those linguistic groups that had received official recognition during the colonial rule and which, in that process of recognition, had standardized their languages by assimilating the local dialects, the linguistic reorganization of the states assured their hegemony over the minority language groups. Although linguistic minorities have been provided constitutional guarantees, particularly with regard to education, various states have pursued discriminatory policies toward them. In addition, once the framework for a culturally pluralistic state was laid out, demands of autonomy and independence by regional groups acquired a certain degree of legitimacy.

These demands have ranged from Assam for the Assamese, Maharashtra for the Maharashtrians, secessionist movements of Khalistan (Punjab), and independence movements for Jammu and Kashmir. At one end of the spectrum, a coincidence of the linguistic, ethnic, and territorial boundaries has given rise to "sons of the soil" movements in which preferences for governmental positions and white-collar jobs in industry and the private sector are demanded for the local community (where *local* is defined as the numerically dominant linguistic group); at the other end, ambitious demands have been made for a separate homeland. While the demands of the dominant local community vis-à-vis "the migrants" have been accommodated, despite the rights guaranteed in the Constitution to move freely throughout the country and to reside and settle in any part of the country, the secessionists' demands have been coercively crushed. In short, the survival of the Indian federation is largely a result of the national leadership's accommodation of the influential regional groups in the national political structure. By avoiding confrontation, the central government "adopted a posture of mediation and arbitration between contending linguistic cultural forces" (Brass, 1990: 151), so long as these did not threaten the territorial integrity of the state.

In addition to the linguistic conflict at the regional level between the dominant language and minority groups, the official language issue has generated protests and violence due to the fact that knowledge of the official language as the language of administration is required for the most sought-after public service jobs. The Indian Constitution makes a distinction between the languages of India and the official language. Although there are 1,652 mother tongues, the Eighth Schedule recognizes 18 languages, effectively creating a language hierarchy. (Recognition of languages by the Eighth Schedule enables these linguistic minorities to receive cultural and educational protection. There have been ongoing struggles on the part of various language groups to be included in the Eighth Schedule. During the past four decades, only four language groups have been successful, increasing the number of recognized languages from 14 to 18).

Article 343 of the Indian constitution, a compromised outcome of one of the Constituent Assembly's bitter disputes, stipulated that Hindi would become the official language of the Indian Union in 1965, and that during the transition period, English should remain the language of the administration. Though Hindi is the language spoken by the largest number of Indians, it is not the language of the majority and it is spoken only in northern and central India. The replacement of English by Hindi implied putting non-Hindi speakers, particularly the educated class, at a disadvantage regarding employment opportunities at the central level and the competitive Indian Administrative Services. The opposition to Hindi came mostly from the South, specifically from Tamil Nadu. In response to violent protests to "Hindi imperialism," Nehru assured the non-Hindi areas that English would continue for an indefinite period. In 1963, the Official Languages Act was passed, which introduced a new concept, the "three language formula," making it mandatory for all schools to teach in three languages: the regional language, English, and Hindi for non-Hindi speakers or a different Indian language for Hindi speakers. After Nehru's death and the changed pro-Hindi leadership in the central government, Tamil Nadu became engulfed in student protests. During late 1964 and early 1965, a large number of deaths (estimates ranged from an official count of 60 to unofficial reports of upward of 300) were caused by police firing at student demonstrations. Two students immolated themselves, Hindi books and signs were burned, and the armed forces were called in to restore order in the state. In 1967, an amendment to the Official Languages Act incorporated Nehru's earlier verbal assurances that Hindi would not be imposed against the wishes of the people and that English would be retained for an indefinite period. It also paved the way for the three language formula as the official policy by allowing all major regional languages, Hindi, and English, as media of examination, a policy which, however, has generally remained ignored by both the Hindi heartland and the South. While the southern states, such as Tamil Nadu, have not made any particular attempt to include Hindi in the school curriculum, the Hindi-speaking states have opted for Sanskrit, the ancient Hindu language, as the "other" language in their educational system. With the exception of Urdu and Punjabi, other languages (such as

Maithali in Bihar, Chattisgarhi in Madhya Pradesh, and several tribal languages), which obtained official constitutional status, have not received either administrative or financial support. English remains the language of communication throughout India, and as Hardgrave and Kochanek point out, its usage as the link language is becoming "increasingly tenuous."

Unofficial Civil Wars: Religion and Caste

On 6 December 1992, a large Hindu mob destroyed the Babri mosque in the city of Ayodhya, in Uttar Pradesh, to reverse the actions of the Mughal dynastic ruler Babar who, according to the local historical legend, had constructed a mosque in 1528 to replace the temple of Ram (Hindu deity celebrated in the *Ramayana* epic). The significance of the demolition of the mosque lies in the shift that ensued in the consensus regarding the national ideological discourse. The conduits for this shift have been the Bharatiya Janata Party (BJP), the Vishwa Hindu Parishad (VHP), and the Rashtriya Swayamsevak Sangh (RSS). No longer were the Hindus willing to accept Nehru's secular normative basis for the nation, and they had begun to explore the possibilities of establishing the Indian nation on the principles of *Hindutva. Hindutva,* or Indianness, holds that there can be only one national identity in the Hindu *rashtra* (nation), that the country can have only one Hindu culture, that there can be many "flowers, but one garland; many rivers, but one ocean." The *Hindutva* concept, in its symbiotic relationship with the notions of *pitrabhumi* (sanctity of the land of one's birth) and *punyabhumi* (holy land), maintains that all differences of ritual, belief, and caste are irrelevant. For *Hindutva,* what matters is the origin in the Hindu nation, which confers on the inhabitants a distinctive identity since, unlike secularism, an imported Western concept that does not emotionally bind people, India's common culture and common civilization are its essential unifying forces. For Hindu chauvinists, *Bharat* (India) is both their motherland and a holy land, both *pitrabhumi* and *punyabhumi.* This has understandably made the Muslims in India, about 100 million in number, 12% of India's population, feel both politically and religiously insecure, all the more so in light of the militancy associated with the *Hindutva* movement. To "regenerate pride" among the Hindu community and to implement the Hindu nationalist agenda, Hindu *senas* (armies) and *raksha dals* (protection/defense groups) emerged. Violent rhetoric became an integral part of the mobilization strategy. Hindu rallies echoed with slogans such as "Jo ham se takrayae ga choor choor ho jayega" (he who challenges us will be crushed to pieces). Leaflets distributed by the VHP clearly evoked the image of divine vengeance seeking Muslim blood, and Sanyasin Rithambari of the VHP gave a call to all men to wake up, "bir bhaiyon jago" (brave brothers, wake up), clench their fists, and defend their nation.

The Shah Bano case and the upper-caste backlash against the implementation of the Mandal Commission report were the catalytic events behind the galvanization of Hindu nationalism. In 1985, India's Supreme Court granted a 73-year-old Muslim woman alimony to be paid by her husband of 43 years after she had been given a divorce in the traditional Muslim manner. This was contrary to Muslim personal law, whereby her husband was not required to pay her a maintenance allowance. This judgment was denounced by Muslim fundamentalists and by the clergy, who raised the alarm that Islam was in danger. This decision, they believed, allowed the state's direct interference with Shariat and paved the way for a uniform civil code. India's Prime Minister Rajiv Gandhi immediately responded to these protests with legislation in the form of the Muslim Women Bill, which removed divorce for Muslim women from the provisions of the current secular laws. The Shah Bano case gave the Hindu nationalist organizations an excellent opportunity to unite Hindus behind the *Hindutva* movement. The government was criticized for appeasing the Muslims who, it was alleged, were allowed to maintain their religious practices, whereas Hindus, the majority community, were being asked to abandon their religious beliefs in the name of "pseudo-secularism." Muslims, the BJP pointed out, would never consider themselves Indians due to their lack of psychological affiliations with Indian culture and civilization.

In 1980, the Mandal Commission (the Backward Classes commission) released its report recommending that 27% of all central government jobs be reserved for the "backward" castes. This, together with reservations for the Scheduled Castes and Tribes, meant that 49.5% of all government positions were reserved. In 1990, Prime Minister V. P. Singh of the coalition National Front government, without consulting either the BJP or the government's Communist partners, declared his intention to implement the report. Upper-caste Hindus, particularly students, who were the major losers in the competitive market for public service jobs, strongly protested the government's decision. In North India, there were several cases of self-immolation. Student demonstrations were met with police fire and police indifference to the wounded in encounters that were captured on tape by Newstrack, a news video program. The BJP, which resigned from the government, saw the reservation policy as creating severe cleavages within the Hindu community. The restoration of the Ram temple at Ayodhya was to provide an excellent opportunity to the Hindu organizations to unify the Hindus by cutting across class and caste lines in the "homogenization of a 'national' Hinduism" (Van Der Veer, 1994: 7). Hindus were asked to close their ranks and assert their identity as the majority community, lest they find themselves "treated like dirt" in their homeland. Malik correctly suggests that "by pitting Ram against Babur, the BJP changed the context of Indian politics. For the majority of Hindus, Ram represents the traditions (*maryadas*) of Hindu culture; now he became a national symbol. Babur, on the other hand, was an invader and conqueror" (1994: 95).

Secessionist Movements

One of the most serious threats to Indian federalism has come from secessionist movements in the tribal regions of the Northeast, in Punjab, and in Jammu and Kashmir. In order to maintain the territorial integrity of the state, the Indian leadership has largely resorted to coercive measures to repress and contain these movements. After the lapse of British paramountcy, the tribal hill people, who had never been under the rule of plains people and with regard to whom the British had followed a policy of isolation, had difficulty reconciling themselves to the fact that after independence, they were to be ruled by the Indian government. Consequently, a large number of "solidarity" movements, such as the Garo National Council, Hmar Association, Kuki National Association, Mizo Union Association, and Naga National Council, erupted in the Northeast to assert autonomy for the tribal regions. Although over the years, the Indian state has responded to these insurgencies by reorganizing the Northeast and granting statehood to Nagaland, Mizoram, Manipur, and Tripura, and by creating legal provisions for the protection of tribal identity and culture, underground guerrilla activity continues.

The Sikh demand for an independent Khalsa state, which can be traced historically to the preindependence period, gained momentum in the 1980s, engulfing the state of Punjab in violence. As reformulated by the Akali Dal (the Sikh party), the earlier demands of the 1973 Anandpur Sahib Resolution for Punjab's limited association with Indian federalism (whereby the central government would retain only the portfolios of defense, foreign affairs, currency, communication, and railways) turned, in 1983, into a call for a full-fledged Khalistan (homeland for the Khalsa–the Sikhs' movement). Some of the underlying causes were Akali Dal's exclusion from political power through an excessive intervention by the central government of Indira Gandhi into regional affairs in Punjab; Indira Gandhi's courting of extremist Sikhs such as Bhindranwale, the leader of the Sikh Student Federation, to neutralize the moderate Akali Dal party; strong feelings in the state that Punjab was being discriminated against in the issue of sharing water resources with neighboring states; the central government's neglect of the state in terms of industrial development; and the widespread fear among Sikhs of assimilation into Hinduism, since the predominant Hindu attitude does not regard Sikhism as separate from Hinduism. In 1983, for the first time in Indian history, a deep gulf emerged between Hindus and Sikhs. Male Hindus and government officials became the targets of extremist Sikh violence. Bhindranwale masterminded terrorism from behind the walls of the holiest of Sikh shrines, the Golden Temple. In June 1984, the Indian government carried out "Operation Bluestar." The Indian army entered the temple, and after a three-day siege, more than 500 people, including Bhindranwale, were killed. The Akal Takhat suffered severe damage. The desecration of the Golden Temple caused great outrage among the Sikh population both within and outside India. On 31 October 1984, Indira Gandhi was assassinated by her two Sikh bodyguards, and this led to the retaliatory mob slaying of at least 2,000 Sikhs by Hindus in New Delhi, Punjab, and Haryana. By 1992, when an electoral process was initiated in Punjab as a result of a series of settlements and accommodations with the moderate Sikhs and an active policy of coercive repression of the militancy, the Khalistan agitation from 1981–91 resulted in the violent deaths of more than 16,000 people.

In 1989, a secessionist movement accompanied by political insurgency began in the predominantly Muslim Kashmir Valley, a part of the Jammu and Kashmir state. Since partition, the state has been a source of constant dispute between India and Pakistan. Jammu and Kashmir had been one of three princely states not to accede to either India or Pakistan. In 1947, the tribesmen of the northwest frontier province of Pakistan took it on themselves to "liberate" Muslim Kashmir. Unable to defend the state, the Hindu Maharaja of Kashmir signed the instrument of accession with India in October 1947. The Indian government accepted the accession on the condition that once law and order were restored, the wishes of the people of Kashmir would be formally consulted in a plebiscite regarding their future. Meanwhile, the Indian Constituent Assembly approved a temporary provision, Article 370, determining Jammu and Kashmir's political relationship with the Indian Union. The article, while restricting the Union's legislative powers to the areas of foreign affairs, defense, and communications, allowed the state government to legislate on residuary powers. In the mid-1950s, the Indian government withdrew its offer of plebiscite to the people of Kashmir, citing two reasons: (1) Pakistani forces had not evacuated the Kashmiri territory (since the tribal invasion, one-third of the state's territory remains with Pakistan), and thus law and order had not been restored to the state, and (2) the Kashmiri population had participated in local and federal elections, and this constituted a de facto endorsement of their association with India.

In terms of recognizing cultural pluralism, the 1950 Indian Constitution had innovatively dealt with Kashmir's requirements for political autonomy in order to preserve the latter's distinct ethnic identity. Through Article 370, the Indian state simultaneously embraced and denied its differences from the Kashmiri society; it recognized the cultural and political identity of the Kashmiri population, yet it asserted that affinities between Kashmir and the Indian state were based on the socialistic and democratic agenda of the Kashmiri nationalist movement against the Maharaja and on a historical concept of cultural identity, *Kashmiriyat,* a secular tradition prevalent in the valley since the fourteenth century. Since 1955, there has been an incremental abrogation of Article 370 and a steady, progressive integration of the state into the Indian polity. In addition, the installation of centrally approved state governments, the repression of democratic opposition, and widespread patronage politics have been responsible for the alienation of the Kashmiri masses. When two nationalist groups, the Jammu and Kashmir Liberation Front and the Hizbul Mujahedeen, led the Azadi (freedom) movement, they enjoyed mass support. Since 1989, an entrenched gun culture has prevailed in the Kashmir Valley. The secular traditions of *Kashmiriyat* have been under serious attack due to the exodus of the Hindu community after a number of Hindus were killed by militant groups in the early years of the movement, and in order to protect their religious identity, a large number of Hindus in the state have openly embraced the Hindu nationalist party, the BJP. The 1991 BJP *yatra* (pilgrimage) to the valley to hoist the Indian flag on Republic Day in Srinagar, the summer capital of the state of Jammu and Kashmir, and the demolition of the Babri mosque have certainly sharpened the Kashmiri Muslim's sense of religious identity. Recently, following the example of Punjab, both parliamentary and regional elections were held in the state, producing a respectable voter turnout of about 40%. The success of the elections is largely attributable to the fatigue of the population with the movement and its accompanying violence (according to an official estimate, since the inception of the movement, more than 20,000 people have been killed). Since the elections, a subdued militancy continues, with occasional bombings and clashes with the armed forces. All the secessionist groups who had refused to participate in the electoral process continue to press their demands for an independent Kashmir.

Conclusion

Will Indian federalism survive and overcome its crisis, emanating from various religious, linguistic, caste, and other identity-based cleavages? One of the major strengths of the Indian polity is its people's commitment to democracy. Although the loyalties of the people to local and regional governments are stronger than to the central government, and although the present elected leaders represent and are primarily accountable to their caste/community/region-based constituencies, the appeal of democracy, as Ashis Nandy suggests, has deepened in India. Politics, defined in terms of local identities and local issues, no longer appears remote to the people. The ICSSR-CSDS-India Today poll after the 1996 election clearly revealed that the democratic system enjoys greater legitimacy today than it did in the past. Most Indians are convinced that their vote matters, and it is the poor and the disadvantaged who defend democracy more vigorously than the elite. As the politics has shifted from the center to the local, Indian federalism needs to redefine itself, and the central government needs to seek a new basis for its legitimacy. Federalism will survive in a multicultural Indian polity, but with a new mandate and with new processes.

Further Reading

Austin, Granville, "The Constitution, Society, and Law," in *India Briefing 1993,* edited by Philip Oldenburg, Boulder, Colo.: Westview Press, 1993

The *India Briefings* are excellent overviews of contemporary politics and the major events of the year.

Brass, Paul R., *The Politics of India since Independence,* Cambridge: Cambridge University Press, 1990

A comprehensive overview of Indian politics, particularly the problems associated with pluralism and national integration.

Chanda, Asok, *Federalism in India: A Study of Union State Relations,* London: George Allen and Unwin, 1965

An institutional history of Indian federalism and a discussion of the constitutional provisions relating to center-state relations.

Frankel, Francine R., *India's Political Economy, 1947–1977: The Gradual Revolution,* Princeton, N.J.: Princeton University Press, 1978

A very detailed analysis of India's political economy and the accommodative politics of the Indian state in reconciling the goals of social equity with a gradual nonconflictual mode of change.

Frankel, Francine, and M. S. A. Rao, editors, *Dominance and State Power in Modern India: Decline of a Social Order,* volume 2, Delhi and Oxford: Oxford University Press, 1989–90

An excellent collection of essays on internal variations of social stratification in India.

Hardgrave, Robert L., and Stanley Kochanek, *India: Government and Politics in a Developing Nation,* 5th edition, Fort Worth, Tex.: Harcourt Brace College Publishers, 1993

A comprehensive introduction to Indian politics and contemporary political events.

Kohli, Atul, *Democracy and Discontent: India's Growing Crisis of Governability,* Cambridge and New York: Cambridge University Press, 1990

A detailed, comparative study of the governability crisis in five districts of the states of Gujarat, Andhra Pradesh, Karnataka, West Bengal, and Tamil Nadu.

Malik, Yogendra K., and V. B. Singh, *Hindu Nationalists in India: The Rise of the Bharatiya Janata Party,* Boulder, Colo.: Westview Press, 1994

A political history of the BJP and an analysis of the circumstances giving rise to the party as a major political force in the 1980s and the 1990s.

Rudolph, Lloyd I., and Susanne Hoeber Rudolph, *In Pursuit of Lakshmi: The Political Economy of the Indian State,* Chicago: University of Chicago Press, 1987

An extensive, detailed analysis of the weak-strong, hard-soft state in which a theory of command and demand politics provides a general framework for explaining the Indian political economy.

Van Der Veer, Peter, *Religious Nationalism: Hindus and Muslims in India,* Berkeley: University of California Press, 1994

Through a focus on the Babri mosque controversy, the relations between religion and politics in a postcolonial context are explored.

Weiner, Myron, and Mary Fainsod Katzenstein, *India's Preferential Policies: Migrants, the Middle Classes, and Ethnic Equality,* Chicago: University of Chicago Press, 1981

A comparative study of ethnicity, preferential politics, and the issue of "foreigners" versus "locals" in the states of Assam, Maharashtra, and Andhra Pradesh.

Reeta Chowdhari Tremblay is associate professor of comparative public policy and South Asian politics at Concordia University in Montreal, Canada, where she is also director of the graduate program in public policy and public administration. Her current areas of research are the secessionist movement in Kashmir, identity based politics and citizenship, and popular culture. She has published extensively in the areas of secessionist movements and public policy.

⌣. ɔter Twelve

National Politics, Regional Politics, and Party Systems

Arun R. Swamy

Over the last decade, observers of India have seen their most cherished assumptions dissolve. In politics, the party that led the country to independence in 1947, the Indian National Congress, appears to have permanently lost its once dominant position and is now one of three principal contenders for power. On the economic front, with active government control of economic decisions largely abandoned since 1991, India is increasingly pinning its hopes for development on market forces.

Many have recognized these trends, but few know what to do with them: both politically and economically, the action in India is increasingly in the country's large states rather than the national capital of New Delhi. Changes in state politics have reshaped the national party system as differences in parties' fortunes across states sustain a multiparty system nationally. Moreover, state government policies, always important for the fate of social welfare programs, are now critical to the success of industrialization efforts, both within their states and nationally.

This chapter examines the linkages between the national and state arenas through a focus on political parties. These linkages involve both the social divisions (cleavages) the parties represent and the policies they promote. However, while the strategies of parties often influence economic development, the issues they campaign on are not always economic ones. Accordingly, this chapter focuses on all factors shaping the party system with the connection to economic development made when relevant.

The first part adopts a bird's-eye view of party competition by providing a brief description of the current national situation and how it differs from the past, an overview of different models of Indian party competition, and a survey of the major social divisions and policy differences that fuel party competition. The second part gives a historical overview of Indian party systems since independence, focusing first on the emergence of opposition to the Indian National Congress, then on national politics after 1971, and finally on regional patterns.

Throughout the chapter, six broad themes emerge: three relating to all-India factors influencing party competition, and three to regional differences. First, party competition in India has been shaped by the adoption of a parliamentary system of government, with a "first-past-the post" electoral system. It is commonly noted that the Congress won huge parliamentary majorities with less than 50% of the vote because individual legislators can be elected by a plurality without a majority. Today, the same system produces wide swings in parties' parliamentary strength with small vote shifts, complicating the ability of parties to form alliances. Second, the evolution of parties other than the Congress has been shaped by the actions of the Congress: Congress policies have determined opposition issues, while the ability of a given party to penetrate new states is often shaped by the performance of Congress in those states. Third, underlying these various opposition efforts, there is a long-term regularity to Indian party competition; this involves a party representing the top and the bottom of society and espousing *protection populism* competing against one representing the middle and espousing *empowerment populism*.

In addition to these all-India factors, there are a number of factors that are responsible for the regionalization of Indian parties. First, differences in the situation of rising, middle-level social groups historically have made it difficult for parties representing these groups to evolve a national strategy. Second, with the erosion of the Congress in northern India, protection populism is also increasingly being espoused by different parties in the various regions, often because the historic absence of party democracy within the Congress has pushed many state-level leaders to break away. Finally, a long-term and long-submerged macroregional conflict, that between north and south, is increasingly shaping party politics. Two possible consequences are the emergence of coalitions of regional parties accompanied by greater demands for decentralization of power or the breakup of the Congress into regional wings, with the southern wing, paradoxically, eventually allying with the Bharatiya Janata Party (BJP).

The Nature of Party Competition in India

From independence in 1947 until the 1960s, there was little party competition to speak of. The Indian National Congress, which led the movement for independence and had name recognition throughout the country, was opposed by many small parties with only local support. The Congress therefore won huge majorities in national and state legislatures while averaging 45% of the vote. After 1967, alliances among opposition parties and the occasional emergence of a single non-Congress party in particular states occasionally deprived the Congress of power in individual states. In 1977, the Congress lost national elections for the first time. However, despite frequent splits in the party, the Congress continued to dominate until 1989. In the elections of 1989, 1991, and 1996, of which the Congress won only the second, and in state elections held during this period, the party saw its share of both votes and seats slip steadily away. In 1996, the Congress received 28% of the popular vote, down from nearly 40% in 1989, and was only the second largest party in parliament. More remarkably, in the two largest states, Uttar Pradesh and Bihar, the party appeared to have vanished entirely.

What has replaced Congress dominance, however, is rather messy because of regional differences. While many states saw two-party systems by the mid-1980s, the parties vary from one state to another. At the national level, the result is a combination of "national" parties with a presence in some states but not others, regional parties limited to a single state in which they are among one of two principal parties, and small local parties, often formed by breakaway factions of one of the others, which are able to affect the outcomes of the elections by depriving one of the larger parties of their winning margin.

Some order is given to this picture by the formation of alliances among the various parties that reduce the principal forces in the present parliament to between three and five. The largest alliance is led by the BJP, a militantly nationalist party that opposes safeguards for minority cultures. The BJP won the largest contingent in parliament in 1996 with a little over 20% of the vote, but it has little presence in the south and east, where it has historically been viewed as a party of the Hindi-speaking north. The second largest alliance in parliament is, of course, that led by the Congress, which retains the widest spread but has been largely eliminated from the two largest states in the Hindi-speaking north and depends on regional partners to win in two southern states. The third largest party, the Janata Dal, has periodically tried to form an alliance, the National Front, with single-state "regional" parties. It champions the interests of farmers, affirmative action programs, and decentralization of power, and it is strong in one northern, one eastern, and one southern state. Owing to the weakness of the links between the Janata Dal and the regional parties, the latter can be treated as a fourth force: concentrated in the south and north-east, the most culturally distinct regions, these regional parties have a sizable presence in the present parliament and have attempted to form a Federal Front. Finally, the Left Front, an alliance led by India's two communist parties, has strength in the southern state of Kerala and the eastern state of West Bengal, and it has also been allied with the Janata Dal since 1991.

After the May 1996 elections, the government was formed by a coalition of the Janata Dal, regional parties, and the Left Front, referred to as the United Front, with support from the Congress, which has almost as many seats in parliament as the entire United Front combined. The situation, however, is unstable, since most of the United Front constituents compete with the Congress in some states. Additionally, the BJP has been making inroads into some of these states. Although in most of these states, the BJP remains a distant third party, it can have an enormous impact by taking away the margin of victory of one of the major parties. A similar effect has been achieved by the Bahujan Samaj Party (BSP), which has been taking votes from Congress in many states (see Table 12.1).

The changes over the last few years have led to a number of scenarios. One expectation, that the BJP, which grew phenomenally from two seats in 1984, would eliminate either the Congress or the Janata Dal, is no longer as likely, since the party has experienced no growth in its vote percentage since 1991 and remains unable to penetrate the south or east. Today, most observers forecast a period of coalition politics and focus on anticipating future alignments. Three kinds of alliance patterns are commonly forecast. One suggests that a three-way contest has emerged between an alliance of the right led by the BJP, one of the center led by Congress, and one of the center-left consisting of the United Front–that is, the Janata Dal, regional parties, and Left Front. Recent events, however, suggest that Janata Dal and regional parties have less in common with the Left Front than with the Congress, especially on the crucial question of whether to liberalize the economy further. The same consideration casts doubt on the second scenario, which suggested two alliances, the BJP and the anti–BJP, with the Congress in the latter. More recently, some commentators have suggested that the Congress, Janata Dal, and regional parties would be able to form a stable centrist alliance, marginalizing both left and right. This, however, faces the difficulty already mentioned: the Congress often competes with the United Front constituents for power.

While the third scenario may succeed the present arrangement in the short run, in the long run, the principal dynamic of Indian politics is a contest such as the one between the Congress and the centrist United Front parties–the Janata Dal and the regional parties. This contest focuses on different approaches to distribution and equity: a basic needs approach, represented at present by Congress, and an emphasis on expanding opportunity for upwardly mobile groups, which is common to the Janata Dal, regional parties, and various parties that have broken away from the Janata Dal. To examine what this means, the various models that have been used to explain Indian politics need to be considered.

Models of Indian Party Systems

The earliest model of Indian politics is the "dominant party" or "one party dominant" system model in which the largest party was viewed as mediating among social conflicts. In terms of policies and ideologies, Congress was described as a "party of consensus," incorporating criticisms from smaller "parties of pressure" in the opposition (Kothari, 1989) and linking different groups to the national government through an elaborate patronage network (Weiner, 1967). In retrospect, it is clear that these early views overstated the case. Differences over policies pushed by the top leadership caused many Congress leaders to form new parties, while the patronage networks used to incorporate lower levels of leadership could not have met the demands of growing numbers of politically conscious voters, since patronage requires spreading scarce resources ("divisible benefits") to individuals. Nonetheless, the model continues to be a reference point for most scholars writing on Indian politics.

Since party competition picked up in 1967, scholars have bemoaned the inability of Indian parties to conform to familiar patterns of competition. While Indian parties often espoused distinct ideologies of the right or left, these positions did not influence the alliances they made. Frequent party splits led many authors to describe Indian parties as mere collections of factions held together by loyalties to individuals and motivated by power alone (Brass, 1984: chapter 4). By the 1980s, many authors described Indian party competition as a politics of "competitive populism" in which parties tried to "outbid" each other in promising more of the same short-term benefits to various groups at the expense of long-term development and nation building (Kohli, 1990, 1994).

If analysts have been unable to find an ideological pattern to Indian politics, they have, at times, discovered some regularities to parties' vote bases. Many researches have found that Congress has built a coalition of the elite social groups and the most marginalized, especially in the largest northern state, Uttar Pradesh (Brass, 1984: chapter 8; Hasan, in Frankel and Rao, 1989–90). During the 1980s, Rudolph and Rudolph (1987) speculated that a coalition between middle-sized farmers and the "backward castes" would emerge as a new "hegemonic coalition." The formation of the Janata Dal in 1988 seemed to confirm this view, but that party, like its predecessor, the Janata Party, broke up, while the BJP emerged as the principal challenger to the Congress. Investigations into the politics of backwardness in different states did produce the most comprehensive recent survey of politics in the major Indian states (Frankel and Rao, 1989–90), but also demonstrated that these patterns varied far more from state to state than the authors had recognized. This recognition was reflected in a third view, which suggested that there was a regularity to Indian party competitions at the state level that is obscured at the national level by the different patterns of state politics (Chhibber and Petrocik, in Sisson and Roy, 1990).

Building on the first two arguments, the "normal" pattern of Indian politics is not "competitive populism" (promising more of the same) but "competing populisms" through which centrist parties use similar rhetoric to describe subtly different policies (Swamy, 1996). Parties espousing empowerment populism depict society as divided between haves and have-nots: they demand broader access to privileges enjoyed by the elite and appeal to people's aspirations for upward mobility. The middle peasant/"backward caste" coalition is one variety of empowerment populism. Other parties tend toward protection populism: expressing concern for the needs of the most vulnerable of the have-nots, they promise social insurance rather than social mobility and build a "sandwich coalition" of the top and bottom of the social structure against the middle. An alliance between the upper and lower castes is one instance of protection populism. Another is a coalition in the southern state of Tamil Nadu, which won support from the very poor and women with minimal social insurance programs (Swamy, 1996). However, focusing on castes as the building blocks of these coalitions underestimates the role of ideology or policy in shaping them. Moreover, by ignoring the different relative situations of middle-status groups in various states, authors have often not realized that Congress coalitions that appear different in various states are actually similar. In particular, the "backward castes" may be included in a protection populist strategy in states where they are a relatively small proportion of the population but not in others. To examine this further, the sources of party competition in social divisions and policy differences between parties need to be explored.

Sources of Party Competition

There are two common ways of describing how parties in a given party system relate to one another–in terms of their policies and in terms of their social support bases. While the two are often held to overlap, much depends on which one treats as coming first. On balance, policy appears to come first (both economic and social policy) because the major policy decisions were made in the 1950s before electoral competition got under way. Opposition parties have often had to find ways to tie together very different grievances about existing policies, with the result that opposition coalitions seldom stay together after they win. Two common themes uniting opposition groups have been the demand for decentralization, and the notion of backwardness.

In economic policy, the theme of decentralization has been used to cover a wide range of demands, from privatization to promotion of small-scale industry. From the 1950s until the recent liberalization, conflicts over economic policy centered on the role of the state in promoting heavy industry. Most parties supported efforts to protect domestic industry and promote industrialization. Restrictions on the areas in which big business could invest were opposed by business groups, while farmers, small businesses, and even socialists opposed the decision to promote investment in heavy industry, arguing that it reduced the capacity of the economy to create new employment. Starting in the 1960s, many of these viewpoints converged around a platform of opposing centralization and economic bigness in the name of Gandhianism. At this point, they often overlapped with demands for greater powers to the states undertaken by culturally distinct linguistic groups.

The partition of British-ruled India into a Muslim-majority Pakistan and Hindu-majority India complicated discussions over the place of regions, religions, and languages in an Indian nation-state. As a general rule, the Congress policy was to attempt to grant cultural groups autonomy in cultural matters without giving them political autonomy. This strategy was pursued toward language policy, the rights of religious minorities, and the related question of how much power to grant provincial (state) governments.

During the 1930s, a tenuous compromise on the language issue would have established Hindi as the "national language" while redrawing state boundaries to coincide with linguistic ones, thereby granting linguistic groups "cultural autonomy." In the 1950s, both measures were resisted, the first by non-Hindi speakers, especially in the south, and the second by Hindi speakers. Eventually, widespread protests in the south led to the creation of linguistic states and, later, an agreement to continue using English as an "associate official language" in the 1960s (Das Gupta, 1970). Struggles over language policy and the establishment of linguistic states have brought into being a number of parties aimed at asserting a distinct regional identity within India, notably in the south and northeast. Regional sentiments, however, have also helped the Communist Party of India–Marxist (CPM) rule for 20 years in the state of West Bengal (Kohli, 1990), demonstrating that regionalism in India is not simply about cultural issues but about economic ones as well. Regional parties have been successful only when they articulate grievances over their state's economic development or other issues, such as language policy, which affect the economic prospects of significant sections of the state. In recent years, regional parties have been at the forefront of demands for greater decentralization of power, allowing states to make their own economic decisions.

This point is even more evident in the case of two other kinds of geographic/cultural constituencies that have become prominent in recent years–movements demanding that smaller states be carved out of the linguistically defined states, and the Scheduled Tribes. The first demand typically arises in areas that are economically less developed, or "backward," frequently because they were administered as part of a "princely state" before independence. The second relates to the aspirations of social groups who were not incorporated into Hindu society as castes because they lived in relatively remote areas. The two are distinct but overlap: in many states, notably the second largest state, Bihar, "backward" areas are also those with large tribal populations. Although historically these constituencies were Congress preserves, the BJP has, ironically, been quicker to champion the demands of breakaway regions for their own states in the belief that smaller states would strengthen the central government.

A somewhat different use of the decentralization theme emerged in the context of religious policy. While the British replaced existing legal practice in many areas, they attempted to allow religious communities to practice their own "personal laws"–primarily, those governing family and relations, including inheritance. This practice was reluctantly continued after independence as a way of assuring minorities that their cultures would be respected. However, the Nehru government did assert the right to reform Hindu law in a number of areas, especially relating to the rights of women (Smith, 1963). This alienated many conservative Hindus in the 1950s and has allowed right-wing nationalists to charge that the majority is being discriminated against.

It is often observed in the context of the personal law conflict that the existence of hereditary social strata among Hindus made changes in Hindu law inevitable. The same circumstance has resulted in an even sharper set of conflicts over affirmative action quotas, or "reservations" for disadvantaged social groups. Here, the sharpest conflicts have not been about reservations for the most disadvantaged groups, the Dalits–also referred to as Scheduled Castes, Harijans, or former "untouchables"–who make up about 12% of the population; rather, they have arisen as the result of proposals to extend reservations to less deprived but still relatively disadvantaged groups, known in legal parlance as "Other Backward Classes" and in political terminology as the "backward castes," or OBCs. Until the 1990s, providing compensatory measures for the "backward" classes was left to state governments, which had a high degree of discretion over both the nature of the preferences–quotas were not obligatory–and over whom to count as "backward." This flexibility in defining "backward castes" has meant that support for reservations can shift when particular groups obtain or lose their entitlement to "backward" class benefits (Galanter, 1984). Moreover, with the decision to enact a single national list of "backward castes" in the 1980s, it was found that in some states, the "backward caste" category included the bulk of traditional peasant or cultivating castes, while in others it did not (Reddy N. Subba, "Mandal Commission Report on Backward Classes: Should It Be Revived or Revised?" paper presented at Workshop on Backward Classes and Reservations, Madras Institute of Development Studies, 3 March 1990). These horizontal conflicts among peasant groups have hampered the ability of the middle peasant "backward caste" coalition to hold.

To understand how peasants could be "backward" in some places and not in others requires a brief digression into the nature of caste. Two common assumptions need to be corrected if any sense is to be made of Indian party politics. One is the image of a single "caste system" composed of four big categories (*varnas*)–Brahmins (Priests), Kshatriyas (Warriors), Vaishyas (Traders), and Sudras (peasants and workers)–with "Untouchables" forming a fifth layer. A second and related misconception is that those who actually worked the land were uniformly under the control of noncultivating high-status castes. In fact, years of anthropological research have demonstrated that the building blocks of Indian society were small communities (*jati*), which have often merged into large caste "clusters" in modern times. While these groups might have been ranked in a local hierarchy, they often made conflicting claims about their own status, and local hierarchies frequently bore little relationship to *varna* categories. This was especially true among cultivating groups. In much of the south, west, and northwest, local power was typically exercised by "dominant castes" consisting of peasant communities that doubled as warriors and even contained royal lineages. (For an overview of these issues, see Kolenda, 1978.) In addition to these historic differences in the status of the peasantry, differences in British policies toward the land in the various provinces strengthened the rights of the cultivating peasantry in some while restricting it in others, so that "peasant" communities have experienced very different opportunities for upward mobility.

This point leads to the last set of important cleavages in Indian politics, those relating to the countryside. During the 1960s, authors believed that the most important conflicts in India would occur within the countryside between the landed and the landless. This did indeed occur in two states, Kerala and West Bengal, where the communists succeeded in mobilizing tenants around the question of land distribution. Elsewhere, however, radical approaches to land distribution found little support.

One reason for the weakness of radical approaches to land distribution is that in much of India, enough peasants owned some land to make them identify their interests with property holders. The hereditary tax collectors (zamindars) whom the British had turned into big landlords were significant only in some parts of the country, and their rights were abolished in the 1950s. However, this still left a large landless population, which the Congress leadership sought to satisfy by imposing ceilings on how much land an individual or family could own. In most areas, these efforts pit a large class of small and medium-sized farmers against the very poor, and the former were a sizable voting block by themselves. Beginning in the 1960s, Congress factions representing newly prosperous farmers left the party in various regions. In the 1980s, farmers emerged as a powerful constituency, demanding higher prices for farm produce and subsidies for inputs into agriculture, arguing that the cities ("India") had been exploiting the countryside ("Bharat," the Hindi name for India).

The shift from conflicts among rural groups to conflicts between the countryside and the city was seen by many analysts as the wave of the future (Rudolph and Rudolph, 1987). This prediction overlooked two facts: the rural poor in all areas depended on wages and did not share the interests of farmers in higher prices, and in many areas, the peasantry were too poor to avail themselves of subsidies to farm inputs. The first division allowed the Congress to appeal to many parties through the use of welfare policies aimed at increasing the consumption of the very poor. The second division, between peasants in richer and poorer regions, led parties representing the former to incorporate the latter through other issues such as "backward castes." It was this strategy that led to the middle peasant/"backward caste" hypothesis, which, as we have seen, fractured over horizontal conflicts between peasant groups in different regions.

A similar conflict over language issues has often prevented parties attempting a middle peasant/"backward class" alliance from establishing a national presence. The continued use of English in education and administration has preserved the advantage of older elite groups in entering professional occupations. Empowerment populists typically propose replacing English with Indian languages in the states but disagree over whether to do this at the national level. Parties in the Hindi-speaking states favor replacing English entirely with Hindi; as we have seen, this is viewed by speakers of the other major Indian languages, each with its own script, as discriminatory.

What this brief survey indicates is that the various lines of division in Indian society cut across each other in an infinite variety of ways, which makes it difficult for parties to find issues that build stable winning coalitions. This general point, and the specific issues noted, will help to make sense of the overview of Indian party evolution that follows.

The Evolution of Party Competition, 1947–97

Although India has had multiparty competition since 1920 and a parliamentary democracy since 1947, the story of modern Indian politics is the story of the Indian National Congress. It was in the forum of the Congress, established in 1885, that elites from the scattered British-ruled provinces of India came to know each other. After 1920, it was the Congress that developed strategies to incorporate new social groups into politics. Thus, the very existence of a national political arena is largely the result of political mobilization by the Congress before independence. Moreover, since the Congress monopolized power during the first 20 years of Indian independence (1947–67) and governed nationally for most of the next 30 years, all Indian parties have been shaped by their relationship to it: many are offshoots from it, all have had to differentiate themselves from it, and regional party systems often reflect the opportunities created by its failure to represent some interests.

The Emergence of Opposition, 1947–71

During the 1950s and 1960s, there were a number of opposition parties to the Congress both on the left and the right. On the right, the Bharatiya Jana Sangh (forbear to today's BJP) championed militant Hindu nationalism while the Swatantra was, outwardly, a more conventional free enterprise party. On the left, in addition to the Communist Party of India, which split into two parties in 1964, there were two socialist parties, which periodically attempted to merge and changed their names accordingly. The differences within the conservative and socialist pairs provide a useful clue to subsequent alignments.

The Jana Sangh, founded in 1951, had its roots in a highly disciplined militant nationalist organization, the Rashtriya Swayamsevak Sangh (RSS). Although the RSS tradition is sometimes thought of by outsiders as a variety of fundamentalism, few knowledgeable observers use this characterization: its harshest critics prefer the term *fascist,* while others simply refer to it as *Hindu nationalist.* This is because reviving Hindu religious practice–as distinct from scapegoating religious minorities–is not a high priority for Hindu nationalists. Hindu nationalists tend to be concerned with the political unity of Hindus, to view religious minorities, especially Muslims, as potential traitors, and to favor the creation of a single national culture. This position led the Jana Sangh to support making Hindi the sole national language, thereby alienating much of the south, to oppose granting minorities separate civil codes, thereby alienating religious minorities, and to resist affirmative action programs based on caste identity, thereby alienating low-caste groups. Historically, Hindu nationalists have favored a highly centralized state (far more centralized even than the Congress supports), have sought protection for domestic industry from foreign competition while opposing regulations on private enterprise, and have pressed for a strong military, including nuclear weapons. Many of these positions continue to be held by the Jana Sangh's successor, the BJP, today (Weiner, 1957; Anderson and Damle, 1987; Graham, 1990; Jaffrelot, 1996; "Why Jana Sangh," in Bhatkal, 1967.)

Unlike the Jana Sangh, the Swatantra was a loosely organized party of leaders who left Congress because of policy differences. They were especially opposed to central economic planning and land reform, and some were also opposed to the changes in Hindu law. These issues were tied together by the slogan "Anti-Statism" (Erdman, 1967). However, unlike the Jana Sangh, the Swatantra favored foreign investment, viewed military spending as wasteful, and, with a strong southern component to the leadership, opposed efforts to make Hindi the national language. (See "Why Swatantra," in Bhatkal, 1967).

These ideological differences were reflected in the two parties' patterns of support. During the 1960s, the Hindu nationalist Jana Sangh had a wide presence in north India but not one deep enough to win many seats. It is generally considered to have drawn principally on Hindu trader groups. By contrast, by recruiting princes who felt threatened by the Congress's assaults on land ownership and traditional Hindu law, Swatantra was able to draw on the traditional allegiances enjoyed by these aristocrats and rapidly established itself as the principal opposition to Congress in a number of states. The support, however, proved ephemeral, much of it eventually going to the Jana Sangh/BJP.

A parallel set of divisions divided the two socialist parties during the 1960s. Outwardly, the conflict turned on whether to accept Congress as a potential ally in building socialism, particularly after Congress declared itself in 1954 in favor of a "socialistic pattern of society." Deeper differences, however, were those that turned on how to define *socialism*. In general, those who favored an alliance with Congress, whom I term *economistic socialists,* were concerned primarily with such economic issues as state ownership of industry, while those who opposed alliances with the Congress placed equal emphasis on social equality. The latter, whom I term *egalitarian socialists,* proposed eliminating the use of English in public life and education (a position that limited their appeal to the north), demanding reservations for the other "backward" classes, and favoring a development strategy oriented toward agriculture and small-scale industry. These issues, paradoxically, made it easier for egalitarian socialists to ally with parties to the right and eventually to merge with agrarian populists drawn from middle peasant communities (Weiner, 1957: chapters 2–5; Brass, 1984: chapter 5; "Why Samyukta Socialist Party," in Bhatkal, 1967).

In 1967, many factors converged to allow opposition parties to experiment with their alliances. The death of two prime ministers had left a leadership vacuum in Congress while two years of drought and food shortages underscored opposition claims that Congress's economic policies ignored the common people. Finally, rising literacy in non-Hindi states and Congress's inept handling of the language policy in 1965 alienated many young voters on both sides of the issue. In many states, Congress lost its majority, and opposition alliances were able to form governments, at times with the help of defectors from the Congress.

Although most of the non-Congress coalitions were short lived, in some states, this period produced enduring changes. Most notable was Tamil Nadu, the one state where a non-Congress party, the regional Dravida Munnetra Kazhagam (DMK), was able to govern for nine years without depending on coalition partners. The DMK was succeeded by another regional party, the Anna Dravida Munnetra Kazhagam (ADMK). In the southern and eastern states of Kerala and West Bengal, communist-led coalitions introduced important changes in land law that helped to expand their support. Finally, breakaway Congress factions in several large states often took dominant peasant groups out of the party. Beginning with the formation of the Bharatiya Kranti Dal (BKD), an agrarian party in Uttar Pradesh (Brass, 1984: chapter 8), this trend culminated with a split in the Congress in 1969. Many state-level leaders, often from dominant peasant communities, who opposed Indira Gandhi's radical agenda broke with her to form the Indian National Congress (Organization) (Congress [O]) and ally with the opposition (Frankel, 1978). These trends, however, were somewhat obscured by Indira Gandhi's landslide victory in 1971.

National Patterns, 1971–85

The period between 1969 and 1975 is generally referred to as Indira Gandhi's "radical phase." Her government took over ownership of the major banks and many other private corporations, imposed new restrictions on investments by large business conglomerates and on imports, and abolished pensions for former princely rulers. However, although Indira Gandhi campaigned in 1971 on the slogan "remove poverty," her efforts in this area were not so much radical as reformist. The two exceptions, efforts to further limit the amount of land one could own and to nationalize the wholesale trade in foodgrains, failed. The long-term legacy of this period was in "poverty alleviation" measures that sought to increase the incomes of the very poor by providing them with cheap credit or employment in public works programs. In addition, Indira Gandhi sought to respond to the demand of egalitarian socialists for "backward" class reservations by identifying herself with the Scheduled Castes and Scheduled Tribes.

The various antipoverty measures undertaken by the Indira Gandhi government faced two important constraints–one political, the other economic. Politically, the implementation of antipoverty programs had to be left to state governments, which led to great variation among states. Economically, the government's resources were squeezed by war in 1971, drought in 1972, and the international oil crisis of 1973. By 1974, the inflation caused by these events resulted in a nationwide railway strike (crushed by the government) and in massive middle-class protests in several northern states. When the protests against her government received a boost from a court order declaring her election to parliament to be void, Indira Gandhi declared a state of national emergency, arrested her opponents, and suspended civil liberties for 18 months.

When the emergency was finally lifted in 1977 and elections were called, the various opposition parties, along with some more breakaway factions of Congress, merged to form the Janata Party and won a landslide victory to match Indira Gandhi's six years earlier. However, unlike Indira Gandhi, who won in all regions, the Janata Party was almost completely shut out from the south, reflecting both the north Indian base of most of its constituents and the relatively effective performance of Congress governments in two key southern states, Andhra Pradesh and Karnataka. The immediate consequence of the Janata Party government was to highlight the inadequacies of a strategy of uniting all non-Congress parties. The party broke up after two years, allowing Indira Gandhi to return to power in 1980. With a second split after the 1980 elections, the Janata Party effectively had three successors: a revived Hindu nationalist party, now named the Bharatiya Janata Party (BJP); a revived agrarian populist/egalitarian socialist party, the Lok Dal; and the Janata Party itself, which contained much of the old Congress (O) as well as many economistic socialists.

The Janata Party did attempt to introduce a number of policy measures that had been proposed for years by the various opposition parties. Reflecting the probusiness heritage of the Swatantra and Jana Sangh, its first budget loosened regulation of economic activity in a number of areas. At the same time, the agrarian populist Charan Singh pushed expanded incentives for agricultural production, new regulations intended to favor small-scale industrial units, and increased funds for local government bodies dominated by middle peasants (Franda, 1979), while an old-time socialist, George Fernandes, imposed new restrictions on foreign investment. Finally, the Janata Party government appointed the Mandal Commission to recommend reservations for the OBCs. Although this commission's report was shelved when Indira Gandhi returned to power in 1980, it was revived at the end of the decade.

While the breakup of the Janata Party helped Indira Gandhi return to power in 1980, this was not the only factor. Indira Gandhi's Indian National Congress (Indira) (Congress [I]) saw its vote share increase from 36% in 1977 to its historic average of about 43%, indicating that many voters in the north had now returned to the party. One major reason was the violent conflict that erupted in many places between the middle and "backward caste" groups on the one hand and Scheduled Caste groups on the other. These enabled Indira Gandhi to reestablish her claim to being the protector of the weakest sections in society.

On her return, Indira Gandhi's Congress (I) Party in the 1980s reinterpreted its "protection populist" stance to respond to new circumstances. Halting steps toward deregulating the economy continued. Exports, in particular, were given a new thrust. In other areas, too, the government appeared to be shifting positions, championing reservations for "backward castes" and adopting a more Hindu stance toward cultural identity. However, while the Congress appeared to champion policies introduced during the Janata interlude, it did so in ways that changed their actual impact.

For example, the Integrated Rural Development Programme (IRDP), intended initially to shift resources toward local bodies in the countryside to use as they wished, was redefined as a program for providing loans directly to poor families, establishing a direct relationship of patronage between the central government and the poor. Similarly, while Indira Gandhi's party began to champion "backward" class reservations in a number of states, the Congress refused to extend reservations to the national level and sought a narrower definition of "backwardness" than the Mandal Commission envisioned. More generally, while the Congress abandoned radical measures aimed at changing the pattern of ownership of productive assets, it expanded the use of welfare programs aimed at directly increasing the poor's purchasing power and access to credit. By the end of the 1980s, the principal proposals of the Congress in the area of welfare policy were employment programs at minimum wages for the rural poor and school lunch programs (which had been introduced successfully in some Congress-ruled states).

Indira Gandhi's second term in power did, however, demonstrate that various social forces opposed to the Congress had come to stay. Farmers' movements became very active, while various fragments of the Janata Party continued to challenge the Congress in some states. In two states, Kerala and West Bengal, a communist alliance, the Left Front, crystallized into a permanent alternative to Congress or, in the case of West Bengal, a permanent governing alliance. In addition, in two southern states where the Congress had won in 1977, Andhra Pradesh and Karnataka, the party lost power in 1983 to a new regional party, the Telugu Desam Party (TDP) in Andhra Pradesh and the Janata Party in Karnataka.

Indira Gandhi's assassination in 1984 interrupted what might have been a resurgence of non-Congress parties. The Congress Party quickly elected her son, Rajiv Gandhi, as its leader, although he had never held office. Elections a month later gave Congress its largest victory ever, with nearly 50% of the vote and over 80% of the seats. However, in 1985 and 1987, non-Congress parties won reelection in Andhra Pradesh, Karnataka, and West Bengal while Congress governments lost in Kerala and Haryana, the former to a communist-led alliance and the latter to the Lok Dal, a farmer-based fragment of the Janata Party.

Rajiv Gandhi's prime ministership (1984–89) facilitated the revival of non-Congress parties in a number of ways. He associated himself openly with the aspirations of the urban middle class, allowing foreign investment in consumer goods for the first time in 30 years and pushing computerization of industry and administration. By 1989, farmers' groups, concerned with their growing indebtedness, were in full revolt. Moreover, his attacks on the politics of patronage and corruption ended, ironically, in his own implication in a major corruption scandal involving imported defense equipment. Finally, Rajiv Gandhi's government opened the way for the rise of the BJP: after overturning a Supreme Court ruling that had struck down traditional Muslim provisions for alimony as inadequate, he apparently sought to placate conservative Hindu opinion by reopening a dispute over the location of a mosque in Ayodhya.

The procurement scandal proved to be the catalyst that allowed various opposition tendencies to unite. The star of Rajiv Gandhi's administration, V. P. Singh, resigned and brought about a merger of the two centrist descendants of the Janata Party of 1977–79, the Lok Dal and the Janata Party. The new party, the Janata Dal, entered into a formal alliance, called the National Front, with regional parties, primarily parties representing middle peasants and "backward castes," and then negotiated with both the Left Front and the BJP to avoid dividing the anti-Congress vote in specified election districts. These elaborate arrangements allowed the opposition alliance to deny the Congress a majority in the 1989 elections, but Congress remained the largest party owing to a solid showing in the south. V. P. Singh formed a government with "outside" support from both left and right. This government lasted less than a year, after which the Janata Dal itself fragmented.

The most important cause of the breakup of the Janata Dal was the issue of reservations for the "backward castes." The National Front campaign had included a promise to implement the recommendations of the Mandal Commission, issued under the Janata Party government of the late 1970s, for establishing a national system of affirmative action quotas for the OBCs. However, as we have seen, many "middle peasant" groups are excluded from the "backward" category. The consequences were dramatic. First, there were riots against the proposal in a number of major cities in the north. Then, the BJP, which opposed the policy but felt that it could not publicly state its stance, pushed for a confrontation over the disputed mosque site at Ayodhya, which provided a pretext to withdraw support from the government. Almost immediately thereafter, the Janata Dal split into two factions. One faction, opposed to V. P. Singh's leadership, contained most of the middle-peasant factions representing caste clusters that formed a minority government supported by the Congress for six months, after which elections were held in which Congress won.

In addition to the conflict over "backward caste" reservations, the Janata Dal government also revealed important conflicts over economic policy. First, farmers' representatives in the government demanded the waiver of all loans to farmers, but this was resisted by others, notably economistic socialists, who favored fiscal conservatism. Further, the period demonstrated that while many economistic socialists in the Janata Dal may have opposed liberalization, most of the other political leaders favored it.

In the 1991 elections, the Congress won enough seats to form a government but was short of a majority. Congress, ironically, had been helped by the assassination of Rajiv Gandhi during the campaign. His assassination, and the fact that Congress returned to power with a voting bloc heavily drawn from the southern states, resulted in the election of P. V. Narasimha Rao as the first southern prime minister. In a number of ways, this development both reflected the growing importance of state-level politicians and shaped the restructuring of Indian party politics. It was the principal state-level southern Congress leaders, those holding or aspiring to the office of chief minister, who rallied southern members of parliament around Rao's candidacy while the near-elimination of Congress from the two largest Hindi-speaking states left potential prime ministerial candidates from those states with no base in the parliamentary party. Subsequently, fears of reviving the predominance of Hindi-speaking politicians may have led Rao to place a low priority on furthering the fortunes of his party in the giant Hindi belt states of Uttar Pradesh and Bihar, with the result that by the 1996 elections, Congress had been reduced to a minor party in these states.

Apart from these regional jealousies, the principal legacy of the Rao years was the dismantling of controls over private economic activity. Although many measures (notably, efforts to sell off state-owned enterprises) were resisted by trade unions and the left, and although the BJP has opposed foreign investment in consumer goods industries, state governments of all parties have been supportive and have begun avidly to seek foreign investment. It is necessary to keep the wide acceptance of some kind of liberalization in mind when we turn to the political developments leading up to and following the 1996 elections.

Regional Patterns and Emerging Configurations

In the 1996 election, the Congress slumped to its worst performance ever. Although it remained the largest party in terms of votes (28%), in terms of seats, it was exceeded by the BJP. A principal reason for this was the exit from the party of important leaders with support in key states. This process began in 1993 over two issues–the failure of the central government to maintain funding for antipoverty programs and Rao's inability to prevent the destruction of the

Ayodhya mosque by a Hindu nationalist mob. It culminated with the exit of most of the Congress state unit in the southern state of Tamil Nadu in a disagreement over the party's choice of regional ally.

The year following the 1996 elections has been a period of flux. An initial attempt by the BJP to form a government failed when the Congress, Janata Dal, regional parties, and left parties all refused to support it. The result was an unusual arrangement in which the Janata Dal again formed a coalition government with regional parties, but one with three important differences from the V. P. Singh government of 1989–90. First, the left parties (with the partial experience of the largest, the CPM), were now part of the governing coalition rather than support from the outside, making this the first national government with communist participation. Second, the Congress, rather than the BJP, supported the alliance from "outside." Finally, regional parties–including one former Janata Dal faction in Uttar Pradesh and a former Congress faction in Tamil Nadu–now had a substantial presence in parliament, together accounting for more members than the Janata Dal.

The government was headed by the country's second southern prime minister, H. D. Deve Gowda of Karnataka, who, unlike his predecessor, neither spoke Hindi nor had any experience of national politics. This arrangement lasted ten months before differences between a new Congress president and Gowda led to the latter's resignation. Several weeks of intensive negotiations produced a new government under foreign minister, I. K. Gujral, a widely respected figure with no political base and no enemies. Remarkably, however, in spite of the fact that this crisis occurred during budget negotiations, and Deve Gowda's finance minister, a southerner and former Congressman, had proposed the most probusiness budget since the 1950s, parties were careful not to allow the budget to collapse as a result of the bickering.

The various reports and interviews that came out of the recent crisis demonstrated the importance of regional factors in two ways that are likely to prove critical to future developments. First, the national strategies of the major parties are increasingly constrained by the fact that any given pair of national alliances competes for power in some states: the relative durability of the National Front-Left Front alliance is due more to the fact that these two alliances do not compete in any state than to ideological similarities. Second, differences between north and south came into the open in negotiations over the choice of a new prime minister. In order to forestall the election of a prime minister with close ties to the Congress, left parties argued for a north Indian leader on the grounds that it was in the north that the BJP needed to be combated. Since a similar argument had been voiced during the election of a new Congress leader in December, it is likely that the question of which region should lead the nation is going to become more salient in future years.

All of these factors suggest that a detailed understanding of regional party systems is critical for any analysis of national politics. While this is beyond the scope of the present chapter, a first approximation is to map out the various kinds of party systems that have emerged in recent years and some reasons for their emergence.

The first pattern to develop was the contest between the Congress and parties of the left, principally the communist party, in the southern state of Kerala and, later, West Bengal. While the success of the left has not been duplicated for reasons already cited, these states have been relatively unaffected by subsequent developments, providing the left with an autonomy in national politics beyond their strength in terms of votes.

Almost simultaneous with the emergence of a Congress-left contest was a contest between the Congress and regional parties, which emerged first in the southern state of Tamil Nadu and the northwestern state of Punjab. Later, the small northern states of Jammu and Kashmir, Assam in the northeast, and the large southern state of Andhra Pradesh joined this group, while Tamil Nadu proceeded to pioneer a third pattern, not yet replicated elsewhere, of a contest between two regional parties. While in Punjab, Assam, and Kashmir, internal ethnic conflicts have played an important role in the emergence of regional parties, in Tamil Nadu, and to a lesser extent Andhra Pradesh, the rise of intermediate social groups was critical.

Around the time the regional parties first became prominent (1967), the seeds of the third pattern were sown. This is the contest between the Congress and the Janata Dal, which characterizes the politics of the southern state of Karnataka, the eastern state of Orissa, and the northern, Hindi-speaking state of Haryana. With the formation of the National Front, in which the Janata Dal allied with regional parties, the further splintering of the Janata Dal, and the apparent incorporation of these splinter groups under the National Front umbrella, this pattern is increasingly merging with the previous one. In one way or another, all of these patterns represent the rise of intermediate social groups, but their splintering also reflects the regional diversity of these groups.

A fourth pattern is provided by states where the Congress and the BJP are the principal parties, including the western state of Gujarat and the northern, Hindi-speaking states of Rajasthan, Madhya Pradesh, and Himachal Pradesh. To this should be added the large, industrial, western state of Maharashtra, where the BJP, in alliance with a regional Hindu nationalist party, the Shiv Sena, has recently come to power.

Finally, there are states where the principal contestants are the Janata Dal, or an offshoot of this party, and the BJP. These states include the two largest and poorest states, Uttar Pradesh and Bihar, although in the latter, the Congress remains a significant force. Here, the most significant development is the collapse of Congress.

To this long list of patterns it should be added that there are a number of states where three parties appear to be significant at present. These include Karnataka, where the BJP has been rising, and Uttar Pradesh, in which a regional "backward caste" party and a regional Dalit (Scheduled Caste) party have divided the anti-BJP vote. A brief summary of the major factors that have produced this situation is as follows.

The first factor is that although the Congress party did not radically alter the distribution of assets in the country, it did implement many programs to ameliorate poverty. Important differences among states in their ability to implement this Congress brand of "pro-poor" policies may account for the differences in the decline of Congress in the 1990s. States where relatively effective performance by the Congress has allowed it to retain a presence include Karnataka and Maharashtra in the 1970s and Gujarat and Madhya Pradesh in the 1980s.

A second factor is the Mandal Commission. The decision to implement the Mandal Commission report led directly to the expansion of BJP support at the expense of the Janata Dal in a number of states. The breakup of the Janata Dal resulted in the exit, almost en masse, of Janata Dal units in Haryana, Rajasthan, and Gujarat, resulting in the demise of the party in these states. In Rajasthan, Janata Dal defectors joined the BJP, while in Gujarat, where they joined the Congress, it was the BJP that saw its vote expand. The pattern just noted is not perfect. In one important state, Uttar Pradesh, owing to personal rivalries, the middle caste/middle peasant faction initially stayed in the Janata Dal while the "backward caste" faction joined the dissidents. This situation, however, was short lived, and Uttar Pradesh now has two regional parties representing these distinct constituencies, each with a base in a different part of the state.

A third factor leading to the present situation is whether individual Congress state units choose to represent dominant caste clusters. Paradoxically, Congress has been weakened in the 1990s by the reentry into the party of factions representing dominant peasant groups. These included Maharashtra, Gujarat, and Karnataka.

Finally, the emergence in northern India of the Bahujan Samaj Party, a party representing the Dalits, has led to the virtual eclipse of the Congress in Uttar Pradesh. Significantly, though, while the BSP sought an alliance with a regional "backward caste" party in Uttar Pradesh, it has recently become allied with the BJP, which represents upper castes in the state, indicating that the logic of a top-bottom coalition remains a powerful force in Indian politics.

Conclusion

For the price of one tank you can fund a thousand clinics; for the price of one tank you can build a thousand schools.

–Ram Vilas Paswan,
Indian Railways Minister, 11 April 1997

During the televised 12-hour debate in the Indian parliament that preceded the fall of Prime Minister H. D. Deve Gowda from office, there was only one exchange that could clearly be said to have addressed questions of economic development. In response to a charge that the Gowda government had neglected defense spending, the railway minister, a representative of the most excluded social groups in Indian society, passionately pointed to the trade-off between spending on defense and social development. In different ways, the railway minister's intervention highlighted how much has changed in Indian politics, and what has not.

What has not changed is the high salience of basic economic questions. In contrast to many researchers who, over the years, have anticipated that universal suffrage in a country such as India would result in ethnic divisions paralyzing electoral politics, the most successful electoral parties in India have always been those that championed bread-and-butter issues. Observers who anticipate the rise of the Hindu nationalist BJP, against whose charge Paswan defended his government, would do well to remember this.

At the same time, however, the fact that Paswan, a Dalit, made his charge from the benches of the Janata Dal, a party built on support from middle-status peasant groups, suggests how much has changed. In the past, the most important Dalit or Scheduled Caste leaders were in the Congress, and the Congress used basic-needs concerns to undercut the demands of upwardly mobile groups while defending an elite vision of nationalism. The collapse of this "sandwich coalition" in the two largest states, both in northern India, with the upper castes moving toward the BJP and the lower castes joining either the Janata Dal or the BSP, has largely eliminated Congress from these states and made it virtually impossible for the party to win a majority in the parliament.

At the same time, horizontal conflicts among middle-status groups owing to regional differences and the strains resulting from the patchwork emergence of opposition to Congress in various states make it unlikely that a single party will replace the Congress. The BJP itself has recognized this by seeking regional allies, suggesting that the short-term future is one of shifting coalitions of regionally based parties.

Further Reading

Andersen, Walter K., and Shridhar D. Damle, *The Brotherhood in Saffron: The Rashtriya Swayamsevak Sangh and Hindu Revivalism,* New Delhi: Vistar, and Boulder, Colo.: Westview Press, 1987

The most recent major study of Hindu nationalism prior to Jaffrelot 1996, with a focus on the RSS.

Barnett, Marguerite Ross, *The Politics of Cultural Nationalism in South India,* Princeton, N.J.: Princeton University Press, 1976

Still the only published book-length work on the rise of regionalism in Tamil Nadu.

Bhatkal, Ramdas G., editor, *Political Alternatives in India,* Bombay: Popular Prakashan, 1967

An extremely useful collection of essays by leaders of the major national parties in 1967 outlining why voters should choose them.

Brass, Paul R., *Caste, Faction, and Party in Indian Politics,* volume 1, *Faction and Party,* Delhi: Chanakya Publications, 1984

A collection of the author's previously published essays, including important works on the socialist movement, factionalism, and agrarian populism in Uttar Pradesh.

——, *The Politics of India since Independence,* 2nd edition, Cambridge: Cambridge University Press, 1995

The most thorough and lucid introduction to Indian politics. Chapters on parties and state and local politics discuss many of the trends reviewed here, including regional party systems.

Brass, Paul R., and Marcus Franda, editors, *Radical Politics in South Asia,* Cambridge, Mass.: MIT Press, 1973

A collection of the editors' essays. Brass's introduction and essay on the socialist party in Bihar are very relevant.

Dasgupta, Jyotirindra, *Language Conflict and National Development: Group Politics and National Language Policy in India,* Berkeley: University of California Press, 1970

Still the most authoritative and thorough work on the evolution of language policy.

Echeverri-Gent, John, *The State and the Poor: Public Policy and Political Development in India and the United States,* Berkeley: University of California Press, 1993

Very important studies of rural employment generation programs in Maharashtra and West Bengal.

Erdman, Howard, *The Swatantra Party and Indian Conservatism,* London: Cambridge University Press, 1967

The major study of the only recognizably conservative national party in Indian political history.

Franda, Marcus, *Small Is Politics: Organizational Alternatives in India's Rural Development,* New Delhi: Wiley Eastern Limited, 1979

An excellent overview of the Janata Party period (1977–79) with emphasis on development proposals.

Frankel, Francine R., *India's Political Economy, 1947–77: The Gradual Revolution,* Princeton, N.J.: Princeton University Press, 1978

The most thorough study of politics and economic policy during the first 25 years of independence.

Frankel, Francine, and M. S. A. Rao, editors, *Dominance and State Power in Modern India: Decline of a Social Order,* 2 volumes, Delhi and Oxford: Oxford University Press, 1989–90

The only comprehensive survey of state politics in all the major states, focusing on the rise of backward castes and middle peasants.

Galanter, Marc, *Competing Equalities: Law and the Backward Classes in Modern India,* Delhi: Oxford University Press, 1984

By far the most important study of the history and evolution of "backward caste" reservations.

Graham, Bruce D., *Hindu Nationalism and Indian Politics: The Origins and Development of the Bharatiya Jana Sangh,* Cambridge and New York: Cambridge University Press, 1990

An excellent study of Hindu nationalism up to the 1960s; complements Jaffrelot, 1996.

Jaffrelot, Christophe, *The Hindu Nationalist Movement and Indian Politics: 1925 to the 1990s,* London: Hurst, and New York: Columbia University Press, 1996

The most recent major work on Hindu nationalism, with excellent material on the strategies of the Bharatiya Janata Party and special reference to the state of Madhya Pradesh.

Kohli, Atul, "Centralization and Powerlessness: India's Democracy in a Comparative Perspective," in *State Power and Social Forces: Domination and Transformation in the Third World,* edited by Joel S. Migdal, Atul Kohli, and Vivienne Shue, Cambridge: Cambridge University Press, 1994

A succinct statement of the author's view that party politics in India have degenerated into "competitive populism."

——, *Democracy and Discontent: India's Growing Crisis of Governability,* Cambridge and New York: Cambridge University Press, 1990

A study of five districts that had been studied by Weiner (1967), arguing that the Congress party had suffered institutional decline since the 1960s.

——, *The State and Poverty in India,* Cambridge: Cambridge University Press, 1987

A landmark comparative study of the relationship between party organization and effectiveness at implementing redistributive policies in Karnataka, West Bengal, and Uttar Pradesh. Argues that land reform legislation requires a disciplined party organization capable of monitoring implementation.

Kohli, Atul, editor, *India's Democracy: An Analysis of Changing State-Society Relations,* Princeton, N.J.: Princeton University Press, 1988

A major collection of essays on the nature of Indian democracy in the 1980s.

Kolenda, Pauline, *Caste in Contemporary India: Beyond Organic Solidarity,* Menlo Park, Calif.: Benjamin/Cummings, 1978

An excellent short survey of the various dimensions of caste.

Kothari, Rajni, "The Congress 'System' in India," in *Politics and the People: In Search of a Humane India,* volume 1, Delhi: Ajanta Publications, 1989

Originally published in *Asian Survey* (1964). A landmark article, laying out the "dominant party system" model of Indian politics.

Lele, Jayant, *Elite Pluralism and Class Rule: Political Development in Maharashtra, India,* Toronto and Buffalo, N.Y.: Toronto University Press, 1981

Study of the dominance of the Maratha caste cluster in the Maharasthra Congress organization. Updated in his piece in Frankel and Rao 1989–90.

Rudolph, Lloyd I., and Susanne Hoeber Rudolph, *In Pursuit of Lakshmi: The Political Economy of the Indian State,* Chicago: University of Chicago Press, 1987

The major work on Indian politics in the 1980s, it advanced the thesis that middle peasants and "backward castes" would come to dominate politics.

Sisson, Richard, and Ramashray Roy, editors, *Diversity and Dominance in Indian Politics,* 2 volumes, New Delhi and Newbury Park, Calif.: Sage, 1990

An important collection of recent essays on party systems, with essays on several states and on the place of various social groups in the Congress coalition.

Smith, Donald Eugene, *India as a Secular State,* Princeton, N.J.: Princeton University Press, 1963

Still the most authoritative and thorough work on policy toward religions. A very useful final chapter on the attitudes of various parties, including the Jana Sangh.

Swamy, Arun R., "Sense, Sentiment, and Populist Coalitions: The Strange Career of Cultural Nationalism in Tamil Nadu," in *Subnational Movements in South Asia,* edited by Allison K. Lewis and Subrata Mitra, Boulder, Colo.: Westview Press, 1996

A study of Tamil nationalism.

Weiner, Myron E., *Party-Building in a New Nation: The Indian National Congress,* Chicago: University of Chicago Press, 1967

The landmark study on how the Congress organization worked at the local level.

——, *Party Politics in India: The Development of a Multi-Party System,* Princeton, N.J.: Princeton University Press, 1957

The first effort at mapping out the major parties in India.

Wood, John R., editor, *State Politics in Contemporary India: Crisis or Continuity?* Boulder, Colo., and London: Westview Press, 1984

An important collection of studies of state politics during Indira Gandhi's last term in office.

Arun R. Swamy received his Ph.D. in political science from the University of California, Berkeley, in 1996. His dissertation, "The Nation, the People, and the Poor: Sandwich Tactics in Party Strategies and Policy Models, India 1931–96," focuses on the sources of national and regional party systems. Dr. Swamy is currently in India doing follow-up research on this topic.

Table 12.1: Percent of Votes (Seats Won) in Fifteen Largest States, 1996 Elections

State (total seats)	*Largest party/alliance*	*Second largest*	*Third largest*
	North (Hindi-speaking)		
Bihar (54)	Janata Dal 31.9* (22)	BJP+Ally† 35.0 (18+6)	Congress 13.0 (2)
Haryana (10)	BJP+Ally† 34.9 (4+3)	Congress 22.6 (2)	Regional† 19.0 (0)
Madhya Pradesh (40)	BJP 41.3 (27)	Congress 31.0 (8)	BSP 8.2 (2)
Rajasthan (25)	BJP 42.4 (12)	Congress 40.5 (12)	
Uttar Pradesh (85)	BJP 33.4 (52)	Regional/UF† 20.8* (16)	BSP 20.6 (6)
	South		
Andhra Pradesh (42)	Congress 39.7 (22)	Regional/UF 32.6* (16)	
Karnataka (28)	Janata Dal 34.9 (16)	Congress 30.3 (5)	BJP 24.8 (6)
Kerala (20)	CPM 21.2* (5)	Congress 38.0* (7)	
Tamil Nadu (39)	Regional/UF+Ally† 52.8 (17+20)	Congress+Ally 26.1 (0+0)	
	East		
Orissa (20)	Congress 44.9 (16)	Janata Dal 30.1 (4)	BJP 13.4 (0)
West Bengal (42)	CPM 36.7* (23)	Congress 40.1 (9)	
	West		
Gujarat (26)	BJP 48.5 (16)	Congress 38.7 (10)	
Maharashtra (48)	BJP+Ally 38.6 (18+15)	Congress 34.8 (15)	
	Other		
Assam (14)	Congress 31.6 (5)	Regional/UF 27.2* (5)	BJP 15.9 (1)
Punjab (13)	BJP+Ally 35.2 (0+9)	Congress 35.1 (2)	BSP 9.4 (3)

* Figures for largest party in alliance only.
† Indicates breakaway faction from Congress or Janata Dal
Note: BJP=Bharatiya Janata Party; BSP=Bahujan Samaj Party; CPM=Communist Party of India--Marxist; UF=United Front; Regional/UF indicates a regional party that joined the United Front
Source: Government of India, *Statistical Report on General Elections 1996 to the Eleventh Lok Sabha,* volume 1, National and State Abstracts, New Delhi: Government of India, Election Commission of India, 1996

Chapter Thirteen

Human Development in Crisis: Investment Failures in Health and Education

A. K. Shiva Kumar

Introduction

India has witnessed several changes since independence in 1947. The adoption of new technologies and the introduction of sophisticated management have rapidly expanded and diversified the country's production base. Between 1951 and 1996, the index of industrial production went up 15 times, foodgrain production increased fourfold, and per capita income more than doubled (Government of India, 1997). In the social sphere, socially disadvantaged communities have benefited from affirmative action, untouchability has lost much of its grip on society, and women today enjoy far greater freedoms than ever before. The country has had a vibrant democracy, and the Constitutional Amendments of 1992–93 have paved the way for democratic local self-governance. Yet 50 years after independence, India continues to be classified as a low human development country (United Nations Development Programme, *Human Development Report 1997,* Oxford: Oxford University Press, 1997).

This chapter analyzes India's achievements and failures in human development and argues that the main reason for the low levels of human development is the gross neglect of basic education and health. A closely related factor has been India's failure to enhance women's capabilities as much as those of men. Without immediate policy measures to augment India's investments in human capital and correct the imbalances that exist, the country's aspirations of recording rates of economic growth comparable to that of the East Asian "tigers" will remain largely unfulfilled.

Investments in Health and Education

India has built up an impressive network of health services (see Table 13.1). Between 1951–92, the number of primary health care centers and subcenters went up from 735 to more than 150,000. During this period, there was a fivefold increase in the number of hospitals, a fourfold increase in the number of dispensaries, and a sevenfold increase in the number of hospital beds. The steady increase in the number of doctors led to a paradoxical situation in which there were more doctors in 1992 than nurses. A doctor in India, on average, serves a population of some 2,440 people. The corresponding figure for Sri Lanka is 7,143, and for sub-Saharan Africa is 18,514 (United Nations Development Programme, *Human Development Report 1997,* Oxford: Oxford University Press, 1997). Besides allopathic medicine, India also has a rich tradition of indigenous medical practices.

India has made substantial investments to improve child nutrition through the Integrated Child Development Services program, noted as the largest program for children anywhere in the world. The services are comprehensive and consist of supplementary feeding, immunization, health checkups, referral services, nonformal preschool education, and nutrition inputs. Today, the program provides care and daily food supplements to nearly 18 million children in India's poorest regions.

Similarly, postindependent India has witnessed a tremendous expansion in the educational system (see Table 13.2). In 1994, the country had close to 817,000 schools, more than 6,400 colleges, and 213 universities. The number of primary schools went up from around 210,000 in 1951 to nearly 573,000 in 1994. Today, close to 95% of the rural population has a primary school within a walking distance of one kilometer. By 1994, there were over 1.7 million primary school teachers and another nearly 2.5 million teachers in middle and secondary schools.

Despite these increases in public provisioning of health and education, the levels of human development in India remain abysmally low.

Human Development in India

The notion of human development in India is based on Amartya Sen's concept of viewing human progress as an expansion of human capabilities (Sen, 1985). In the capability framework, attention is paid to the functionings that a person can or cannot achieve. Functionings refer to the many valuable things that a person can do or be. This would include, for instance, escaping avoidable mortality, being healthy, being well-nourished, and so on. As Sen puts it, "Capability reflects a person's freedom to choose between different ways of living. The underlying motivation–the focusing on freedom–is well captured by Marx's claim that what we need is replacing the domination of circumstances and chance over individuals by the domination of individuals over chance and circumstances" (Sen, 1989).

People's achievements critically depend on the available opportunities as well as on the freedoms that they enjoy to make choices. The Human Development Reports published by United Nations Development Programme (UNDP) since 1990 have strongly advocated capability expansion as the goal of development. These reports point out that people are the real wealth of a nation. Human progress is not about income expansion and commodity production. Instead, development must be viewed as a process of enlarging people's choices. According to the 1990 UNDP, "The most critical of these wide-ranging choices are to live a long and healthy life, to be educated and to have access to resources needed for a decent standard of living. Additional choices include political freedom, guaranteed human rights and personal self-respect." In the human development framework, the constituents of human capital (such as good health, education, skill, etc.) are not regarded merely as inputs that contribute to expanding output, but they are primarily critical elements of human life. These elements have an intrinsic importance as well as an instrumental significance. In other words, to be educated, healthy, and well nourished are valuable in themselves over and above the fact that they can contribute to an expansion of output (Drèze and Sen, 1995).

The Human Development Reports have also introduced the Human Development Index (HDI) that attempts to capture three essential components of human life–longevity, knowledge, and basic income for a decent living standard. Longevity and knowledge correspond to the formation of human capabilities, and income is seen as a proxy measure for the choices that people have in putting their capabilities to use. The HDI contains three indicators: life expectancy to measure longevity, adult literacy and enrollment as indicators of educational attainment to measure knowledge, and an appropriately adjusted real gross domestic product (GDP) per capita (in purchasing-power-parity adjusted dollars) to serve as a surrogate for command over resources needed for a decent living (United Nations Development Programme, *Human Development Report 1997,* Oxford: Oxford University Press, 1997). Even though the concept of human development is larger than the measure, the index has the virtue of incorporating human choices other than income. Table 13.3 presents basic data on human development and the HDI for India and other countries and regions of the world.

Life expectancy is not only an indicator of the quantity of life, but of its quality as well. Increases in longevity reflect progress along multiple fronts, including the incomes and earnings of people, the prevention and control of diseases, the increase of knowledge and awareness, the availability of safe drinking water, and the provisioning and efficacy of health services. Between 1951–95, life expectancy at birth in India nearly doubled, to 61 years, and infant mortality was halved, to 74 deaths per 1,000 live births. Remarkable progress has been recorded in immunizing children and in eradicating guinea worm. Close to 85% of the population is reported to have access to safe drinking water. Still, India's life expectancy at birth is 11 years lower than Sri Lanka's. Close to 2.2 million infants die each year, and most of these deaths are avoidable. Some 100,000 to 125,000 Indian women die from pregnancy-related causes, accounting for almost 25% of annual maternal deaths worldwide (UNICEF, 1995).

India also has a long way to go in improving the nutritional status of its children. The country has achieved near self-sufficiency in foodgrain production, it has built up a good safety stock of foodgrains, and famines have been virtually eliminated. Yet some 60 million children below four years of age remain moderately to severely malnourished. Equally shocking is the fact that the proportion of underweight children below five years of age–50% in India–far exceeds the levels of malnutrition (31%) reported in sub-Saharan Africa (United Nations Development Programme, *Human Development Report 1997,* Oxford: Oxford University Press, 1997).

Even more glaring than India's shortfall in health is its dismal performance in basic education. Literacy nearly tripled during 1951–91; however, today, India's adult literacy rate (51%) is lower than that of sub-Saharan Africa (56%). It is only marginally higher than the average of least-developed countries (48%), and significantly lower than that of China (81%) and Sri Lanka (90%). India accounts for almost a third of the world's illiterate population. Nearly a third of India's children in the age group six to 14 years are currently out of school. Only about two-thirds of the children reach grade five of primary schooling, and of those completing grade five, many cannot even read or write a simple sentence.

In 1994, India's real GDP per capita expressed in purchasing-power-parity (PPP) adjusted dollars was PPP US$1,348. This was lower than the average for sub-Saharan Africa and was comparable to the income levels reported by Bangladesh (PPP US$1,331). India's per capita income was 50% of the levels reported by China. It was 40% of the per capita income levels reported by Sri Lanka and Indonesia, and only 12% of the Republic of Korea's per capita income. All these countries with higher income levels today also have lower rates of illiteracy.

India ranks 138th out of 175 countries for which the HDI is computed. Its HDI value (0.446) is just about half the value of East Asia's HDI (0.881) and 40% lower than China's HDI. Among the most important factors contributing to the country's low level of human development are India's extremely low levels of achievements in health, nutrition, and basic education.

Interstate Disparities and Differentials

National aggregates mask significant achievements as well as glaring disparities between states and communities. Table 13.4 presents data on human development indicators for the most populous Indian states. Several types of inequalities characterize India's development.

Between states, Kerala, for instance, reported an infant mortality rate (IMR) of 16 deaths per 1,000 live births in 1995. Only 39 countries in the world–and all of them by far richer–reported a lower infant mortality rate. On the other hand, Orissa reported an IMR of 103 per 1,000 live births in 1995, and there were only 24 countries that had a higher rate of infant mortality. The life expectancy of a girl born in Kerala today, about 74 years, is 20 years more than that of a girl born in Uttar Pradesh. Less than 20% of adult women were illiterate in Kerala in 1991. On the other hand, more than 80% of women were illiterate in Bihar, Rajasthan, and Uttar Pradesh.

Between rural and urban areas, human development indicators for urban areas are better than those reported for rural areas. For instance, life expectancy at birth in rural areas (59.4 years between 1989–93), where 76% of India's population resides, is much lower than in urban areas (64.9 years). Literacy rates in urban areas (73%) are significantly higher than in rural areas (45%).

Between communities, achievement levels among communities classified as belonging to Scheduled Castes and Scheduled Tribes are lower than the rest of society. According to the National Family Health Survey 1992–93 (International Institute for Population Sciences, 1995), the under-five mortality rate among Scheduled Castes was 149 deaths per 1,000 live births–almost 33% higher than the rate in the rest of the population. In 1991, only 24% of Scheduled Caste women and 18% of Scheduled Tribe women were literate. Literacy rates among rural women belonging to Scheduled Tribes was as low as 4% in Rajasthan and 9% in Andhra Pradesh (Census of India, *Final Population Totals: A Brief Analysis of Primary Census Abstract: Paper 2 of 1992,* New Delhi: Office of Registrar General and Census Commissioner, 1992).

Topping the Indian states in terms of human development is Kerala, with an HDI value of 0.603, comparable to the levels of human development achieved by China. At the bottom of the scale are Madhya Pradesh (HDI value of 0.349) and Uttar Pradesh (HDI value of 0.348). These two states rank along with Madagascar, Nepal, Rwanda, and Senegal as among the regions with the lowest levels of human development in the world.

Women's Achievements and Gender Inequality

Women fare worse than men by most social indicators. India is one of the few countries where there are fewer women than men–927 females per 1,000 males–a reflection of serious gender inequality and discrimination. This contrasts with the situation in the industrial world, where there are on average 1,040 women for every 1,000 men. Whereas typically life expectancy at birth for women, due to their biological advantage, exceeds that of men by five years, in the case of India, the differential was less than a year during 1989–93. In four states–Bihar, Madhya Pradesh, Orissa, and Uttar Pradesh, which account for nearly 30% of the country's population–life expectancy at birth for women was lower than that of men. Such discrimination against girls and women is also reflected in the strikingly lower levels of literacy (39%) among women aged seven and above than among men (64%). Employment opportunities are also unevenly distributed between women and men. The work participation rate among men is 52%, whereas in the case of women, it is 23% (Census of India, *Final Population Totals,* New Delhi: Office of Registrar General and Census Commissioner, 1991). Apart from the fact that much of the work that women do goes unrecognized, they are also often paid lower wages for comparable work. In the political sphere, too, women are not fully represented. They accounted for some 7% of elected Lok Sabha members of parliament. In 1996, there were women representatives to the Lok Sabha from only 13 out of India's 32 states and Union Territories.

An innovative contribution of the 1995 Human Development Report (United Nations Development Programme, *Human Development Report 1995,* Oxford: Oxford University Press, 1995) has been the introduction of the Gender-related Development Index (GDI). The GDI concentrates on the same variables as the HDI but focuses on both the inequality between men and women as well as on the average achievement of all people taken together. In other words, the GDI is the HDI adjusted for gender equality. The construction of the GDI for Indian states reveals not only the extent of gender inequalities within India but also the country's poor performance vis-à-vis other nations of the world (Shiva Kumar, 1996). At the top of the list of Indian states is Kerala, with a GDI value of 0.597. Uttar Pradesh is at the bottom, with a GDI value of 0.310, next to Benin (see Table 13.4). Looked at differently, the GDI value for Uttar Pradesh is only half that of Kerala. There are only 13 countries in the world with lower GDI values than Bihar and Uttar Pradesh. Twice as many people, living in abysmal conditions of human deprivation, live in Uttar Pradesh and Bihar (combined population of 225 million in 1991) than in the 13 countries that had lower GDI values.

Learning from Experience

Investments in health and education are critical. In addition to the levels of expenditure, it is equally important to understand how well these translate into health and education outcomes. In terms of public expenditures, for instance, India spends less on health than many other countries (see Table 13.5).

India's public expenditures on health went up from 0.3% of gross national product (GNP) in 1960 to 1.8% of India's gross domestic product (GDP) in 1990. This was the average level of spending by the least developed countries but considerably lower than the levels of spending by many of the "good health at low cost countries," such as China (2.1%) and Costa Rica (5.6%). However, public expenditure on education compares favorably with many other countries and regions of the world; as a share of GNP, it increased steadily from 1.2% in 1950–51 to 3.8% in 1993–94. This is higher than the average of 3.6% of GNP spent by developing countries and 2.6% spent by China (United Nations Development Programme, *Human Development Report 1997*, Oxford: Oxford University Press, 1997). Yet in both health and education, India has a long way to go in terms of ending deprivations.

The patterns of India's investments in health reveal several biases. These have contributed in large measure to the low levels and to the glaring disparities in achievements. Much of the investments have been in curative health and located in urban areas. Of the 13,692 hospitals in the country in 1993, 69% were located in urban areas, in which 24% of the population resides (Government of India, 1994). Similarly, nearly 80% of hospital beds are located in urban areas. Less than a third of India's 600,000 villages have a primary health care center or subcenter located in them. According to the National Family Health Survey 1992–93 (International Institute for Population Sciences, 1995), the proportion of fully immunized children aged 12 to 23 months was 51% in urban areas and only 31% in rural areas.

Private spending is estimated to be 75% of total health expenditures. Such a high proportion of private spending suggests not only that the burden of disease remains high but also that most families do not have adequate access to government facilities that are often provided free or for a nominal charge. Apart from the limited public provisioning, there is also the question of quality of services. Government health care centers in many parts of India are known for their inefficiencies and poor quality of service. This is not to suggest that private health care services, especially for the poor, is of good quality. Rural private health care providers frequently exploit the ignorance of the poor. In the absence of any effective regulation, besides charging high fees for wrong treatments, many private health care providers also do not adhere to norms of hygiene and public safety.

India has the largest number of malnourished children in the world. It was once believed that Indian children do not normally grow as fast or as large as children in other countries. This is, however, not true. Studies have conclusively established that global standards of height and weight apply to Indian children as well, provided they have access to proper food, health care, and attention (Nutrition Foundation of India, *Growth Performance of Affluent Indian Children–Under 5s,* New Delhi: Nutrition Foundation of India, 1991).

It has been customary to regard income poverty as an important factor contributing to India's malnutrition, but statistical analyses show that it is reasonable to attribute only half of the malnutrition to income poverty per se (UNICEF, 1995). It has often been argued that the high levels of malnutrition reflect a serious shortage of food. It is true that if children do not get food to eat, they will be malnourished, but the amount of food that a child needs is very little, especially between the ages of six to 18 months, when children are often malnourished and remain so thereafter. The real solution to malnutrition lies in more care rather than more money or more food. Even at birth, one in every three children born is of low birth weight, the result of malnutrition in the womb. Pregnant mothers need to be cared for and given proper food and rest. Similarly, children need to be cared for. In the initial months of life, breast milk is often adequate to provide effective nutrition, but after six months, it becomes necessary to supplement breast milk with solid, mushy foods in order to ensure that the child does not become malnourished. However, infants at this age cannot eat by themselves. They need to be fed small quantities frequently throughout the day. The National Family Health Survey also reveals that only 31% of children aged six to nine months receive breast milk and solid or mushy foods. Apart from ignorance, often the mother or older sibling lacks the time and patience to feed young babies (Vulimiri Ramalingaswami, Urban Jonsson, and Jon Rohde, commentary on nutrition in UNICEF, *The Progress of Nations,* New York: UNICEF, 1996 and UNICEF India, 1995).

Article 45 of the Constitution of India (1950) enjoins that "the State shall endeavor to provide, within a period of ten years from the commencement of this Constitution, for free and compulsory primary education for all children until they complete the age of fourteen years," yet the intention to provide free and compulsory primary education for all remains an elusive goal. India is among the few countries in the world where primary education is not compulsory. A consequence of this has been the perpetuation of the practice of child labor. Commenting on this feature, Weiner writes:

> Modern states regard education as a legal duty, not merely as a right: parents are required to send their children to school, children are required to attend school, and the state is obligated to enforce compulsory education. . . . The State thus stands as the ultimate guardian of children, protecting them against both parents and would-be employers. This is not the view held in India. Primary education is not compulsory, nor is child labor illegal.

As with health, biases in educational investments have contributed to the low levels and inequalities in achievements. In the initial years of India's independence, the government concentrated on promoting higher education and on setting up heavily subsidized colleges and universities. Elementary education remained neglected. Between 1951–81, for instance, the number of colleges went up from 578 to 7,577, and the number of universities from 28 to 123. During this period, the number of primary schools did increase, and between 1950–90, expenditure on elementary education went up from 0.46% of GNP to 1.72%. However, this expansion was grossly inadequate to provide universal education to a growing population. In 1991, there were some 329 million illiterate people in the country aged seven and above–more than the number of illiterates, 305 million in 1981. In the First Five-Year Plan (1951–56), 56% of total plan resources for education were allocated to elementary education. This proportion fell to 30% during the Fourth Five-Year Plan (1969–74) and only now has begun to increase. The Eighth Five-Year Plan (1992–97) allocated 43% to elementary education–still lower than the levels earmarked during the first plan (Government of India, 1993).

Not only are dropout rates in primary schooling (grades one through five) high (36%), the levels of educational achievements of children are also extremely low. A National Advisory Committee appointed by the Government of India in 1993 pointed out that in addition to the physical burden, manifest in the burden of carrying a schoolbag, which, on average, weighs 4 kilograms in most public schools in the cities, learning has become monotonous and joyless. The committee reported that "both the teacher and the child have lost the sense of joy in being involved with the educational process. Teaching and learning have both become a chore for a great number of teachers and children."

Primary schooling is almost entirely financed by the government. However, most government school buildings are in poor physical condition. Typically, classrooms are overcrowded and teachers lack the basic facilities of blackboards, drinking water, toilets, and teaching aids. The pupil-teacher ratios are high, and in many rural schools, the teacher has to engage in multigrade teaching. At the same time, irregular functioning of schools and teacher absenteeism affect the motivation of school children. Even though primary education is said to be offered "free," the costs of schooling–uniforms, textbooks, notebooks, etc.–are high even in government schools and add up to substantial amounts, especially for income-poor families. As a result of all these factors, even though demand for basic education is strong and parental motivation is high, both children and parents do not find it attractive to send their children to school.

Learning from Kerala

Examining interstate differentials points to the remarkable success achieved by Kerala in promoting health and education. Kerala's achievements are significant compared to the performance of the country or Uttar Pradesh–the state with the lowest HDI (see Table 13.6). Despite being a low-income state (like Uttar Pradesh), Kerala has recorded impressive gains in life expectancy and literacy. This is the outcome of conscious policy choices that the state made in order to invest in people's health and education. Today, Kerala spends more on health and education than the average Indian state (and Uttar Pradesh). However, it is not a question of spending alone; the dispersal of health and educational services has been more widespread and equitable in Kerala than in Uttar Pradesh. For instance, some 57% of Kerala's hospital beds are located in rural areas as against 21% for India. Kerala has achieved universal primary schooling and a zero dropout rate in primary schools.

In addition to the levels and patterns of investment in health and education, several other factors have contributed to Kerala's human development. Drèze and Sen and Ramachandran, for instance, point to the role of public action in promoting an equitable expansion of opportunities relating to elementary education, land reforms, the position of women, and the provisioning of health services. This has been backed in Kerala by strong political participation and activism, public monitoring, and community involvement.

The Road Ahead

In 1996, the Government of India set itself the goal of eradicating income poverty by the year 2005. In order to realize this goal, it has identified three key factors: (1) ensure a GDP growth rate of at least 6% per annum over the next ten years; (2) ensure provisioning of at least seven basic minimum services–universal access to safe drinking water, 100% coverage of primary health care centers, universalization of primary education, public housing assurance to all shelterless, deserving families, extension of the midday meal throughout all primary schools, road connectivity to all villages and habitations, and streamlining the public distribution system targeted to families below the income poverty line; and (3) ensure that the income-poor and the socially disadvantaged groups receive special attention and priority.

To a large extent, such assurances have been repeatedly made by the central government since the First Five-Year Plan was launched in 1951. Political consensus has also been built up for poverty eradication, but surely India has to learn from its past experience and do things differently if these goals are to be realized. The question is, what should be some of the changes in policy thinking and formulation?

First, India must recognize and capitalize on the strong complementarity that exists between economic expansion and improvements in the quality of people's lives. In 1960, India had a life expectancy of 44 years and an adult literacy rate of 34%. Conditions in Botswana were not very different, with an average life expectancy of 45.5 years and an adult literacy rate of 41% (United Nations Development Programme, *Human Development Report 1996,* Oxford: Oxford University Press, 1996). In 1960, Botswana had a lower level of real GDP per capita (PPP US$474) than India (PPP US$617). By 1993, the situation had changed. Sixty-eight percent of Botswana's adult population was literate as against 51% in India. During this period, Botswana increased its per capita income to PPP US$5,220–almost four times higher than India's per capita income of PPP US$1,348 in 1994. Similarly, in 1960, South Korea and India had similar levels of per capita income. By 1994, South Korea's income was nearly eight times higher than India's. This increase in income between 1960 and 1994 coincided with a period when adult illiteracy in South Korea fell from 46% to 2%. Again, if China, Indonesia, and Thailand have all achieved and sustained higher levels of per capita incomes than India, it is because they have a much higher level of literacy and basic education than India has. If human poverty has to be eradicated, India must invest in its people–in their health and education.

Second, India needs to strike a balance in its development. This balance is not on the economic front alone–between receipts and expenditures, imports and exports, and savings and investments. A balance is needed between economic growth and an expansion of social and political opportunities, between the development of physical infrastructure and social infrastructure. At the same time, several imbalances need to be corrected–between men and women, rural and urban areas, and socially disadvantaged communities and the rest of society.

Third, there is the issue of resources for investing in human capital. Clearly, more financial resources are required if all children have to attend school, if all villages must have access to a primary health care center, if all communities must have access to safe water, if all pregnant mothers have to be assured of safe motherhood. Additional resources could be mobilized by improving the tax-to-GDP ratio and ensuring a growth rate of 6% to 8% per annum, by eliminating subsidies to the rich, by cutting the losses of public enterprises, and by reducing defense spending. But there is also a need for improving the quality and efficacy of services, for correcting imbalances in public expenditures, for plugging leaks and reducing wastage, and for ensuring greater efficiency in spending.

Fourth, the state needs to play a more proactive role in promoting basic health and education than it has in the past. In the public sector, increased provisioning has to be matched by a dramatic improvement in capacity utilization and quality. In the private sector, too, efficiency has to be promoted through an appropriate regulatory framework.

Fifth, the persistent gaps in the levels of achievements between men and women need to be narrowed and overcome. Opportunities must be created and expanded for women to gain equal access to basic health care, education, and other essential social services. At the same time, opportunities must be created to enable them to participate more fully in economic and political decision making.

Finally, human development has to be participatory. It must be planned and managed locally by people whose lives are affected by it. Communities must participate actively to shape programs and to ensure that opportunities are expanded and that the benefits are shared equitably. For this, structures of local self-governance must be strengthened, and people's participation needs to be encouraged.

India has the potential for rapid economic growth. The economic reforms introduced by the Government of India in 1991 have focused on expanding economic opportunities. Liberalization, deregulation, and the opening up of the economy to foreign investments and global competition are beginning to transform the economic scene, but India's ability to realize this economic potential will depend critically on how quickly and effectively it tackles the crisis in human capital. India can achieve and sustain economic and human progress only if every effort is made to eradicate illiteracy, make primary education compulsory, and ensure equitable access to good quality health care for all its people.

Further Reading

Drèze, Jean, and Amartya Sen, *India, Economic Development, and Social Opportunities,* Delhi: Oxford University Press, 1995

Offers an excellent analysis of India's record in ending endemic deprivation since independence and contains a very useful discussion of the role of public action in addressing the problem.

Government of India, *Economic Survey 1994–95,* New Delhi: Government of India, Ministry of Finance, 1995

—, *Economic Survey 1996–97,* New Delhi: Ministry of Finance, 1997

These annual publications report on the country's economic performance during the year. They describe recent economic trends and major policy changes.

—, *Education for All: The Indian Scene,* New Delhi: Government of India, Department of Education, Ministry of Human Resource Development, 1993

Traces the progress and achievements in education since 1947, lists the many failures, and discusses recent initiatives of the GOI for achieving universal elementary education.

—, *Health Information of India 1994,* New Delhi: Government of India, Ministry of Health and Family Welfare, 1994

Presents detailed statistics on health in India.

Ramachandran, V. K., "On Kerala's Development Achievements," in *Indian Development,* edited by Jean Drèze and Amartya Sen, New Delhi: Oxford University Press, 1997.

A good review of Kerala's development record. Identifies the main reasons for Kerala's extraordinary success in human development.

Sen, Amartya, *Commodity and Capabilities,* Amsterdam: North-Holland, 1985

Elaborates on and discusses the human capabilities framework.

—, *Development as Capability Expansion,* United Nations: Journal of Development Planning, No. 19, 1989

Discusses the policy significance and implications of treating an expansion of human capabilities as the goal of development planning.

Shiva Kumar, A. K., "UNDP's Gender-Related Development Index," in *Economic and Political Weekly* [Bombay] 31, no. 14 (1996)

Constructs the Gender-related Development Index for the most populous Indian states using the methodology presented in the 1995 Human Development Report.

UNICEF, *The Progress of Indian States,* New Delhi: UNICEF India, 1995

Looks at the performance of different Indian states by means of several indicators of children's well being, including survival, health, education, gender equality, nutrition, and child labor.

Weiner, Myron, *The Child and the State in India,* Princeton, N.J.: Princeton University Press, 1991

Argues that one main reason for the persistence of child labor in India is the absence of compulsory primary education. The author explores the reasons why India has failed to enact legislation that makes primary education compulsory.

World Bank, *India: Policy and Finance Strategies for Strengthening Primary Health Care Services,* report no. 13042-IN, Washington, D.C.: World Bank, 1995

Analyzes the primary health care system in India and suggests ways of strengthening both the delivery and efficacy of health services in the country.

——, *India: Primary Education Achievement and Challenges,* report no. 15756-IN, Washington, D.C.: World Bank, 1996

A comprehensive review of primary education in India; looks at some of the conditions that need to be fulfilled if India is to achieve universal compulsory education by the year 2000.

A. K. Shiva Kumar is a member of the Human Development Report team in New York. He lives in New Delhi, works closely with UNICEF and UNDP in India, and is a visiting faculty member at Harvard University, where he teaches economics and public policy.

Table 13.1: Expansion of Health Services in India, 1951–92

	1951	*1961*	*1971*	*1981*	*1991*	*1992*
Medical colleges	28	60	98	111	128	146
Hospitals	2,694	3,094	3,862	6,804	11,174	13,692
Dispensaries	6,515	9,406	12,180	6,751	27,431	27,403
Community health centers	-	-	-	217	2,071	2,193
Primary health centers	725	2,565	5,112	5,740	20,450	20,719
Subcenters	-	-	28,489	51,405	130,984	131,378
Hospital beds	117,178	230,000	348,655	569,495	810,548	na
Doctors	61,840	83,756	151,129	268,712	394,068	410,875
Dentists	3,290	3,582	5,512	8,648	10,751	11,300
Nurses	16,550	35,584	80,620	154,280	340,208	385,410

Source: Ministry of Health and Family Welfare, quoted in Government of India, *Economic Survey 1996–97,* New Delhi: Government of India, Ministry of Finance, 1997

Table 13.2: Expansion of Educational Services in India, 1951–94

	1950–51	*1960–61*	*1970–71*	*1980–81*	*1990–91*	*1993–94*
Schools (in 000s) of which:						
Primary	209.7	330.4	408.4	494.5	560.9	572.9
Middle	13.6	49.7	90.6	118.4	151.5	155.7
Secondary	7.4	17.3	37.1	51.6	79.8	88.4
Colleges of which:						
General	370	967	2,285	6,421	4,862	5,339
Professional	208	852	992	1,156	1,765	1,125
Universities	28	45	93	123	184	213
Teachers (in 000s)						
Primary schools	538	742	1,060	1,363	1,616	1,703
Middle school	86	345	638	851	1,072	1,080
Secondary school	127	296	629	926	1,335	1,405

Sources: various issues of the *Annual Report* of the Department of Education, Ministry of Human Resource Development, Government of India

Table 13.3: Human Development in India and Selected Regions of the World

1994	*Life expectancy at birth (years)*	*Adult literacy rate (%)*	*Real GDP per capita (PPP$)*	*Human Development Index (HDI)*
A. Regions of the world				
Developing countries of which:	61.8	69.7	2,904	0.576
Sub-Saharan Africa	50.0	55.9	1,377	0.380
South Asia	61.4	49.7	1,686	0.459
Arab states	62.9	54.7	4,450	0.636
Latin America/Caribbean	69.0	86.2	5,873	0.829
East Asia	71.5	96.2	9,429	0.881
Least developed countries	50.4	48.1	965	0.336
Industrial countries	74.1	98.5	15,986	0.911
B. Selected countries				
Bangladesh	56.4	37.3	1,331	0.366
India	61.3	51.2	1,348	0.446
China	68.9	80.9	2,604	0.626
Indonesia	63.5	83.2	3,740	0.668
Sri Lanka	72.2	90.1	3,277	0.711
Botswana*	65.2	68.0	5,220	0.741
Republic of Korea	71.5	97.9	10,656	0.890

Note: GDP=gross domestic product; PPP$ = purchasing-power-parity adjusted dollars
*Relates to 1993;
Source: United Nations Development Programme, *Human Development Report 1997*, Oxford: Oxford University Press, 1997

13.4: Human Development Index (HDI) and Gender-related Development Index (GDI) for Indian States

	Population (millions)	*Female-male ratio*	*Life expectancy at birth (years)*		*Adult literacy rate (%)*		*Share of earned income (%)*		*Per capita state domestic product (Rs./year)*	*Human development index (HDI)*	*Gender-related development index (GDI)*
			Females	*Males*	*Females*	*Males*	*Females*	*Males*			
	1991	*1991*	*1990–92*		*1991*				*1991–92*		
	(1)	(2)	(3)	(4)	(5)	(6)	(7)	(8)	(9)	(10)	(11)
1. Kerala	29	1,036	74.4	68.8	80.6	91.7	12.4	87.6	4,618	0.603	0.565
2. Maharashtra	79	634	64.7	63.1	44.2	74.4	29.4	70.6	8,180	0.523	0.492
3. Gujarat	41	934	61.3	59.1	41.8	70.4	26.8	73.2	6,425	0.467	0.437
4. Himachal Pradesh	5	976	64.2	63.8	35.5	64.4	37.5	62.5	5,355	0.454	0.432
5. Punjab	20	882	67.5	65.4	41.8	60.5	5.9	94.1	9,643	0.529	0.424
6. Karnataka	45	960	63.6	60.0	37.7	65.3	25.4	74.6	5,555	0.448	0.417
7. Tamil Nadu	56	974	63.2	61.0	35.8	65.0	21.4	78.6	5,078	0.438	0.402
8. West Bengal	68	917	62.0	60.5	42.8	69.3	8.0	92.0	5,383	0.459	0.399
9. Andhra Pradesh	67	972	61.5	59.0	27.3	52.4	27.2	72.8	5,570	0.400	0.371
10. Haryana	16	865	63.6	62.2	27.0	64.3	7.0	93.0	8,690	0.489	0.370
11. Assam	22	923	53.8	54.6	33.9	62.4	23.7	76.3	4,230	0.379	0.347
12. Orissa	32	971	54.8	55.9	29.0	62.5	19.1	80.9	4,068	0.373	0.329
13. Madhya Pradesh	66	931	53.5	54.1	24.3	56.6	25.4	74.6	4,077	0.349	0.312

13.4: Human Development Index (HDI) and Gender-related Development Index (GDI) for Indian States (Continued)

	Population (millions)	*Female-male ratio*	*Life expectancy at birth (years)*		*Adult literacy rate (%)*		*Share of earned income (%)*		*Per capita state domestic product (Rs./year)*	*Human development index (HDI)*	*Gender-related development index (GDI)*
			Females	*Males*	*Females*	*Males*	*Females*	*Males*			
	1991	*1991*	*1990–92*		*1991*				*1991–92*		
	(1)	(2)	(3)	(4)	(5)	(6)	(7)	(8)	(9)	(10)	(11)
14. Rajasthan	44	910	57.8	57.6	17.5	52.7	23.0	77.0	4,361	0.356	0.309
15. Bihar	86	911	58.3	60.4	18.2	55.3	21.8	78.2	2,904	0.354	0.306
16. Uttar Pradesh	139	879	54.6	56.8	20.6	53.6	12.9	87.1	4,012	0.348	0.293
INDIA	846	927	59.4	59.0	33.9	62.4	23.2	76.8	5,583	0.423	0.388

Note: 1. Data pertain to 16 of India's most populous states. States have been arranged in descending order of GDI. 2. The methodology for computation of HDI and GDI are described in United Nations Development Programme, *Human Development Report 1995,* Oxford: Oxford University Press, 1995, and in Shiva Kumar, 1996.

Sources: Columns 1 and 2 from Census of India, 1991; Columns 3 and 4 from Sample Registration System quoted in Drèze and Sen, 1995; column 5 and 6 are estimates derived from 1991 census figures of total literacy quoted in Shiva Kumar, 1996; column 9 is from Government of India, 1995; columns 7,8,10, and 11 are from Shiva Kumar, 1996.

Table 13.5 Public Expenditure on Health and Education, 1980–92

	Public expenditure on health		*Public expenditure on education*	
	as % of GNP 1960	*as % of GDP 1990*	*as % of GNP 1980*	*as % of GNP 1993–94*
Developing countries of which:	0.9	2.1	3.8	3.6
Sub-Saharan Africa	0.7	2.5	5.1	5.5
South Asia	0.6	1.4	4.3	3.5
of which India	0.3	1.8	2.8	3.8
Arab states	0.9	2.9	4.1	-
Latin America/Caribbean	-	2.4	3.7	3.6
East Asia	0.9	2.3	2.9	3.3
of which China	1.3	2.1	2.5	2.6
Least developed countries	-	1.8	3.1	2.8

Source: United Nations Development Programme, *Human Development Report 1997*, Oxford: Oxford University Press, 1997

Table 13.6: Differentials in Social Sector Provisioning and Achievements—Kerala, Uttar Pradesh, and India

	Year	*Source*	*Kerala*	*INDIA*	*Uttar Pradesh*
Human development index	1991	Shiva Kumar, 1996	0.603	0.423	0.348
Gender-related development index	1991	Shiva Kumar, 1996	0.565	0.293	0.388
Per capita state domestic product (Rs.)	1994–95	Economic Survey, 1996–97	6,983	8,281	5,331
Annual growth in state domestic product	1980–95	Economic Survey, 1996–97	3.56%	5.05%	3.85%
Political participation:					
Voter turnout rate in Lok Sabha elections	1996	Election Commission			
Females			70%	57%	47%
Males			72%	66%	55%
Health:					
Life expectancy at birth	1989-93	SRS	72.0	59.4	55.9
Urban			71.8	58.0	55.0
Rural			72.8	64.9	60.4
Infant mortality rate	1995	SRS	16	74	103
Total fertility rate	1992–93	IIPS (1995)	2.00	3.39	4.82

Table 13.6: Differentials in Social Sector Provisioning and Achievements—Kerala, Uttar Pradesh, and India (Continued)

	Year	*Source*	*Kerala*	*INDIA*	*Uttar Pradesh*
% of villages with health care center or subcenter	1992–93	IIPS (1995)	96.2	30.3	20.1
% of births assisted by health professional	1992–93	IIPS (1995)	89.7	34.2	17.2
% of births delivered in health facility	1992–93	IIPS (1995)	87.8	25.5	11.2
% of children 12-23 months fully vaccinated	1992–93	IIPS (1995)	54.4	35.4	19.8
% of children under 4 years who are underweight	1992–93	IIPS (1995)	28.5	59.0	53.4
% of children 6-9 months receiving solid/mushy foods	1992–93	IIPS (1995)	69.3	31.4	19.4
% of households with no toilet facilities	1992–93	IIPS (1995)	29.1	69.7	77.1
% of hospital beds in rural areas	1993	Government of India, 1994	57	20.5	5.5*
Government health expenditure per capita (Rs.)	1991	World Bank, 1995	72	58	32
Education:					
Literacy rate for people aged 7 and above (%)	1991	Census of India, 1991	89.8	52.2	41.6
Excess of male over female literacy (% points)	1991	Census of India, 1991	7	25	29
% attending school (females aged 6–14 years)	1992–93	IIPS (1995)	94.8	58.9	48.2
Drop-out rates in primary school	1992–93	World Bank, 1996	0	36	20
Ratio of enrolled students to teachers	1993	World Bank, 1996	47	49	42

Table 13.6: Differentials in Social Sector Provisioning and Achievements—Kerala, Uttar Pradesh, and India (Continued)

	Year	*Source*	*Kerala*	*INDIA*	*Uttar Pradesh*
Education budget as % of State Domestic Product	1992–93	World Bank, 1996	7.5	4.0	4.0
Education as % of total budget expenditure	1992–93	World Bank, 1996	28.1	22.6	20.9
Per capita education expenditure (Rs.)	1992–93	World Bank, 1996	368	260	176

SRS=Sample Registration System; IIPS=International Institute for Population Sciences
*Relates to 1986

Chapter Fourteen

Indian Economic Reforms: Popular Perception and Public Debate

Girijesh Pant

In 1991, a new era of Indian economic history began. Having started with a mixed economy model that envisaged a complementary rather than a competitive relationship between the public and private sectors, that provided the state rather than the market the commanding power to allocate resources, and that gave equity a premium over efficiency, India reached a stage where the logic of evolution demanded a departure from the policies that could not accelerate the pace of the economy beyond the "Hindu growth rate." Like many other developing countries, India decided to adopt a market-based regime by initiating a macroeconomic stabilization and structural adjustment program, which are popularly described as the economic reforms.

Commentators on Indian political economy trace the beginning of these reforms to the days of Indira Gandhi, when she came back to power after a short spell of political wilderness in the early 1980s. At that time, she needed a new agenda to make a fresh start, particularly when the magic of *garibi hatao* (abolish poverty) lost its appeal. Rajiv Gandhi, who succeeded her, further crystallized and articulated the content of the reforms as a means to take India into the twenty-first century. Both Indira Gandhi and Rajiv Gandhi, however, were hesitant to make a big leap forward. It was the fragile and minority government of P. V. Narasimha Rao that took the bold step to embark on the journey of the rocky road to market.

With the advantage of hindsight, it can be argued that the initiatives of the Gandhis, despite their limited nature, did help Rao create political consensus within the party in favor of reforms by projecting them as a continuation of earlier policies. But at the same time, Rao needed his own agenda to look distinct and different. As a result, he took full advantage of the economic crisis of the time. According to the government assessment, India was faced with the prospect of defaulting on its international commitments. Access to the external commercial credit market was completely denied, India's international credit rating had been downgraded, and the international financial community's confidence in India's ability to manage its economy had been severely eroded. Consequently, India could only borrow against the security of its gold reserves by physically transporting the gold abroad (Government of India, 1995). The crisis was projected as a catastrophe, and the reform was characterized as if There Is (was) No Alternative–the TINA syndrome.

During five years of reforms (1991–96), the Indian economy did fairly well. It broke the barrier of slow growth by making an average of 6% during 1991–96. According to the economic surveys of 1994–95 and 1995–96, growth of the real gross domestic product (GDP) at factor cost, which had fallen to a mere 0.8% in the crisis year of 1991–92, recovered within a year to reach 5.1% in 1992–93, 5% in 1993–94, 6.3% in 1994–95, and 6.5% in 1995–96. The foodgrain stock, which had come down to 168 million tons in 1991–92, went up to a little over 190 million tons in 1994–95. Exports experienced a boom for three consecutive years, with annual growth averaging 19% in 1993–94 and 1994–95, and going up to 24% in the following year. Above all, the foreign reserves, which were barely at US$1 billion in 1991, have been maintained at over US$20 billion since then.

These achievements were widely acclaimed both inside and outside the country. India was even projected as the emerging economic power by the International Monetary Fund (IMF) and the World Bank. However, despite this image building, when Rao went to the polls to seek endorsement for his success, he got the worst deal: the Indian National Congress suffered a crushing defeat. Clearly Rao's government could not convert the economic success of his administration into electoral gains, and this requires some explanation.

Sometimes facts do not speak for themselves. Perceptions or misperceptions dominate the public discourse, and public opinion does influence the electoral verdict. In a democracy, any government that ignores the popular mood is bound to pay heavily. Rao's government committed this mistake: it did not realize the need to educate the people on the meaning of reform. Rao was a poor communicator; eloquence was not his strong point. His government did not remember that even leaders as charismatic as Jawaharlal Nehru and Indira Gandhi went to the people before launching any new scheme. Through his public speeches, Nehru taught the illiterate Indian masses the meaning of socialism, the need for science and technology, the advantage of big dams, and the relevance of steel by using local idioms and vocabulary. He prepared them to forgo the pleasure of current consumption to reap the benefit in future years, as did Indira Gandhi when she sought bank nationalization. The state-led model thus never looked as borrowed from the Soviet Union or drawn from Fabian Socialism as the market reforms appeared to be dictated by the IMF and the World Bank. It is not surprising that people saw in the reforms the return of the East India Company of the colonial period; perceptions are conditioned by historical experience. The people did not protest Rao's policies because they were tired of the populist slogans and practices, but they were not sure of the new policies either. As Nelson puts it, "the losers know who they will be, the gainers are much less certain" (J. M. Nelson, "How Market Reforms and Democratic Consolidation Affect Each Other," in *Intricate Links: Democratisation and Market Reforms in Latin America and Eastern Europe,* edited by J. M. Nelson et al., New Brunswick: Transaction Publishers, 1994: 9). This is not to argue that economic reforms were the deciding factor in the election of 1996; many other issues, from caste to religion, played their role. But from the perspective of the market-democracy paradigm, the Indian experience is an interesting case to be studied. India is a 50-year-old parliamentary democracy; it does not have to address the issues of dual transition. For India, the challenge is democracy with the market on the commanding height.

During the first phase of reform (1991–96), with the unfolding of the package, a number of issues came into the public debate, including the pace of reform, its sequencing, political sustainability, the impact on poverty and equity, regional imbalances, the nature of competition, leveling of the field, etc. In this chapter, an attempt has been made to capture some of the leading issues that were debated during the first phase of reforms and the societal response they evoked. This study is limited to the main issues debated by the three major constituencies who were the representative voices in the Indian public debate of that period: the public sector, the small producers, and the corporate sector.

Privatizing the Public Sector

Most public sector undertakings in India are reported to be running at a loss and are thus said to be a drag on the economy. In 1956, there were 21 units in the state sector; their current number is estimated at 1,300. According to the World Bank, these units manage 55% of the capital stock and over 25% of the nonagriculture GDP (The World Bank, 1995). These units are engaged in a wide variety of economic activities (railways, power, telecommunication, steel, oil and gas, petro chemicals, banking, insurance, tourism, etc.). Mostly over staffed and inefficient, these units have been surviving on government support. Quite often, these are seen as political constituencies rather than as commercial enterprises.

Any scheme aimed at rationalizing the resource utilization of an economy would find these public-sector undertakings a liability to be dispensed with. The Indian reform strategy, therefore, envisages privatization of these units, but the experience of various countries suggests that this has been one of the thorny components of the reform package: "State owned enterprise reform can cost a government its support base. Consequently, politicians everywhere carefully weigh any change in state owned enterprise policy, naturally preferring policies that benefit their constituencies and help them remain in office over policies that undermine support and may precipitate their removal" (The World Bank, *Bureaucrats in Business: The Economics and Politics of Government Ownership,* New York: Oxford University Press, 1995). It is even recognized by the World Bank that unless the following conditions are met, state-owned enterprise (SOE) reforms cannot be implemented.

- *Political desirability.* The benefits to the leadership and its constituencies must outweigh the costs of the reforms.
- *Political feasibility.* The leadership must be able to enact the reforms and overcome opposition.
- *Credibility.* Promises that the leadership makes to compensate losers and protect investors' property rights must be believable.

Resistance to reforms traditionally comes not only from labor but also from management and the political elite, who wield power by virtue of being members in the decision-making bodies. India's experience is not very different. It is argued that "the internal rate of return that the business politics generates in India is pretty high" (Joshi and Little, 1994). Here, public enterprises come in handy. Public enterprises are used to create and provide jobs for the relatives of politicians (often at higher than market wages) or as parking places for selected members of the political party who could not be accommodated in the cabinet. This explains why privatization has low ranking in the reform strategy of the Indian government.

The Industrial Policy Statement announced in July 1991 outlines the policy in terms of restructuring the portfolio of investments, the equity (ownership) pattern of public-sector enterprises (PSEs), and the quality of interface between the government and the PSEs, the boards of directors, the sick enterprises, the safety net, and the policy environment. It has also reduced the areas reserved for the public sector from 17 to six (defense products, atomic energy, coal and lignite, mineral oils, rail transport, and minerals specified in the schedule to the Atomic Energy Order of 1953). According to the *Economic Survey 1995–96,* between 1 July 1991 and December 1995, the record of privatization shows that equity in only 40 Central Public Enterprises were sold during this period. The quantity of equity sold was less than 10% in the case of 19 units, between 10% to 20% in the case of seven, between 20% to 30% in the case of six, between 30% to 40% in the case of seven, and between 40% to 50% in the case of one. The policy seems to have three objectives: to mobilize resources to reduce the fiscal gap, to gauge the mindset or assess the resistance of labor, and to satisfy the global community, especially the IMF and the World Bank. This means that the controlling ownership and management remains with the government, and the policy did trigger a debate at various levels.

On implementation of the Industrial Policy Statement, the government faced criticism for its failure to mobilize resources in accordance to its own estimates. It was expected that in the budget of 1994–95, Rs. 40,000 crores would be mobilized, but only Rs. 2,330 crores could be obtained in the first round. Most of the units brought to the market were profit-earning enterprises, such as Container Corporation of India, Indian Oil Corporation, Mahanagar Telephone Nigam, Oil and Natural Gas Commission, etc. Clearly, there were not many takers. What is more alarming, as pointed out by the Comptroller and Audit General of India, the government sold the shares of these units not only below the market rate but at the reserve price, thus incurring a loss of Rs. 3,400 crores (*Financial Express* [Bombay], 24 June 1995). The weak response from the market was attributed to two main problems: (1) the timing was wrong because the share market at that time was having a shortage of funds, and (2) the image of PSU was not good enough to fetch high prices in the market. Big business argued that the very partial nature of the divestment was responsible for the poor response. According to the president of the Confederation of Indian Industries (CII), partial divestment does not lead to change to management or provide autonomy to the management, and therefore it does not improve the prospects of the company. Divestment needed to be undertaken in more than 50% of the equity so that the units could be reorganized (*Hindustan Times* [New Delhi], 1 March 1995).

While big business was demanding a larger scale of divestment, labor opposed it. It should be pointed out here that two-thirds of the organized labor force is employed by the public sector, and their unions have been quite powerful. The story of UP State Cement Corporation Limited may be cited here as an example.

The government decided to sell the unit because it had accumulated a loss of Rs. 15 million. Dalmia, a private group, agreed to buy 50% of its equity and take over the management to run it as a joint venture. The employees opposed the move and went to court. A petition was filed prohibiting the government to convert the unit into a private enterprise. The government decided to hand over only 48% of the equity, keeping 2% until the court decided the case. But that was not acceptable to labor. The unit experienced labor unrest and strikes. This unrest became violent when police were called in to contain it, leading to the deaths of nine workers. The situation became increasingly dramatic when the government was replaced by another government, which decided to cancel the agreement. Certainly this cannot be taken as a representative case, but it does illustrate the kind of resistance that the scheme could generate (Bajaj, 1994: M121).

The banking and insurance sector is another example in which the government under the pressure of the employees has not been in a position to take the reform forward. On 8 April 1994, about 150,000 employees in the banking and insurance sector held a joint strike, forcing the government to defer its decision (*Times* [New Delhi], 9 April 1994).What is significant is that in opposing the move, the officers, managers, and white-collar class put up a collective front.

The government policy to bring the multinationals into the public sector with a view to break monopolies was generally welcomed but became an issue of debate for what was described as its discriminatory nature. In the case of the power sector, it was pointed out that the kind of counter guarantees and power purchase agreements that were signed forced the local units to compete on unequal grounds. Such a move would not add to their strength but would instead kill them, it was argued. Enron was cited as a case in which a foreign firm was awarded a power project on favorable terms. According to the Parliamentary Standing Committee on Industry, the policy of providing a guarantee of 16% return, payable in foreign exchange, was a very generous offer. Moreover, the per unit power cost of the project had been on the high side. It was high because the capital cost of the project, given by Enron, was very high. It was estimated that the power cost from this plant would either raise power prices to levels that would affect growth adversely or raise subsidies to levels that would render the budget of the State Electricity Board of India (SEBI) unviable. The estimates given by the National Confederation of Officers Association of Central PSUs, the capital cost of the project was phenomenal and the deal amounted to "surrendering to the Cartel." The question was raised that if counterguarantees were given to the foreign companies, then why should BHEL (Bharat Heavy Electricals Limited) and NTPC (National Thermal Power Corporation), the public sector undertakings who have to recover Rs. 1,000 and Rs. 2,000 crores respectively from SEBI, not be given similar treatment (*Financial Express* [Bombay], 21 October 1994). Interestingly, the US Undersecretary of State for South Asia, Robin Raphel, in her visit to India, repeatedly made it clear that without counterguarantees, US business would be reluctant to come to India.

The issue of discrimination against the PSU was also raised in the parliament and the press, who pointed out the case of the Oil and Natural Gas Commission (ONGC)–a public sector unit in the oil and gas sector. The government commissioned report "Hydrocarbon Perspective 2010" became controversial because of its distinct bias in favor of the private sector, particularly in asking for handing over the most profitable discoveries and proven blocks to foreign investors and recommending that the pipeline investment made by ONGC be taken as national assets to be used by the private sector as well. This was opposed by ONGC because it wanted to receive its spending on the pipeline. The commission did not oppose foreign participation but has been asking that the foreign partner also share part of the exploration cost. Indian policy in the oil sector was described as privatizing profit and nationalizing losses (*Economic Times* [New Delhi], 4 February 1995).

The privatization process in labor surplus economies such as India's cannot ignore the issue of employment. The public sector in 1991 provided jobs to 18 million people compared to 7 million in the organized private sector; however, if small-scale units are also added, then it is estimated at 24 million. The total of 42 million in the organized sector compares with 54 million in the nonorganized sector. These figures are for the nonagriculture sector. It is calculated that since the 1980s, the private sector has not been creating additional jobs despite increased investment. Therefore, with the privatization of the public sector, the prospect of additional employment generation cannot be rated high. According to one estimate, 500,000 to 800,000 jobs were lost since the reforms began in the organized sector (Parthasarthy, 1996: 1859–69).

The necessity to retrench labor was found in the logic of competition. It was argued that cost per unit of production was much less in the case of the nonorganized sector, so companies were increasingly using labor in the nonorganized market: according to Banerjee-Guha, Procter and Gamble argued that the cost of Rs. 122 per unit worker in the Kalwa plant to manufacture Vicks was too high compared to the per unit labor cost of Rs. 53 to make the same in the disorganized sector (Banerjee-Guha, 1996: L-23). The trade unions have been trying to oppose such moves, but in the context of reforms, they have not been very successful because the mainstream political parties are committed to reforms. Moreover, with the declining jobs in the organized sector, the trade unions have been losing strength.

Apparently liberalization has put the unions on the defensive. The growth of independent trade unions outside the umbrella of political trade unions can be attributed to emerging political culture under the market regime. Consumerism has made the labor force more materialistic than it was before, and the unions are satisfied with palliatives and donations from time to time. Management does not want to negate unions but rather create a form of business union (Sadri et al., 1994). The Indian corporate sector has been putting pressure on the state to redefine the labor laws by declaring the Exit Policy. The state has been experimenting with various methods. It has announced a safety net–the creation of the National Renewal Fund (NRF)–to provide funds for training and redeployment, compensation for employees who are affected by restructuring or closure of industrial units, both in the private and public sectors, and the provision of funds for employment generation schemes both in the organized and nonorganized sector. The voluntary retirement scheme (VRS) is one such device. In 1992–93, 38,531 workers were covered under the scheme, but their number has decreased to 30,953 in 1993–94. According to a study of the scheme in the Durgapur industrial area of West Bengal, although workers are willing to accept compensation in some cases, there are many instances of coercion from management to accept the deal, and the amount of VRs utilized primarily for nonproductive purposes, thus reducing the chances for retired workers to begin any self-employment venture (Khasnabis and Banerjea, 1996: L-70).

Protest of the Marginal Groups: Global Encounters Local

The public debate on reforms has been quite vibrant with regard to the consequences of liberalization on small producers. Small-scale production has been a vital component of the national economy not only as a source of employment but also as a process of dispersing economic activities in order to ensure maximum exploitation of latent resources both human and material. Given the potential of the sector to earn foreign exchange, the government has of late been encouraging the private sector, both Indian and foreign, to participate in traditional sectors such as the export of handicrafts, cottage goods, and marine resources. With the increasing participation of companies in these sectors, production has registered big jumps due to the new channels of marketing introduced there. As noted in the *Economic Times,* "Progress has affected all three areas–the product mix, the techniques used and the workers–their composition and source" (*Economic Times* [New Delhi], 5 March 1995). It has brought prosperity to the small towns where these units are located but most often by passing over the townspeople. For some, it also has led to uprooting from the ancestral home and moving to the city for white-collar jobs, and for others, it has posed the question of fighting for survival. The marine sector illustrates the second case.

India has a vast coastline. Aquaculture has a very rich potential for earning precious foreign exchange. Since one of the objectives of the reforms is to pile up the reserves no matter how they are accumulated, the Government of India has been keen to exploit marine resources to build up the reserves. It has prepared a blueprint for what it calls the "Blue Revolution," and the World Bank is supporting and financing it. The deep-sea fishing policy (DSF) announced by the government has provided liberal incentives in different forms to the Indian private sector, in collaboration with foreign companies, to exploit these resources. High profit and government incentive have led a large number of companies to enter the field, but environmentalists have argued that aquaculture could destroy ground water resources and agricultural land because of salinization, and waste from the farms pollutes the ocean and shoreline, leading to a decline in catch. Thus it could destroy the livelihood of local fishermen. Moreover, it was found that on the strength of superior technology, trawler owners have been taking over the locals' share: "the trawlers are not only violating the monsoon trawling or night trawling ban in the territorial waters, they now want to establish total control over the whole area" (Mukul, 1994: 475). In other words, a zero sum situation was emerging, creating conflicts between the local fishermen and the companies. Many fishermen who have borrowed from the banks have been facing serious crises in the absence of adequate returns. According to one account from a village called Anjengo, in Kerala, south India, "The trawlers and their men attack and destroy our boats and nets in the sea. Sometimes they capture our boats and nets. Nowadays we go fishing with fear and apprehension in our hearts and whenever we hear a sound or see the moving trawlers, we immediately rush back to the seashore" (Mukul, 1994: 475).

Though the reports of tension and conflict have been in the news, the situation acquired serious dimensions in November 1993. As the account goes,

> It was raining. We were fishing in our small boats with gillnets. We did not hear the sound of the trawlers which suddenly surrounded us and people attacked us with stones, knives and sticks. They captured our boats and destroyed our nets. Paniadima, Stephen and Vincent jumped into the sea to save their lives. But Antony and two other fisher folk were taken to Shaktikullangara and kept hostage by the trawler-owners. All of them were seriously beaten up and three were hospitalised. (Mukul, 1994: 475)

On 8 November, attempts were made to get them released with the help of the local authorities, but these failed. This incident has shaken the fishermen. On 28 November, fishermen, church leaders, and Kerala Swathantra Matsya Thozhilali Federation's activists gathered together to form a village-level association. In the perception of these activists, though sporadic conflict between the locals and the mechanized trawler had been taking place for sometime, it was the announcement of the new policy that had emboldened the companies: "When the government is aggressively pursuing the deep-sea fishing policy and starting many new joint ventures to exploit the fishes in the coastal areas, the battle in the sea is bound to aggravate" (Mukul, 1994: 475). The association called for a fishing *bandh* (strike) on February 1994 in all the coastal belts of the country and declared, "If the government is not enforcing the laws, we will make our laws and enforce them on our strength" (Mukul, 1994: 475). Alarmed by these developments, the government decided to form a fact-finding committee with a proviso to suggest remedial measures. The Murari Committee report provided new ground rules protecting the interests of the local fishermen.

Popular protest against the DuPont Nylon plant in Goa is another example of the people's struggle against the invasion of corporations on their locale. The multinational was given permission to start a synthetic nylon factory. Local activists found out about the polluting impact of the plant and decided to oppose it. The local village government was informed by local activists about the dangerous consequences. A meeting was called with the company in which elected members of the local government were invited. "Local opponents presented documents that showed that adipic acid and hexamethylene diamine (HMDA), the primary chemical used in the nylon 6,6 process, were classified as hazardous substances by the US government. The panchayat also raised the objection that the 250,000 liters of water required daily to run the factory would drastically lower the water table and make the local agriculture impossible" (Cohen and Sarangi, 1995: 563). All five panchayats voted against DuPont, and DuPont admitted that it had not taken adequate measures regarding four critical areas (ground water protection, waste water treatment, solid waste recycling, and air pollution control) and promised to take remedial measures. But by this time, popular opinion became hostile to the plant. Despite the promise of creating 650 jobs, of which 80% would be reserved for the locals, the plant could not get popular support. On the contrary, an opposition movement was launched that was peaceful in the beginning but became violent as the state decided to support the company. On 18 January 1995, protesters held a rally in front of the Goa assembly and demanded the project be scrapped. "They accused the prime minister of India and chief minister of Goa as public relation agents for American multinational" (Cohen and Sarangi, 1995: 563). According to one account, on the morning of 23 January, a busload of American experts from DuPont, accompanied by three police jeeps, were met by 70 protesters, mostly women and children sitting on the road leading to the factory. When the women refused to let the bus pass, the police advised the Americans to return to the state capital. Back in Panjim, DuPont officials met the chief minister and demanded more aggressive action.

By 4:30 P.M., two busloads of police returned to the scene and fired on the protesters, killing one. The villagers retaliated by burning police jeeps. A complete strike was called against the police action, and later, the factory premises were occupied by the opposition. Eventually, the company decided to leave and move the plant to another state of India but on much better terms.

These two episodes, though of extreme kinds, suggest that in a conflict between a company and the people, the state is likely to support the former. With the help of a civil society, the marginal section, the victims of the reforms, can protect their interests. Intervention by civil society can even improve the terms and conditions of contracts between the state and the multinational companies.

The Corporate Debate

Liberalization brought up a number of new issues for the corporate sector. The Indian corporate sector registered a significant jump in their profit and their turnover increased by 23.11% in the first half of 1994–95, compared to their performance during the first half of the previous year. The increase in profit after tax has been "explosive": 87.6% compared to 72% in the preceding year. Of the 100 top companies, the first ten had sales revenue up by 24.8% against the average of 19.6%. Even after adjustment for inflation (10.8% in 1993–94), these companies grew nearly five times as fast as the index of industrial production, which grew by 3%. The top ten also captured larger shares of the sales, suggesting the importance of scale (*Business Standard* [Calcutta], 7 March 1995). The survey also pointed out that Indian companies prefer diversification within the domestic market to going to overseas markets. Possibly these companies found it more easy to compete in the domestic market. The high capital cost and limited unit capacity have been additional factors deterring them from going overseas. Moreover, liberalization opened many new areas in the domestic market where returns were found to be fairly high. But a few companies did expand abroad, particularly in the pharmaceutical sector, where the low cost made them competitive. The *Business Standard* survey also pointed out that instead of competing with foreign firms, Indian corporations often ended up forging strategic alliances with them. This has happened across the board, ranging from fast-moving consumer goods to power, telecommunications, and financial services.

The process of merger, acquisition, strategic alliance, and hostile takeover of Indian companies made the corporate debate lively by bringing many new issues to the agenda. With a number of market leaders losing ground to the multinationals and facing threats to their survival, demands for leveling of the playing field gained prominence. According to the findings of the Indian Institute of Management, there has been an accumulation in merger and acquisition since liberalization and a dramatic increase, amounting to almost 50%, in the share of such mergers and acquisitions accounted for by transnational companies (*Industrial Growth: Trends and Production,* Calcutta: Indian Institute of Management, 1994). Thus, concern was expressed about protecting Indian companies against the threat of deindustrialization. Interestingly enough, the Indian prime minister showed concern about it as well. Speaking at the annual session of the Federation of Indian Chambers of Commerce and Industry (FICCI), the leading chamber of commerce, he recognized the need to contain the adverse impact of foreign companies on the economy and to preserve the core of Indian business in Indian hands. However, his emphasis was different when he addressed the centenary meeting of the Confederation of Indian Industries, another chamber representing the interests of foreign capital, where he promised to foreign companies to provide the most competitive environment possible and emphasized the irreversibility of the reform process.

Increasing domination of foreign brand names in the Indian market emerged as another leading concern in the debate. According to one survey, the top ten brand names and 20 of the top 30 are foreign-owned, and in a number of joint ventures, Indian partners were being forced to be in the minority. Some of the companies decided to join hands to restrict the takeover. In the field of color television production, BPL and Videocon together account for 44% of the market, and they have decided to bid jointly for the takeover of a public sector unit, Uptron. Their concern was not as much to preserve their share in the present market as it was to check the Japanese companies showing interest in the market. It should be pointed out that both these companies collaborate with foreign companies, but they were worried that the promising market might induce their collaborators to enhance their share and change the character of the company. It was observed that the premium of brand name and penetration pricing might lead to rearrangement of the market share if not an imminent shakeout. The new alliance among the Indian companies was not confined to the private sector alone. A few public sector units also took initiatives in this direction, even with the private sector. In the telecommunications sector, Tata Telecom (private sector) and Bharat Electronics (public sector) decided to explore the possibility of working together to combine the strengths of the two organizations in order to face the competition. Such combinations are being formed not only to face multinationals but also to lobby with the government. The state owned IPCL (Indian Petro Chemicals Limited) and Reliance, who enjoy a virtual monopoly in petrochemicals, have jointly put up the case against lowering the custom duty because it reduces their profit margin.

The takeover of Indian companies by foreign companies by means of market manipulation also became hot issues in the debate. In its 12th report, the Parliamentary Standing Committee on Industry has shown concern about this phenomenon. It has raised the issue of the price at which shares have been transferred to parent firms. Reportedly, one multinational firm acquired shares to raise its equity from 40% to 50% at the premium of Rs. 50 on a share of Rs. 10 while the market value of the share was Rs. 615. Similarly, a pharmaceutical firm managed to buy preferential shares at a price of Rs. 110 against the ruling market price exceeding Rs. 1,000 per share. Discounts have amounted to 80% in the case of a foreign engineering firm and 70% for a firm specializing in electrical and electronic products. Having picked up shares at discount prices in large numbers, the companies have taken over some of the leading Indian companies. The most startling illustration is the acquisition of the dominant Indian soft drink producing firm by an international giant, which has given the latter access to 60% of the market share.

The hostile takeover of Indian companies was sometimes projected as a battle between foreign and national capital. This could be seen in the Indian Tobacco Company (ITC) versus British-American Tobacco (BAT) case. British-American Tobacco decided to enhance its share in ITC, which ranks fifth in terms of turnover among the Indian top ten companies, to dictate its future production line. This was in contradiction to the plan of the chairman. When the difference between the two could not be resolved by negotiation, BAT indicated its desire to have a new man as the chairman. To defend his position, the ITC chairman projected the debate in terms of a takeover of an Indian multinational by a foreign interest. Some of the Indian dailies even compared the episode with the return of the East India Company. Though the Indian chairman could manage to save the situation with the help of financial institutions, the issue did add a new dimension to the debate. It was inevitable that with liberalization, foreign companies would like to enhance their presence in the market. The Indian corporate sector ought to develop its own strength by building up research and development to negotiate with the multinational corporations (MNCs) for better terms. Another view was that the state cannot remain a silent spectator to the jostling out of India companies. In fact, there was a perception among the industrialists, popularly described as the

"Bombay Club," that the government has been favoring the multinationals (*Financial Express* [Bombay], 2 January 1995).

The prospect of strengthening of oligopolistic tendencies in the Indian market has also been an issue. It cannot be denied that the protective environment promoted rent-seeking tendencies in the Indian economy. It was expected that with the opening up of the economy, the number of players would go up, correcting the distortions. Consequently, the market would be moving toward perfect competition. But the evidence during this period does not support this conclusion. In fact, with merger and acquisition, the number of players is going down. This came to the forefront when the Supreme Court handed down a verdict in favor of a merger between TOMCO (Tata Oil Mills Company) and HLL (Hindustan Lever Limited). The verdict raised the issue of the validity of such mergers. It was pointed out by the parliamentary committee that such mergers have led to a situation in which a subsidiary of a multinational corporation is going to control 70% of India's toilet market and 30% of its detergent market.

Sustaining the Reform Politically

Clearly, as India moves ahead with its reform agenda, it is going to face many new questions and issues. One major issue is going to be that of sustaining the reforms politically. The literature on reforms admits the necessity of a suitable political climate and a strategy of political management to implement them. As J. Waterbury puts it, even in authoritarian systems, structural adjustment programs require careful political management of the interests that contribute to the regime's coalition. Some resources have to be used for political purposes, and because resources are scarce, such utilization appears wasteful in terms of economic efficiency. However, a regime that alienates most or all of its coalition partners will not be able to pursue the adjustment process (Waterbury, 1989: 55). The slow pace of reform in many countries can be explained by the high hidden political cost, particularly when the regimes initiating reforms have been discredited for their failed performance.

In the case of India's reforms, they are to be implemented within the parameter of parliamentary democracy, and therefore, influences and pressures from diverse interest groups are going to impinge on the pace, content, and direction of reform. It should be recalled that the Nehruvian model India pursued after its independence contained varied societal interests, and its conception started long before independence. It provided a fine symbiosis between the economic interests of the classes and the political aspirations of the masses. The national vision of a socialistic pattern of society thus was not conceived to bring socialism to India but to create confidence among the masses for a system that was geared to the interests of the classes. As a confidence-building measure, the state periodically initiated a few popular measures, such as the poverty alleviation program. These equity-oriented measures were needed to legitimize the policies. The symbiosis, however, started feeling the pressure of erosion because the economy failed to grow to accommodate the quantitative and qualitative changes made by the demands of the society. While quantitative changes were the result of demography, qualitative changes were the consequence of a burgeoning middle class estimated to be about 200 million. The new middle class in the age of global communication became a critical factor in pointing out the limitations of the Nehruvian model. In addition, Indian polity witnessed the rise of a new political class, a product of vote-bank electoral politics, the "Mandalites" as described in the Indian political lexicon. This is a caste coalition of the lower castes who have been socially, politically, and economically marginal in the Indian power structure. However, since independence, these castes have gained political strength and the clout to play a critical role in the number game of electoral politics. Traditionally, they have been the vote bank of the Indian National Congress Party. Disillusioned by the Congress Party's performance and interest preferences, they have emerged as an autonomous political class that underlines the political erosion of the Nehruvian model and the disjunction between polity and economy.

The crisis of 1991 provided the ruling class a unique opportunity to discard the earlier model on legitimate grounds. But unlike the earlier model, which could synthesize diverse societal interests into a national vision, the new model has yet to develop a corresponding perspective. Rao's government did not care to sensitize different segments of society, nor did it try to articulate the neoliberal reforms in terms of a future vision. Rao couldn't solicit the necessary legitimacy or confidence to be returned to power; perhaps he was depending on the magic of trickle-down theory. However, in the 1996 election, none of the mainstream political parties contested Rao's reforms with an alternative economic agenda. The electoral verdict thus cannot be interpreted as a rejection of market reforms. The government that came to power is dominated by the Mandal class, which is not averse to consumerism. Consequently, the policy of the government on reforms is no different than that of Rao's government.

In contrast to the earlier model, in which the centralized state evolved the synergy between class interest and mass aspiration under a neoliberal model, the reforms have to be accomplished by a minimalist and disempowered state. It should be pointed out here that the state in the postcolonial societies is perceived as the arbiter, hence central in the conflict-resolution process. It is not a residual entity, as perceived in the West. Retreat of the state can thus pose the problem of credibility for the reforms. The fact that in these societies, an imperfect state is preferred to an imperfect market cannot be ignored in redefining state-market relations. In the context of the required adjustments, the state became at once the subject and object of change.

It is clear from the debate that the imperfect market lacks both the strength and the legitimacy to undertake the reforms on its own. The space vacated by the state cannot be taken by the market. This brings the third actor onto the stage–the civil society. The strength of the civil society lies in its very nature of being indigenous. This is not the place to elaborate on the role of civil society in market reforms; however, a distinction needs to be made between the sponsored civil society and the civil society that grows from the soil. It is important to note that in the era of reforms, a number of sponsored nongovernmental organizations (NGOs) have become visible in India. These organizations quite often distort rather than enrich the public debate.

From the preceding account of public debate and popular perception, it is clear that market reform in India cannot be carried out by ignoring the interests of the local economy and their impact on the social habitat. India is a parliamentary democracy; it is a society of many cultures, religions, and ethnic identities, and it is an economy that has yet to integrate 30% to 35% of its population into the mainstream. A successful reform strategy cannot be premised by ignoring these realities, and hence, the issues of poverty and inequality cannot be pushed to a later date, "the second generation reforms." These will have to be an integral part in each and every step of the process. Neoliberal reforms are not socialist reforms, but they have to be embedded into the society and articulated in terms of a national vision in which the local diversities, specificities, and distortions are not dismissed as marginal issues.

Further Reading

Bajaj, J. L., "Divesting State Ownership: A Tale of Two Companies," in *Economic and Political Weekly* 29, no. 35 (1994)

Banerjee-Guha, Swapana, "Dividing Space and Labor: Spatial Dynamics of Multinational Corporations," in *Economic and Political Weekly* 31, no. 24 (1996)

Bhagwati, Jagdish N., and T. N. Srinivasan, *Indian Economic Reform,* New Delhi: Associated Chambers of Commerce and Industry of India, 1993

A small monograph from the two leading economists that provides a brief appraisal of the reform process in India. This study was undertaken on behalf of the Ministry of Finance of the Government of India.

Cohen, Gary, and Satinath Sarangi, "People's Struggle against DuPont in Goa," in *Economic and Political Weekly* 30, no. 13 (1995)

Government of India, *Economic Survey 1994–95,* New Delhi: Government of India, Ministry of Finance, 1995

——, *Economic Survey 1995–96,* New Delhi: Government of India, Ministry of Finance, 1996

The *Economic Survey* is published by the Ministry of Finance every year before the budget is passed. It is the government's assessment of the health of the economy and is treated as the primary source of data.

Joshi, Vijay, and I. M. D. Little, *India: Macroeconomics and Political Economy 1964–1991,* Washington, D.C.: World Bank, 1994

This book covers the Indian economy from 1964 to 1991 in the framework of the new political economy, the years in which Indira Gandhi was shaping the politics and the economy of the country to the days of Rajiv Gandhi. Traces the genesis of the 1991 reforms to these years.

Khasnabis, Ratan, and Sudipti Banerjea, "Political Economy of Voluntary Retirement," in *Economic and Political Weekly* 31, no. 52 (1996)

Mukul, "Traditional Fisherfolk Fight New Economic Policy," in *Economic and Political Weekly* 29, no. 9 (1994)

Parthasarthy, G., "Unorganised Sector and Structural Adjustment," in *Economic and Political Weekly* 13, no. 28 (1996)

Sadri, S., et al., "Are Trade Unions in India Withering Away?" in *Indian Journal of Labour Economics* [New Delhi] 37, no. 1 (1994)

Discusses the fate of trade unions in the reform environment of India.

Waterbury, J., "The Political Management of Economic Adjustment and Reform," in *Fragile Coalitions: The Politics of Economic Adjustment,* edited by Joan M. Nelson et al., Oxford: Transaction Books, 1989

Concerns the political economy of economic reforms in developing countries. The volume as a whole draws on the experiences of many countries and touches on some of the fundamentals, such as the role of the state, limitations of the market, and the constraints of implementing reforms in developing societies.

World Bank, *Economic Developments in India: Achievements and Challenges,* Washington, D.C.: World Bank, 1995

Dr. Girijesh Pant is chairman of the Centre for West Asian and African Studies in the School of International Studies at Jawaharlal Nehru University, New Delhi.

International Political Economy

Chapter Fifteen

Changing Regimes in Technology Transfer and Intellectual Property in India

Sunil K. Sahu

Introduction

On 15 April 1994, India, along with 110 other nations, signed the Uruguay Round of multilateral trade negotiations of the General Agreement on Tariffs and Trade (GATT). The Uruguay Round results have created a new multilateral trading system and the World Trade Organization (WTO), which replaced GATT on 1 January 1995 with an elaborate institutional mechanism–similar to the ones in the United Nations and the World Bank–to oversee a comprehensive set of rules and disciplines in world commerce and act as the judge of economic globalization, including trade in agriculture, textiles, services, and intellectual property, in addition to the traditional area of merchandise trade covered under GATT. The new trade organization will have far-reaching effects on the trade policy of member countries, as it has brought an end to the era in which "the laws of international trade took place along a country's border, at its customs stations and its ports" (David E. Sanger, "Playing the Trade Card," in the *New York Times,* 17 February 1997, Y27).

By signing this treaty, India has committed itself to a new intellectual property regime that was controversial from the beginning, especially in its provisions related to patent protection. India's participation in this trade treaty will significantly affect the country's existing patent law (1970) and its technology regulation policy (1969). Moreover, this will have serious implications for the pattern of industrial and technological development in India and the role of the Indian private sector versus that of the foreign multinational corporations in India's future development.

In India, the reaction to the signing of the treaty was generally negative. The opposition parties and intellectuals charged the Rao government, which had signed the treaty in Marrakesh in April 1994 and subsequently ratified the new transitional legal framework through an ordinance, with "abject sellout and surrender of the country's sovereignty," and they threatened to carry the battle against GATT to the streets unless the government pulled out of this multilateral trade deal "fashioned by developed nations for boosting their sagging exports." In particular, the criticism has centered around its intellectual property accord, formally referred to as the trade-related intellectual property rights (TRIPs).

The vast body of literature on intellectual property, particularly patents, in developed countries suggests that increases in intellectual property protection generate research and development activity sufficient to offset the social cost of the limited monopoly granted to patentees, copyright holders, and other owners of intellectual property. However, similar research is lacking for developing countries (Siebeck, 1990). This chapter analyzes the impact of TRIPs on technology transfer and intellectual property rights in India by (1) evaluating the import of technology, the cost of technology transfer to India, India's science and technology policy, and research and development (R and D) activities, (2) analyzing the case of the Indian pharmaceutical industry, especially the effects of the 1970 patent law on the growth of the Indian sector and its technological capability, and (3) discussing the prospects of the pharmaceutical industry in the post–GATT regime.

Technology Transfer and Intellectual Property Rights: Why, What, and How

Technological change plays an important role in the development of a country. In the developing nations, there has been a shift in emphasis in the last two decades toward the role of technology in the development process. In economic terms, technology emerged as the key resource and input–even more important than capital and labor–for industrial growth and development. The ability to industrialize through the mere transfer of already existing technology from the developed to the developing nation, rather than independent invention, has thus been emphasized in the development efforts made by Third World nations. Therefore, the academic and policy debate has increasingly focused on the role of modern technology in the development process in the Third World (United Nations Industrial Development Organization, *Technological Self-Reliance of the Developing Countries: Towards Operational Strategy,* in Development and Transfer of Technology Series, no. 15, Vienna: United Nations, 1981: 5).

Technology transfer is the utilization of an existing technique in an instance where it has not previously been used. It refers to the sale or transfer of technology in its embodied forms (tools and machinery) as well as its disembodied forms (know-how and know-why of processes and products). The multinational corporations (MNCs) of the Western nations and Japan, the main suppliers of technology to the Third World, are responsible for 80% to 90% of such transfers. Though there are several modes of technology transfer–joint venture and licensing to unaffiliated companies, for example–the most important channel of such transfers has been foreign direct investment by MNCs.

While the salience of technology in the development process is widely recognized by the developing nations, they face a technology market that is typically oligopolistic, especially for the more recently developed technologies. Thus, there is a controversy regarding the issues of sources, ownership, appropriateness, and (especially) costs and benefits of technology to the developing nations. In the 1970s and early 1980s, developing nations vigorously argued–even demanded in the United Nations and other international forums–that the MNCs of the advanced industrialized nations should provide them the industrial technology they needed either free or at a nominal price because the marginal cost of such transfers was insignificant once a technology had been invented.

The MNCs, on the other hand, took a different view; they considered the technology–a proprietary knowledge–as an economic entity, not a free or social good, which was to be bought and sold internationally for the purpose of making profits. Faced with the efforts of a number of developing countries–notably Argentina, Brazil, India, and Mexico–to impose new regulations for improved control over the technology transfer process, the MNCs sought to restrict the free flow of technologies to the Third World by imposing patent obligations and intellectual property rights requirements (IPRs) on the developing countries.

It is important to note that the conventional thinking as reflected in the literature on IPRs in the pre-1970s period emphasized that developing countries "should be exempt from any international patent arrangement" (E. Penrose, *The Economics of International Patent Systems,* Baltimore: Johns Hopkins University Press, 1951: 233) as they gained nothing from granting foreign patents since they themselves did little patenting abroad and received nothing for the price they paid for the use of foreign inventions. The negative attitude toward IPRs grew stronger in developing countries in the 1970s, and they sought to establish, in the context of the North-South debate on the New International Economic Order (NIEO), a Code of Conduct on the Transfer of Technology. Though unsuccessful in their efforts, mainly because of the opposition of the United States (H. P. Kunz-Hallstein, "The United States Proposal for a GATT Agreement on Intellectual Property and the Paris Convention for the Protection of Industrial Property," in *Vanderbilt Journal of Transnational Law* 22, no. 2 [1989]), many developing countries enacted technology transfer and patent protection legislations in the 1970s.

By the mid-1980s, developing nations moderated their stand on technology transfer and intellectual property issues in response to the changing international environment–the shift in the pattern of world trade in the 1980s and 1990s (exports of North and South, with the exception of African countries, largely consisted of similar goods) and the resultant adjustment challenges; the recognition of the importance of knowledge-based, high technology industries for developing countries; and the fear of the use of trade laws by developed nations, in particular, the Special 301 provision of the US 1988 Trade Acts. Furthermore, developed nations, especially the United States, made trade concessions dependent on intellectual property protection. The developing countries, therefore, responded by bringing about a significant change in their intellectual property law after the mid-1980s. Mexico, for example, enacted new regulations in January 1990 that weakened its previous legislations (of 1972 and 1982) on technology transfer, trademarks, and patents (United Nations, *Intellectual Property Rights and Foreign Direct Investment,* New York: United Nations, 1993: 11).

The TRIPs Agreement

By the late 1980s, leading industrial nations began to view intellectual property as a new basis of comparative advantage as they recognized that there had been a gradual erosion in their traditional areas of production, especially in the manufacturing industry in the newly industrializing countries (NICs). The large multinational firms in the developed countries–for example, US corporations such as IBM, Pfizer, and Microsoft, which had large intellectual property portfolios–were successful in convincing their law makers and political leaders (and it coincided with the views expressed by many economists) that with the growth of intangible goods in international trade, intellectual property, which can be copied and transmitted across national boundaries with relative ease, will increase the trade-distorting effects of counterfeit products. They argued that the presence of such products, estimated to cost firms from industrialized countries about US$60 billion annually, was a potential obstacle to further trade liberalization (see United Nations Publications, no. E.90.II.A15).

This concern was reinforced by the growing awareness of the loss of commercial opportunities in many industries in developed nations–computer software and microelectronics, entertainment, chemicals, pharmaceuticals, and biotechnology, for example. Such concerns led industrial nations, especially the United States, to insist on the inclusion of IPR protection in the Uruguay Round trade negotiations. Though intellectual property featured almost as a footnote on a crowded agenda at the start of the Uruguay Round in December 1986, eight years later, TRIPs emerged as one of the major breakthroughs of the GATT negotiations. Because of the substantive and procedural protection given under the treaty, the TRIPs agreement, which sets intellectual property standards in the areas of copyrights, geographic indications, industrial designs, layouts of integrated circuits, trade secrets, and patents, is the most important international agreement on intellectual property in this century.

The TRIPs agreement integrates the existing systems of IPR protection as provided under the Berne Convention, the Paris Convention, and the International Convention for the Protection of Performances, Producers of Phonograms, and Broadcasting Organizations and compliments them in those areas where there was no international consensus. Among other things, the treaty

> Grants all intellectual privileges that exist for nationals to foreign members–the concept of national treatment or reciprocity (art. 3);
>
> Applies the concept of "most-favored nation" to the area of intellectual property rights (art. 4);
>
> Prevents discrimination in the international transfer of technology (art. 8);
>
> Requires all members to grant patents for any inventions, whether a product or process, as long as the innovation is an "inventive step" and is capable of "industrial application" (the term of patent protection is 20 years from the filing date); and
>
> Requires that the signatory countries make "fair and equitable" enforcement procedures under their national law (Chaudry and Walsh, 1995: 84).

These and other provisions of TRIPs are designed to set the substantive standards to which every member country is required to adhere regarding copyrights, trademarks, geographical indications, industrial designs, and integrated circuits. Trade-related intellectual property rights have thus brought about a fundamental shift in the international intellectual property regime "away from the country-based approach to the multilateral approach" (McDorman, 1994: 120). This norm-creating aspect of the TRIPs agreement is fundamentally different than the objective of the existing international regime on intellectual property. Therefore, developing nations were opposed to the accord, but they grudgingly accepted it due to the fear that the consequences of refraining from acceding to the new GATT treaty would be far worse. In particular, they feared the US trade retaliations invoking the Special 301 provision of the 1988 Trade Act against countries refusing to obey the GATT codes on IPR. It was also, in part, a response to the perceived fear that the United States would suspend the Generalized System of Preferences (GSP) privileges–an important consideration for many developing nations. (For example, about 14% of India's exports to the United States were accorded GSP status in the 1990s.) It is not surprising, therefore, that developed nations succeeded in extracting many concessions from the developing nations leading to the signing of the Final Act of GATT's Uruguay Round.

Technology Transfer and the Indian Experience

The four-and-a-half decades of planned development in India have brought about a substantial diversification in its industrial base, with the consequence that it can now produce a wide range of industrial products. Over the years, India has achieved almost total self-reliance in basic and capital goods industries, which now account for about half of the total value added in manufacturing. In various sectors–such as mining, irrigation, power, transport, and communication–indigenous capacities have already been established to the point of "virtual self-sufficiency so that further expansion can be based primarily on indigenous equipment" (Government of India, 1985: 167). India's growing technological capability has enabled it to be the leading exporter of technologies among the developing nations (Lall, 1982). Yet the government is concerned that "in large areas of economic activities, relatively obsolete, cost-ineffective technology continues to be applied, the pace of scientific and technological innovation remains unimpressive, and the adoption of the available scientific and technological knowledge is tardy" (Government of India, 1985: 319).

India's acquisition of a high degree of technological capability has been accompanied by technological obsolescence and inefficiency in many of its industries. This paradoxical situation can be explained by analyzing India's record on technology imports, the cost of technology imports, science and technology policy, and its research and development (R and D) efforts.

Technology Imports

It is generally true that India has achieved a greater degree of self-reliance in the manufacture of a wide range of consumer goods through a government policy that blocked foreign technology in many industries crucial to economic development, but its industrialization strategy of developing an integrated industrial structure in basic and heavy industry has resulted in importing technology since the beginning of the Second Five-Year Plan in 1955–56. India's adoption of a strategy of import-substitution–to protect the domestic industry, particularly in consumer goods–further led to its dependence on foreign technology. Its industrial strategy involved extensive collaboration in order to import technology for the metallurgical, capital goods, transport, consumer durable, agricultural exports, engineering, and pharmaceutical industries.

Even though the Indian government has followed a strict policy toward multinational corporations, especially from the late 1960s until the early 1980s, the number of foreign collaboration agreements approved by the government is quite large. As Table 15.1 indicates, between 1957 and 1977, the government approved 4,932 foreign agreements (both licensing and direct investments), of which the public sector accounted for approximately 5% and the private sector 95%. The number of foreign collaborations approved between 1974 and 1984 totaled 4,667. After the liberalization of India's import policy under Prime Minister Rajiv Gandhi, the rate of approval reached 740 in 1984 and 1,024 in 1985, with a moderate decline between 1986 and 1989. The average number of foreign collaborations approved per year thus went up from 270 in the 1970s to 660 in the early 1980s. The rate of approvals was particularly high during the second and third plan periods–the years of the "big push" to industry–and in the 1980s when a new thrust was given to industry. However, when industrial growth decelerated from 1966 to 1982 (see Isher Judge Ahluwalia, *Industrial Growth in India,* Delhi: Oxford University Press, 1985), there was a substantial drop in the number of technical collaborations approved. During 1985–90, the total number of new agreements approved was 5,203 (Bhagavan, 1995: M2). Therefore, the number of technical collaboration agreements entered into by India has been viewed by some analysts as a vital sign of growth and industrialization rather than as a depressor of economic growth (Nayar, 1983: 2:107).

India's technical collaboration agreements are heavily concentrated in technologically intensive fields of manufacturing such as capital goods and advanced intermediates (chemicals). In a detailed study of India's technical collaboration agreements, Balasubramanyam has shown that of all the agreements approved by the government between 1957 and 1970, more than 50% were related to electrical and nonelectrical machinery, machine tools, and transport equipment (Balasubramanyam, 1984: 150). If we take into account all the foreign collaboration agreements (6,959) approved by the government between 1957 and 1982, the industrial distribution shows a similar pattern: industrial machinery and machine tools accounted for 40%, electrical equipment and electronics 19%, chemical and pharmaceutical equipment 15%, transport and construction equipment 9%, and technical consultancy only 1% (Lall, 1985: 50). In the industry distribution, the Reserve Bank of India (RBI) survey for 1977–78 to 1980–81 also found that "machinery and machine tools alone accounted for 45% of the companies with purely technical collaboration agreements" (*Foreign Collaboration in Indian Industries: Fourth Survey Report,* Bombay: Reserve Bank of India, 1985: 129–30).

Since the list of government approvals includes renewals, amendments, and agreements that have already expired, the total number of agreements approved by the government, while it is indicative of the growth trend in foreign technical collaboration, may be misleading because it does not give a correct account of the number of agreements in force. Moreover, the list does not tell us the number of approved agreements that were never taken up. In a survey of 1,815 collaboration agreements approved by the government between 1975 and 1981, it was found that 63% of them were never implemented (Indian Investment Center, *Changing Forms of Foreign Investment in India,* New Delhi: Indian Investment Center, 1982: 24). "If this could be generalized to the universe of agreements," says Sanjaya Lall, "only some 2,600 agreements (including those with equity investment) materialized in 1957–82, and about 1,200 pure technical ones in 1969–82" (Lall, 1985: 51).

It is also important to distinguish between different types of technology transfer agreements, especially direct foreign investment (DFI) versus pure technical collaboration agreements. The RBI surveys, which attempt to study the agreements in force rather than those approved, suggest that in the 1960s, there was a higher proportion of agreements by foreign subsidiaries and foreign minority companies than by Indian companies. A study by Balasubramanyam confirms that "the will to transfer technology on the part of the foreign firms was much less in the case of technical collaboration agreements than in the case of direct investment and joint ventures (Balasubramanyam, 1973: 129).

While India encouraged technology transfer through DFI until the mid-1960s (partly because of the foreign exchange crisis of the late 1950s), in the late 1960s, it opted for technology import via licensing. Though the government standardized the procedure of technology import in the early 1950s and gave approval for technology imports, specific policies on technology transfer were formulated between 1965 and 1968. The technology import policy included five important features:

1. Royalty ceilings were prescribed for various industries.
2. The standard permitted duration of agreement was reduced from ten to five years. Renewals were not allowed unless they involved more advanced or different technology.
3. The only export restrictions that were permitted were to countries where the technology exporter had had subsidiaries, affiliates, or licensees.
4. The use of a technology supplier's trade mark was not allowed in India.
5. No restrictions were allowed on the technology importer's right to sell or sublicense the technology (Ashok V. Desai, "Indigenous and Foreign Determinants of Technological Change in Indian Industry," in *Economic and Political Weekly* special no. [1985]: 2,086, and National Council of Applied Economic Research, *Foreign Technology and Investment,* New Delhi: National Council of Applied Economic Research, 1971).

The government thus tried to regulate the transfer of technology in order to keep the cost of technology import down. India was the first developing nation that took restrictive measures to discourage technology imports via private capital investment or minority foreign capital participation. Enos has estimated that between 1969 and 1979, DFI and joint ventures accounted for only 14.7% of India's technology acquisitions (Enos, 1989).

The implementation of government policies brought significant results: the duration of agreements registered a sharp decline, and the share of outright sales of technology increased sharply. In his study of technological change in Indian industry, Desai has found that whereas 36% of the agreements in 1951–67 were for ten years, in 1977–80, ten-year agreements had come down to 3%, and the "share of outright sales of technology rose from 13 to 29 percent" (Desai, "Indigenous and Foreign Determinants of Technological Change in Indian Industry," in *Economic and Political Weekly* special no. [1985]: 2,086).

The RBI data for the 1970s, presented in Table 15.2, reveal that in the 1970s, there was a substantial increase of more than 20% over the previous decade in the number of purely technical collaboration agreements signed between Indian public and private firms and technology-supplying companies. In the wake of India's liberalization of the economy in the 1980s, foreign equity participation remained low. In 1985–86, for example, of the 133 companies that entered the capital market to set up various projects, only 16 (12.7%) had foreign equity participation (*Economic Times,* 6 November 1986).

Between 1969 and 1982, India had, in fact, approved equity flows of only US$80 million–half of which did not materialize. A total of US$40 million in direct foreign investment in 14 years compares favorably with the record of other newly industrializing countries in the 1970s: a net inflow of US$14 billion in Brazil, US$7 billion in Mexico, US$648 million in South Korea, and US$1.5 billion in Argentina (Lall, 1985: 49). Another measure of comparison is the stock of direct foreign investment as a percentage of the gross national product (GNP) and stock of the gross domestic investment (GDI) in a country. In 1977–78, these stocks were comparatively low in India–2.1% and 1.1% respectively–as against 4.8% and 2.6% in Argentina, 6.6% and 4.2% in Brazil, 5.6% and 3.3% in Mexico, and 3.1% and 2.4% in South Korea (Federation of Indian Chambers of Commerce and Industry, "Some Questions Related to Import of Technology and Domestic R and D," New Delhi: Federation of Indian Chambers of Commerce and Industry, 1986). Thus, compared to other NICs, India's imports of technology have indeed been small. However, the technology import situation underwent significant change in the wake of the policy of economic liberalization adopted by the Rao government in 1991.

Cost of Technology Imports

There is a vast body of theoretical and empirical literature on multinationals as exporters of technology to developing nations through DFI. Most studies suggest that multinationals prefer this mode of transfer. In the literature on technology transfer, a distinction is made between the direct and indirect cost of technology import. The direct cost usually consists of royalties, technical know-how fees, profits, dividends, and interest payments–whether technology is transferred through DFI or through a purely technical collaboration agreement. Indirect cost refers to the phenomenon of import dependence and the implications of technology import on balance of payments, industrial structure, equity, and technological environment. Though it is possible to quantify the direct cost, the issue of indirect cost does not easily lend itself to quantitative verifications. A quantitative assessment of real costs is thus difficult to make, especially at the national level.

The data on private sector companies' remittances on account of profits, dividends, royalties, technical fees, and interest payments between 1956–57 and 1981–82 are presented in Table 15.3. The total remittances by the Indian companies show a steady, gradual increase from Rs. 29.7 in 1957–58 to Rs. 398.85 in 1981–82 (for recent years, the figures are even higher). The drastic increase in technical fees from Rs. 43.97 in 1979–80 to Rs. 270.70 in 1981–82 is due mainly to the sophisticated technology acquired by Indian companies in recent years. Mature technologies are more costly to obtain. Since the royalty rate was low after the mid-1960s–1.8% to 3.0% after tax–the technology suppliers started charging higher technical fees (Subrahmanian, 1986). However, an empirical study of 211 technology-importing firms, using the data collected by the National Council of Applied Economic Research (NCAER), has found that in a large majority of collaborations, there was very little increase in lump-sum payments between the late 1970s and early 1980s. Ghayur Alam has found that "while the average lump-sum payments made by 615 collaborations during 1977–79 was Rs. 16.3 lakhs, during 1980–83 it had increased to Rs. 17.46 lakhs" (Alam, 1985: 2,075). His study reveals that out of 1,459 agreements, only 73 had involved payments of Rs. 1 crore each, which accounted for 40% of the total outgo. It can therefore be hypothesized that the large increase in lump-sum payments in the early 1980s (shown as fees in column 5, Table 15.3) was due to the import liberalization policies of the early 1980s. Import liberalization has resulted in a "few, but well publicized instances, where technology payments have been larger than those in the past" (Alam, 1985: 2,075). Furthermore, the table suggests a significant decline in the remittance of profits by foreign firms in the late 1970s and early 1980s, due mainly to the dilution of equity by subsidiaries in response to the Foreign Exchange Regulation Act (FERA) of 1973.

India's direct cost of acquiring technology when compared to other NICs appears to be rather low. Lall has made a comparison of foreign licensed technology in India (based on the data collected from the 433 largest private sector and 203 largest public sector companies) with that in Korea, Brazil, and Mexico (based on the data from the *World Development Report,* 1979). He has calculated, after deflating licensing payments by manufacturing value added, that in 1979, Mexico had the highest dependence on foreign licensed technology (2.7%), followed by Brazil (1.9%), South Korea (1.1%), and India (0.7%–0.8%).

The contribution of imported technology to production value has been found to be fairly high in India (Lall, 1985). According to a recent study, whereas India remits 0.32% of its GNP toward the import of foreign technologies, the contribution of foreign technologies to the production value is about 8.3% of the GNP (V. Govindarajulu, "India's S and T Capability," in *Economic and Political Weekly* 25, nos. 7 and 8 [1990]: M39–40).

In the import of capital goods, India was also found to be least dependent among the NICs: while India's import was US$1.6 billion in 1978–79, for Brazil it was US$4.4 billion (1980), for Mexico US$3.5 billion (1980), and for South Korea US$5.3 billion (1979)(Lall, 1985: 52–53). However, after the adoption of the new economic policy in 1991, the Rao government brought about significant policy change in many areas. For example, the ceiling on lump-sum payments was raised to Rs. 1 crore, royalties for domestic sales and exports were fixed at 5% and 8% respectively, and the collaboration duration was extended to ten years. But there is a paucity of studies analyzing the impact of such policy change on the Indian economy due to the absence of aggregate data on the subject in the postliberalization period.

Science and Technology Policy

The importance of science and technology in national development was recognized by the nationalist leaders as early as 1939, when the Indian National Congress appointed a National Planning Committee chaired by Jawaharlal Nehru. It was emphasized in India's First Five-Year Plan (1951–56), which devoted a whole chapter to science and technology, stressing the importance of scientific and industrial research in India's development.

The emphasis of the first plan was on mastering the latest technology to increase productivity. For Nehru, the role of science in India's development was crucial. He talked of promoting a "scientific temper"; that is, the popularization of a scientific outlook among the people. A Ministry of Scientific Research and Natural Resources was created (Nehru kept the portfolio to himself, which added to the importance of the ministry), and the government gave priority to the establishment of an infrastructure for research and development.

This infrastructure consisted of a chain of national laboratories formed by the Council of Scientific and Industrial Research (CSIR, formed in 1942); establishment of the department of Atomic Energy, the Department of Science and Technology, and the Department of Scientific and Industrial Research, as well as expansion of the science and technology departments of the universities; the creation of the Indian Institutes of Technology (IIT); and the provision of adequate resources to meet the requirements of these organizations (Rahman, 1980).

Though the Second Five-Year Plan (1956–61), which outlined the strategy of developing basic and heavy industry in the public sector, also emphasized scientific research, the Scientific Policy Resolution (passed by the parliament in 1958) recognized the need for the creation of science and technology manpower and emphasized the training of scientific personnel. The institutions established in the ensuing years to train science and technology manpower have produced impressive results. In about 25 years, India has succeeded in creating the third largest pool of technically skilled manpower in the world, next only to the United States and Russia: in 1982–83, India had 322,000 graduate engineers, 464,000 diploma holders in engineering, and 2 million science graduates and post-graduates–a total of 2.8 million technically qualified people, not counting medical and agricultural graduates (Government of India, *Draft Fifth Five-Year Plan 1974–79,* New Delhi: Government of India, 1974; Long, 1979).

It should, however, be noted that the emphasis in India's science and technology policy was on the creation of a scientific infrastructure. The *Draft Fifth Five-Year Plan* (1974–79) recognized for the first time the weakness in India's approach to science policy and pointed out that the earlier plans had lacked a technology policy. Though the five-year plans had underscored the importance of science and technology in India's development, in none of the plans was the relationship between science and technology clearly brought out.

The fifth plan had thus identified the central problem in India's science and technology policy–the country's inability to apply scientific knowledge successfully to increased productivity in the industrial sector. India's first Science and Technology Plan (1974–79) drew attention to the socio-economic and political aspects of the import of technology. The absence of integration between development planning and technology planning was found to be the main shortcoming of the science and technology policy in India (see V. V. Bhatt, *Development Perspectives: Problems, Strategy, and Policies,* New York: Pergamon Press, 1980: chapter 6). Accordingly, the government announced the Technology Policy Statement (TPS) in January 1983.

The basic objectives of the TPS are the attainment of technological self-reliance, the development of indigenous technology, and the efficient absorption and adaptation of imported technology appropriate to national priorities and resources (*Economic Times,* 4 January 1983). The statement carefully made a distinction between self-reliance and self-sufficiency. It maintained that India will continue to import know-how selectively, taking into account its needs, capabilities, and the time required to develop indigenous technology. Though the TPS did not lay down modalities for its implementation, the government has since taken steps to formulate a simplified, liberalized scheme to encourage technology imports for modernizing identified thrust industries. For the first time, the government addressed the problems of inefficiency, technological obsolescence, and the lack of international competitiveness in the industrial sector.

The specific measures taken by the government include (1) selective enhancement of royalty rates and lump-sum payments, (2) modification of rules with regard to the period of agreement, (3) enhancement of limits of imports made for promoting technological upgrading and modernization under the Technical Development Fund, (4) import of select machinery and equipment "not domestically available" under Open General License (OGL)–that is, import without specific license–"for creation of new capacities in the export sector as well as to upgrade technology and improve the quality of products," (5) fiscal incentives to encourage in-house R and D by the private sector firms, and (6) delicensing of several industries.

The government made a few significant changes in its policies following the announcement of the TPS. Recognizing that the excessive regulation and control of the economy–popularly known as the "permit-quota raj"–throughout the 1960s and 1970s had resulted in slow growth in the industrial sector (Ahluwalia, 1985; Bardhan, 1984), technological lagging, and industrial products not being internationally competitive, the government decided to introduce a series of measures of trade liberalization. The liberalization of the import regime, which began in the early 1980s, gave attention particularly to the import of technology. The strategy of a liberal technology import was based on the belief that if government simplified rules and procedures and liberalized inputs of foreign exchange to the Indian private sector for acquisition of imports of know-how, designs, consultancy, capital goods, etc., it would upgrade Indian products to international standards.

A stable, long-term import-export policy was adopted in 1985 when the government decided to formulate import and export policy every three years instead of annually. The entry of foreign technology was, however, made easy in the 1991 New Industrial Policy, which gave automatic approval for technology agreements and foreign investments (and foreign equity participation up to 51%) for most industries. Therefore, India has substantially strengthened its trade liberalization policy in the 1990s.

Research and Development (R and D)

India has recognized the importance of in-house R and D for the absorption of imported technology and for the development of new processes and products. To encourage the setting up of R and D centers recognized by the Department of Science and Technology (DST), the government started a scheme in 1974 that provides a number of fiscal incentives–for example, R and D expenditures being 100% income tax deductible. Such incentives have had a positive effect on the growth of R and D units in the private sector: the number of industries with recognized R and D centers has risen from 106 in 1973 to 930 in 1986. (Of these 930, 89 belonged to the public sector.) There has been a steady increase in R and D manpower employed by in-house R and D units. In 1975–76, about 13,000 R and D personnel were employed by about 400 units. During 1981–82, the number increased to over 41,000 for about 750 units, and by 1986 the manpower for 930 units had reached 45,000 (*Economics Times,* 25 February 1987).

India's expenditure on R and D and related science and technology activities has grown rapidly since 1951, from the annual average of Rs. 4 crores during the First Five-Year Plan period to Rs. 1,507 crores during the Seventh Five-Year Plan period to the expected annual average of Rs. 4,000 crores in the Eighth Five-Year Plan period.

The national expenditure on science and technology activities has increased from Rs. 4.68 crores in 1951 to Rs. 173.37 crores in 1971 to Rs. 1,003.45 crores in 1981 to Rs. 3,303.55 crores in 1988–which represents 0.02%, 0.47%, 0.66%, and 1.10%, respectively, of the gross national product (GNP). India's allocation of resources to R and D and science and technology activities, measured in terms of the percentage of the GNP, is higher than other NICs, with the exception of South Korea, and is comparable to many developed countries, such as Canada, Australia, and Italy (Patel, 1989). The difference in per capita R and D expenditure between India and most industrialized countries, however, is enormous. For example, India's US$2.78 per capita R and D expenditure in 1984 was 57 times less than the average of US$159 per capita for developed countries (see V. Govindarajulu, "India's S and T Capability," in *Economic and Political Weekly* 25, nos. 7 and 8 [1990]: M37–38, and Lester C. Thurow, "Maintaining Technological Leadership in a World Economy," in *MRS Bulletin* 14, no. 4 [1989]: 43–48]. Furthermore, India spent only 23% of its total R and D resources on industrial R and D compared to 50% or more spent on nuclear energy, space, and defense research. In 1982–83, for example, India spent Rs. 286 crores out of Rs. 1,237 crores on industrial research, which was a mere 0.63% of the sales turnover. The level of expenditure has also differed significantly between public and private sector companies. According to the RBI Fourth Foreign Collaboration Survey, the average annual difference in R and D expenditure between the private and public sector companies was 1 to 4.

A number of empirical studies on technology transfer have suggested that the companies that did their own R and D got a better return on their technology import. According to one study, the companies that did their own R and D "unpackaged their technology requirements and imported only those components they could not generate economically or fast enough, they were better informed about the technology market before entering it as buyers, and they received greater benefit from technology imports in terms of their own product and process development" (Desai, 1980: 75). The government has recognized the importance of in-house R and D in accelerating meaningful absorption and adaptation of imported technology in the private sector and has sought to bring production and R and D sectors in the country closer. Accordingly, in August 1986, it announced four stringent conditions for foreign collaboration agreements for technology payments of over Rs. 2 crores. First, it is now obligatory for Indian entrepreneurs wishing foreign collaboration to involve competent R and D personnel from within the enterprise (or from any other competent R and D institution in the relevant area) in the process of technology acquisition from the negotiating stage. Second, the Indian entrepreneur will have to submit a time-bound program for technology absorption/adaptation/ improvement (TAAI) within six months of the issue of the foreign collaboration approval. Third, it is compulsory for technology units registered with the Department of Scientific and Industrial Research to set up in-house R and D facilities or to enter into long-term consultancy agreement with any relevant R and D institution in the country within two years of receiving the foreign collaboration approval letter. Finally, the collaboration agreement to be executed by the Indian party and the foreign collaborator will not deny an indigenous R and D institution, identified for the purpose of examining the TAAI plans, any access to the production unit of the Indian enterprise (*Economic Times,* 29 August 1986).

The government unfortunately has not strictly adhered to these conditions in technology transfer negotiations. The data on R and D in industry available from the Ministry of Science and Technology suggest a general tendency toward stagnation in R and D efforts of the private sector in the 1980s: "R and D expenditure as a percentage of sales turnover of firms in the private sector as a whole has declined on an average from 0.78 per annum during the period 1975–76 to 1979–80 to 0.68 in the period 1980–81 to 1986–87 (Subramanian, 1993: 17). However, there seems to be some agreement among policy makers and academic analysts that India should raise its R and D- and science and technology-related expenditure to as high as 2% of the GNP. A higher allocation of resources to R and D would indeed modernize India's industrial sector and make its products internationally competitive in a global economy.

The Indian Patent Regime and the Pharmaceutical Industry

Patent protection in most developed countries is granted for periods of 16 to 20 years–16 years in Great Britain and 17 years in the United States. In the pharmaceutical field, however, a distinction is made between process patents and product patents; whereas a product patent is for a newly invented end product, a process patent protects a new method of manufacturing an existing product. While the United States and Great Britain permit both, Switzerland and Canada allowed only process patents until 1995. Italy, from 1939 until 1978, permitted no pharmaceutical product or process patents at all. It is important to note that in most countries with strong patent systems, the law either omits compulsory licensing, as in the United States, or provides for licensing only under extreme conditions.

Patent laws in the Third World have undergone significant changes in the 1960s and 1970s. The drug industry is considered vital in maintaining the health of a nation, but there has been a virtual monopoly of multinationals over this industry. In order to remedy this situation, many Third World countries–notably Brazil, Mexico, Colombia, and India–sought to revise their patent laws in the late 1960s and early 1970s. Consequently, the patent protection in these countries has been very weak until the passage of the Uruguay Round.

The Indian Patent Act, drawn up by a joint committee of the two houses of the

Indian parliament in 1970, was an important step taken by the government to break the monopoly of drug multinationals. The act, which replaced the old Indian Patents and Design Act of 1911, lowered the period of validity of patents in general from 16 to 14 years (and of patents in the field of food, drugs, and medicines to a period of seven years) and raised the scales of fees payable for renewing the patent. An important feature of the new law is the provision that grants patent protection only to processes and not products. The rationale was that product patents allowed companies to gain a monopoly market by combining drugs and chemicals in different formulations. Indeed, the legislation, inspired by the reports of various enquiry committees, diluted patent rights in an effort to promote local development work, process research, and manufacture. Other important provisions of the act were

1. To broaden considerably the grounds for the issue of compulsory licensing
2. To give the controller wide powers in determining the terms of settlement
3. To allow the unrestricted use of patented inventions by the government for its own purposes
4. To provide for the automatic endorsement of patents in the field of foods, drugs, and medicines and patents for the methods or processes for the manufacture or production of chemical substances with the words "licenses of rights" after a period of three years from the date of sealing of the patent (see Amiya Kumar Bagchi et al., "India Patents Act and Its Relation to Technological Development in India," in *Economic and Political Weekly* 19, no. 7 [1984]: 287–302, and S. Vedaraman, "The New Indian Patent Law," in *IDMA Bulletin* 10, no. 8 [1984]: 115–24).

In addition, the law made mandatory a worldwide search of patent literature to establish the novelty of a product or process; in the past, patents had been granted for processes outdated elsewhere. The 1970 law therefore provides a very weak patent protection in India.

The provisions of the act were thus far-reaching and reflected the thinking of policy makers in the late 1960s, particularly Prime Minister Indira Gandhi. Addressing the 34th World Health Assembly in Geneva in May 1982, she reiterated the stand she had taken more than a decade before: "The idea of a better-ordered world is one in which medical discoveries will be free of patents and there will be no profiteering from life and death" (*IDMA Bulletin* 15, no. 25 [1984]: 391).

The West, particularly the United States, expressed its displeasure with the Indian law from the time it was enacted. During the Reagan and Bush administrations, the United States demanded that the Indian government strengthen its patent protection laws. In May 1989, the Bush administration retaliated by excluding India from the trade "priority countries" listed under the Super 301 provision of the 1988 Omnibus Trade and Competitive Act (*New York Times,* 2 June 1989). In the 1980s, the government of India took the stand, in periodic policy statements, that it did not want to revise the 1970 patent law. However, the government's policy changed under Prime Minister Rao when his government adopted the policy of economic liberalization in 1991 and later signed, despite popular opposition, the GATT accord in 1994.

Though the analysts and policy makers do not agree on the pros and cons of the Patent Act, and the debate still continues, it is a widely accepted view that its enactment has contributed significantly to the growth of the national sector–both public and private–in the Indian pharmaceutical industry. Dr. Y. K. Hamied, the managing director of CIPLA, a leading Indian company, has described it as "the most significant milestone in [India's] aim to achieve self-sufficiency and self-reliance in the drug industry." Indeed, the decade following the implementation of the Patent Act in 1972–73 witnessed unprecedented progress and development of bulk drug manufacture by both the private (Indian) and public sector drug companies. Using indigenously developed technology, the national sector of the industry began to produce a number of bulk drugs and their formulations, such as ampicillin, amoxycillin, erythromycin, ethambutol, metronidazole, propranolol, and trimethoprim. (In India, unlike some other countries–China, for example–burden of proof of infringement is on the patentee, even for a process patent; this is practically impossible to prove in the case of imported items.) According to Hamied, by 1984, the contribution of the national sector of the industry had reached 65% of the drug formulations and 83% of the bulk drug production in the country, and this sector also contributed to over 65% of the exports of drugs and pharmaceuticals from India. In 1983–84, this sector produced Rs. 295 crores' worth of bulk drugs as against Rs. 60 crores by the foreign sector and Rs. 1,000 crores' worth of formulations as against Rs. 760 crores by the foreign sector.

The growth of the national sector was further accelerated by the new drug policy (NDP) announced by the Janata government in March 1978. The NDP, which was based primarily on the recommendations of the Hathi Committee Report, divided drugs into three groups for purposes of reserving items for production by various sectors. Whereas the production of 17 essential drugs was reserved for the public sector and the production of 27 items was reserved for the Indian sector–public and private–64 items were open for licensing to all sectors, including the foreign sector.

The NDP imposed restrictions on the growth and expansion of FERA companies. Those firms not manufacturing bulk drugs and those producing "low technology drugs" were required to bring down their foreign equity holdings to 40%; however, those foreign companies producing "high technology" drugs were allowed to retain foreign equity in excess of 40% subject to dilution formulas linked to expansion projects. In the case of product-mix, the FERA companies were required to maintain a ratio of 1 to 5 in the production of bulk drugs to formulations, whereas the ratio for Indian companies was 1 to 10. Furthermore, FERA and Monopolies and Restrictive Trade Practices Act (MRTP) companies were required to make 50% of their bulk drug production available to nonassociated formulators (the ratio being 40% and 30% for the public and Indian companies respectively).

The Patent Act (1970), especially its provision of compulsory licensing, and the policy of sectoral reservation in the NDP (1978) did contribute significantly to the growth and expansion of the national sector, both public and private. Once the new patent law, which only allowed process patents, went into effect in 1973, it became possible for Indian companies to manufacture the patented drugs legally by paying 4% royalty on sales to the patent holder or to manufacture the patented drug themselves through a different process. Many companies decided to develop their own processes through in-house R and D instead of paying the royalty; they did so by simply altering the molecular structure of a drug. The introduction of the system of "drug canalization" almost eliminated the problem of transfer pricing so widespread in the pharmaceutical industry.

By 1984, multinationals had been dislodged (to use a term from Dennis Encarnation's study) from their dominant position to being a junior partner in the industry. In that year, the national sector's contribution to the total production of formulation and bulk drugs had reached 65% and 83% respectively. The range of products manufactured by the Indian sector also became sophisticated, and its contribution to the growth of bulk drug production was significant. The private national sector seems to have benefited most from the protective policies of the government. In less than a decade, it reached maturity and became competitive with the foreign sector in a highly research-intensive industry. Many private-sector Indian firms acquired the technological capability to manufacture bulk drugs from the basic stages. The Indian sector of the industry–companies such as Ranbaxy, Cipla, Cadila, Alembic, Lupin, Torrent, Sarabhai–has achieved near self-sufficiency in the production of bulk drugs and has emerged as the leading world producer of bulk drugs such as ciprofloxacin, dextrapropoxyphene, ethambutol, ibuprofen, norfloxacin, sulphamethoxazol, and trimethoprim.

India's pharmaceutical export to developed and developing countries has grown sharply in recent years; between 1985–86 and 1994–95, exports grew 14 times from Rs. 140 crores to over Rs. 2,000 crores (B. K. Keayla, *New Patent Regime,* New Delhi: National Working Group on Patent Laws, 1996). India exports a large number of bulk drugs–ampicilinem diazepam and ethambutol, for example–and formulations. The recent changes in Eastern Europe and the former Soviet Union have offered new opportunities to the industry. For example, Indian pharmaceutical companies are crowding to Russia to tap its vast potential and to cash in on the strategic advantages that India enjoys in this market. More than six Indian drug firms–Ranbaxy, Cadila, Dr. Reddy's, Sol Pharma, and J. B. Chemicals–have opened offices in Moscow, some of them manned by resident directors and general managers. Indian exporters enjoy a price advantage over their competitors (mainly Turkey and Egypt)–for example, a strip of ten tablets of the anti-ulcer drug Zantac will cost US$27 in Europe against US$1 for the Indian equivalent. These private exports to Russia were worth Rs. 300 crores in 1994.

The Pharmaceutical Industry in the Post-GATT Regime

The issue of the impact of the strengthened IPR regime on the Indian pharmaceutical industry has been a subject of debate and controversy in India, both before and after the signing of the GATT accord. The critics view TRIPs, especially their provision of product patents, to be a zero-sum game as they believe that it will inevitably make the multinational companies dislodge the national sector's dominance, which owes its phenomenal growth in the 1970s and 1980s to the Patent Act of 1970. The granting of product patent, the critics maintain, would mean that the Indian pharmaceutical companies could no longer produce, through their own novel processes, the drugs still under patent in other countries. They also argue that the post-GATT regime will lead to higher consumer prices for drugs, larger foreign exchange outflow due to large imports, and smaller employment generation in the medium and small sectors of the industry due to lower domestic production.

On the other hand, the supporters of the treaty, the government, and drug multinationals and their industry association Organization of Pharmaceutical Producers of India (OPPI), maintain that the critics' apprehension of a rise in price is unfounded as the government can still clamp on price controls in the event of unreasonable price rises. They argue that a large number of drugs will go off patent in the next five years, leaving only 7% to 8% of the Indian market covered by patents by the year 2000. According to one estimate, the share of patented products may not exceed 25% of the total market in the year 2025. Moreover, the compulsory licensing provision under TRIPs will allow the Indian government to protect public interest. The supporters of the GATT accord also point to the provision in TRIPs that gives the Indian industry a pipeline period of ten years (until 1 January 2005) in which it should find some alternative strategy to stay healthy.

The emerging strategy of many Indian pharmaceutical companies in the post-GATT era is to enter into strategic marketing alliance and collaborative research and access to the latest technology through forming joint ventures with foreign multinationals (*Business World,* 3 April 1996: 102). Leading Indian pharmaceutical companies are preparing themselves for massive changes in their business environment as a result of the thrust toward globalization and the rapidly integrating world pharmaceutical scenario. For example, Lupin Laboratories group, a Rs. 600-crores company with an integrated technology strength in synthesis, fermentation, biotechnology, and herbal medicine, has adopted the strategy of expansion in technology and upgrading the quality and selection of products for both Indian and foreign markets. It has set up a manufacturing facility in Thailand and is looking for opportunities in the United States and elsewhere. In particular, the company is making a sustained effort toward expanding exports to the developed markets of its highly potent antibiotic cephalosporin, which had a worldwide sales of US$8 billion in 1990 and has since been increasing at the rate of 6% annually.

At present, only a few companies–for example, Sun Pharma, Wockhandt, Dr. Reddy's Lab–have made investments in basic research. The Indian drug industry has not yet shown substantial interest in indigenous research and development. On an average, a company in India spends about 2% of its turnover on R and D compared to 15 to 17% in developed countries–despite a series of incentives, such as 100% one-time relief on R and D expenditure, weighted tax deduction of 125% for sponsored research in national laboratories, and the special scheme of the Department of Science and Technology (DST). In fact, the Ministry of Chemicals and Fertilizers is expected to come up with new incentives for inspiring the private sector to invest in R and D and face competition from multinationals.

India's pharmaceutical industry will witness radical changes during the first decade of the new IPR regime. A number of Indian companies, expecting to get a share of the expanding domestic and foreign markets, are investing crores in the production of bulk drugs and formulations and the research and development of new drugs. The industry expects to achieve 3.5% of the global market in pharmaceutical drugs, estimated at Rs. 36,000 crores, by the turn of the century (*Times,* 21 February 1996). Since Indian producers have in the past shown their capacity to adapt, developing the skills required for a particular situation, there is no reason to doubt their ability to withstand the challenge of globalization. In the last 25 years, India has been successful in adapting, assimilating, and improving on imported technology. The growth in the Indian pharmaceutical industry in the next decade will most likely be dominated by the export of generics, the fastest-growing segment of the industry.

Conclusion

The Indian government's efforts to develop science and technology infrastructure and its restrictive policy frame for the transfer of technology has allowed India to develop a technological capability not found in any other NIC. India has acquired its industrial technology through a large number of technical collaboration agreements. In the first two decades of independence, the government has followed a liberal policy, and DFI was the most important mode of technology transfer. In the late 1960s, the tight regulation of technology import began and lasted until the late 1970s. To make the country technologically self-reliant, the government preferred a pure technical collaboration agreement to DFI, which allowed the development of indigenous technological capability. The strict regulations regarding technology transfer did keep the cost of technology imports down, but the late 1970s witnessed the beginning of a period of relaxation of regulations. India's approach to reforms in the 1980s was incremental, but it changed qualitatively in 1991 under Prime Minister Rao, when his government accelerated the process of liberalization by radically liberalizing the import regime, industrial policy and foreign investment, and the trade and exchange rate policy.

India's Patent Act, which lasted for a period of about 22 years, contributed significantly to the growth and expansion of the national sector–both public and private–in the pharmaceutical industry. As a result, India's pharmaceutical industry is the most advanced in the Third World. Through the government policy of sectoral reservation and compulsory licensing, India achieved total self-sufficiency in the production of formulations, and in bulk drugs it has acquired the know-how for sophisticated process technologies. India's signing of the GATT accord, especially its provision of product patent, will have a far-reaching effect on the pharmaceutical industry. The Indian firms will have to face the challenge of the new patent regime, with the gradually increasing proportions of patented drugs along with progressive delicensing of drug production and decontrol of drug prices carried on by the government. Given the state of the industry, serious and efficient Indian players will, over time, grow big enough to have their own full-fledged R and D capable of introducing new patented drugs.

The pharmaceutical industry is in for a major restructuring as it passes through a period of flux and adjustment under the post-GATT regime. At this stage, we can discern a few trends. First, a number of Indian companies are opting for backward integration, as many bulk drug manufacturers are moving into the manufacture of drug intermediates; this will give the companies a lower cost. Second, there has been a tendency in the industry to scale up operation to achieve economies of scale. Third, the fragmented industry will move toward consolidation into bigger units where return will outweigh investment. In this game, multinationals and large Indian companies will be the major beneficiaries. Fourth, Indian firms are eager to collaborate with medium- and small-sized foreign companies to get access to markets in other countries. Fifth, Indian companies are moving toward exploring markets for generic products whose patents have expired. Finally, the strong patent protection is not likely to have a negative effect on the Indian pharmaceutical industry in terms of its growth, profitability, technology transfer, and availability of drugs, though it will have negative welfare implications for the masses–for example, a likely increase in the price of drugs.

A stronger IPR protection will allow for the transfer of sophisticated technology at a rapid rate as it will increase the technology supplier's willingness to transfer new technology because of the reduced fear of illegal imitation. This may help the modernization of India's industry at a time when many industries in India are faced with technological obsolescence. It may also lead to more DFI, though the relationship between intellectual property protection and DFI–and increased R and D and capital formation–has not yet been conclusively shown. Nevertheless, the impact of the new IPR regime will vary from industry to industry and will depend on other government policies, such as the policy on foreign collaboration and DFI and fiscal concessions for R and D in firms.

Acknowledgment

The author collected data and conducted field research in India for this project during his sabbatical leave from DePauw University in the spring of 1996. He is thankful to the Faculty Development Committee and the President's Discretionary Fund of DePauw University for their support of this study. He would also like to thank Professor Robert Newton for his computer-related help.

Further Reading

Alam, Ghayur, "India's Technology Policy and Its Influence on Technology Imports and Technology Development," in *Economic and Political Weekly* 20, nos. 45, 46, 47 (1985)

Analyzes the influence of government policies between the 1950s and 1980s on the technology imports and technology development activities of Indian firms and argues that while the government policies have been successful in regulating certain quantitative aspects of technology imports, their success in promoting technological activities within Indian industry has been limited.

Balasubramanyam, V. N., *The Economy of India,* London: Weidenfeld and Nicolson, and Boulder, Colo.: Westview Press, 1984

A nontechnical introduction to India's economic performance and policies from independence until the early 1980s. This book also provides a succinct review of critical policy questions, such as the then pervasive nature of import and industrial controls and regulation and their effect on efficiency, growth, and poverty, and the debate between proponents and opponents of greater outward orientation of the economy.

——, *International Transfer of Technology to India,* New York: Praeger, 1973

Based on the empirical study of 20 Indian firms, this monograph focuses on India's foreign technical collaboration agreements. It examines the feasibility of technical collaboration agreements as instruments of technology transfer in a relatively pure sense and analyzes the lessons to be learned from the Indian experience in the field of technical collaboration agreements.

Bhagavan, M. R., "Technological Implications of Structural Adjustment: Case of India," in *Economic and Political Weekly* 30, nos. 7 and 8 (1995)

Chaudry, Peggy E., and Michael G. Walsh, "Intellectual Property Rights: Changing Levels of Protection under GATT, NAFTA, and the EU," in *Columbia Journal of World Business* (Summer 1995)

Analyzes the recent changes in the laws protecting intellectual property rights in the GATT and two regional trade blocks–the European Union (EU) and the North American Free Trade Agreement (NAFTA). The article also examines the north-south controversy by focusing on the current problem of patent piracy in the pharmaceutical industry and the effect these changes have on this sector.

Desai, Ashok V., "The Origin and Direction of Industrial R and D in India," in *Research Policy* 9, no. 1 (1980)

Based on a survey of 84 firms in 1972 and 1978, this article analyzes the trends in industrial R and D in India in the 1960s and 1970s. The article shows that there has been a rapid rise in R and D expenditure and a shift in its composition toward in-house corporate R and D and away from R and D in government laboratories.

Encarnation, Dennis J., *Dislodging Multinationals: India's Strategy in Comparative Perspective,* Ithaca, N.Y.: Cornell University Press, 1989

Analyzes the changing relations among multinationals, the state, and local enterprises in India in the first four decades of its independence. By comparing India with other NICs, especially Brazil and South Korea, the study shows how Indian enterprises and states greatly improved their bargaining power relative to multinationals and succeeded in dislodging the latter from their preeminent position.

Enos, J. L., "The Transfer of Technology," in *Journal of Asia-Pacific Economic Literature* 3, no. 1 (March 1989)

Discusses the modes and types of technology transferred from the developed to the developing countries as evidenced by case studies from Asia and the Pacific. The article also identifies current issues in technology transfer such as costs, restrictions on availability, and measures of success.

Government of India, *Seventh Five-Year Plan,* Delhi: Government of India, Planning Commission, 1985

——, *Sixth Five-Year Plan 1980–85,* Delhi: Government of India, Planning Commission, 1981

Lall, Sanjaya, "Technological Learning in the Third World: Some Implications of Technology Exports," in *The Economics of New Technology in Developing Countries,* edited by Frances Stewart and Jeffrey James, London: Westview Press, 1982

——, "Trade in Technology by a Slowly Industrializing Country: India," in *International Technology Transfer: Concept, Measures, and Comparisons,* edited by Nathan Rosenberg and Claudio Frischtak, New York: Praeger, 1985

McDorman, Ted L., "Unilateralism (Section 301) to Multilateralism (GATT): Settlement of International Intellectual Property Dispute after the Uruguay Round," in *International Trade and Intellectual Property: The Search for a Balanced System,* edited by George R. Stewart, Myra Tawfik, and Maurine Irish, Boulder, Colo.: Westview Press, 1994

Nayar, Baldev Raj, *India's Quest for Technological Independence: The Results of Policy,* New Delhi: Lancers, 1983

Patel, Surendra J., "Main Elements in Shaping Future Technology Policies for India," in *Economic and Political Weekly* 24, no. 9 (1989)

Rahman, A., "Evolution of Science Policy in India after Independence," in *Science and Society,* edited by A. Rahman and P. N. Chowdhury, New Delhi: Center of R and D Management, CSIR, 1980

Siebeck, Wolfgang E., editor, *Strengthening Protection of Intellectual Property in Developing Countries: A Survey of the Literature,* Washington, D.C.: World Bank, 1990

Provides a review of the vast literature–theoretical and empirical–on economics of patents and other instruments of intellectual property. The article points out that the vast majority of studies focus on industrial economies, and similar research is lacking for developing countries; therefore, the author proposes a research agenda that includes, among other things, and assessment of intellectual property protection in developing countries.

Subramanian, K. K., "Technological Capability under Economic Liberalization: Experience of Indian Industry in the 80s," in *Sectoral Growth and Change,* edited by Yoginder K. Alagh et al., New Delhi: Her-Anand Publications, 1993

Explores the direction in which economic liberalism in the 1980s, compared to the earlier period of planned economic development and protectionism, affects India's technological capability. The author argues in favor of a relatively free-market environment for technology import accompanied by positive state intervention to encourage domestic R and D by Indian firms.

—, "Technology Import: Regulation Reduces Cost," in *Economic and Political Weekly* 21, no. 32 (1986)

Sunil K. Sahu is an associate professor of political science at DePauw University in Greencastle, Indiana. He is the author of the book *Technology Transfer and Development in the Third World* (forthcoming) and has published numerous articles in scholarly journals.

Table 15.1: Foreign Collaboration Proposals Approved by Government

Year	*Number of Approvals*
1948–55	284
1956	82
1957	81
1958	103
1959	150
1960	380
1961	403
1962	298
1963	298
1964	471
1965	296
1966	202
1967	182
1968	131
1969	135
1970	183
1971	245
1972	257
1973	265
1974	359
1975	271
1976	277
1977	267
1978	307
1979	267
1980	526
1981	389
1982	591
1983	673
1984	740
1985	1,024
1986 (Nov.)	600

Sources: *Foreign Collaboration in Indian Industry: Survey Report*, Bombay: Reserve Bank of India, 1968; *Foreign Collaboration in India: Second Survey Report*, Bombay: Reserve Bank of India, 1974; *Foreign Collaboration in Indian Industry: Fourth Survey Report*, Bombay: Reserve Bank of India, 1985; *Business Environment*, 1986; *Hindu*, 8 November 1986

Table 15.2 Foreign Technical Collaboration Agreement in Force, 1961–80: Classification of Companies

	1961–64		*1964–70*		*1977–80*	
	Number	*%*	*Number*	*%*	*Number*	*%*
Subsidiaries*	144	13.4	167	15.2	49	7.2
Minority foreign capital participation	445	42.3	489	44.5	300	44.6
Pure technical collaboration (Indian companies, private and public)	462	44.0	442	40.3	424	63.2
Total	1,051	100.0	1,098	100.0	673	100.0

*A subsidiary is a company incorporated in India but having a majority (more than 50%) holding in equity capital by a single foreign company; minority participation involves such holdings of 50% or less. Companies that have no foreign company equity capital participation but have technical collaboration agreements fall into the pure technical collaboration group.

Sources: *Foreign Collaboration in Indian Industry: Survey Report,* Bombay: Reserve Bank of India, 1968; *Foreign Collaboration in Indian Industry: Second Survey Report,* Bombay: Reserve Bank of India, 1974; *Foreign Collaboration in Indian Industry: Fourth Survey Report,* Bombay: Reserve Bank of India, 1985

Table 15.3: Remittances Abroad by Indian Private Sector Companies 1956–57 to 1982–83 (in Rs. Crore)

Year	*Profits*	*Dividends*	*Royalties*	*Technical fees*	*Interest*	*Total remittances*
1	2	3	4	5	6	7
1956–57	19.40	7.10	1.20	-	2.70	30.40
1957–58	17.40	8.80	0.90	-	2.60	29.70
1958–59	20.00	8.30	1.30	-	5.20	34.80
1959–60	16.40	11.70	1.80	-	6.20	36.10
1960–61	18.90	12.60	2.50	-	7.60	41.60
1961–62	12.40	18.50	2.40	-	7.20	40.50
1962–63	19.40	21.50	3.60	-	4.50	49.00
1963–64	12.80	18.80	4.60	-	10.80	47.00
1964–65	15.60	22.00	4.40	3.60	6.20	51.80
1965–66	13.50	19.40	2.95	6.98	-	42.83
1966–67	14.47	28.77	5.13	10.43	-	58.80
1967–68	15.95	32.70	4.32	14.68	-	67.65
1968–69	12.96	30.25	4.78	17.97	12.73	78.69
1969–70	12.72	31.41	5.80	13.05	9.28	72.26
1970–71	13.12	43.48	5.23	20.63	12.80	95.26
1971–72	9.94	38.87	5.86	13.90	12.13	80.70
1972–73	15.54	39.08	7.33	11.33	15.60	88:88
1973–74	21.91	37.51	6.21	14.08	16.27	95.98
1974–75	7.19	18.46	8.46	12.56	36.70	83.37
1975–76	20.36	24.84	10.49	25.66	24.65	106.00
1976–77	19.39	48.47	15.88	37.80	25.11	146.65
1977–78	10.13	68.01	19.50	28.14	22.70	148.48
1978–79	10.24	54.35	12.65	36.31	26.63	118.50
1979–80	14.37	50.92	0.53	43.97	25.22	144.01
1980–81	12.10	55.92	8.88	104.93	22.32	204.15
1981–82	12.16	58.92	15.99	270.70	41.08	398.85

Source: Economic Intelligence Service, *Basic Statistics Relating to the Indian Economy,* volume 1, *All India,* Bombay: Center for Monitoring Indian Economy, 1985; K. V. Swaminathan and S. Varadarajan, "Cooperation in High Technology," unpublished paper, 1984, cited in Nagesh Kumar, "Technology Policy in India: An Overview of Its Evolution and an Assessment," in *The Development Process of the Indian Economy,* edited by P. R. Brahmanand and V. R. Panchamukhi, Bombay: Himalaya Publishing House, 1987: 469–70

Chapter Sixteen

Regional Organizations for Trade and Security: SAARC, ASEAN, APEC

M. J. Vinod

Introduction

Regional cooperation is the product of the realization that national development objectives can be best pursued through collective efforts. Conceptually, regional cooperation reflects an association of states within a region to promote common objectives, to meet common needs, and to resolve common problems. Regional cooperation may very widely in terms of form, scope, and size. Today, 120 countries, accounting for about 82% of world trade, are members of various types of regional trade arrangements. These include associations of nations that work in the form of free-trade areas, trade preference associations, customs unions, and common markets.

This chapter analyzes regional organizations in the Asia-Pacific region from the perspective of the nexus between trade and security. Three such organizations are examined: the Association of Southeast Asian Nations (ASEAN), the Asia-Pacific Economic Cooperation (APEC), and the South Asian Association for Regional Cooperation (SAARC). The concept of security is going through a process of reconceptualization. During the Cold War, the focus was clearly on the military dimensions of security. We now see a shift away from military security to human security. The relationship between trade and security assumes significance within this broader framework of security and defines the working of regional organizations such as ASEAN, APEC, and SAARC.

Regional Cooperation in South Asia: The South Asian Association for Regional Cooperation (SAARC)

It took almost three decades for countries in South Asia to recognize the need to address common problems through cooperation and coordination. Informed by the amazing strides made by the neighboring ASEAN countries through cooperative endeavors, South Asian states could no longer afford to overlook such arrangements. The rationale for regional cooperation in South Asia includes the following: the region is one of the poorest in the world, the detrimental aid and trade climate leaves the countries in the region with little choice but to cooperate, and the benefits that could accrue from regional cooperation cannot be ignored.

The present thrust toward regional cooperation among the South Asian states is a relatively new phenomenon. The late president of Bangladesh, Zia ur Rahman, advocated this idea for quite some time, first broaching it with Indian political leaders during a visit to India in December 1977. He also discussed it with his counterparts in Pakistan, Nepal, and Sri Lanka in 1977, 1978, and 1979, respectively. The formal origins of

SAARC can be traced back to a proposal made by Zia ur Rahman in May 1980 in which he called for a summit meeting of the leaders of the seven South Asian states. The summit he visualized would "explore the possibilities of establishing a framework for regional cooperation" (Muni, 1984: 30–31).

Initially, both India and Pakistan were rather hesitant to accept the idea. This hesitancy was more a reflection of their respective perceptions of national interests, especially in relation to each other. Their divergent perceptions of national security can be traced to the days of the partition of the Indian subcontinent in August 1947. Since then, the two countries have fought three bitter wars. However, they did not reject the idea altogether; they sought to modify the proposal in light of these considerations. The idea went through much discussion and debate before the seven South Asian states decided on a charter for the organization. The smaller countries in the region–Sri Lanka, Nepal, Bhutan, and Maldives–straightaway accepted the proposal.

In terms of its origin, composition, and aims, SAARC is rather unique. India occupies an especially significant position in the region both in terms of its size and location. On any criteria of comparison–population, area, resources, economic development, armed forces, etc.–the asymmetry between India and the rest of the SAARC countries becomes rather obvious. Hence, it was considered appropriate to opt for a modest beginning in regional cooperation (Pran Chopra, 1986: 73). All the seven South Asian states–India, Pakistan, Sri Lanka, Bangladesh, Bhutan, Nepal, and Maldives–also share many common values that are rooted in their social, ethnic, cultural, and historical traditions.

The South Asian Association for Regional Cooperation was formally established on 8 December 1985. The formation of SAARC marked the beginning of a new endeavor in the region, and the organization has created a new set of relations on a regional basis. The preamble of the charter adopted at the Dacca summit in 1985 asks member states to avoid raising all "contentious, bilateral, political issues" and focus on development, peace, amity, stability, and progress in the region. The rule of unanimity has been consciously agreed upon in view of the existing political realities in the region. This in effect gives all members a veto power. Clause D of Article 1 of the charter states that the major objective of SAARC is to "contribute to mutual trust, understanding and appreciation of one another's problems."

The institutional framework provided for in the SAARC charter consists of

- Annual summit meetings of the heads of state or government
- A council of ministers consisting of the foreign ministers of the member states, which is largely responsible for formulating policies and reviewing the progress of cooperation
- A standing committee of foreign secretaries responsible for overall monitoring and coordination of programs, approval of projects, mobilization of regional and external resources, and identification of new areas of cooperation
- Technical committees responsible for implementation, coordination, and monitoring of the programs in their respective areas of cooperation

The SAARC secretariat has been functioning in Kathmandu since January 1987. It has been regularly organizing workshops, seminars, and training programs by officials and experts in the SAARC states. The secretariat, along with the regular meetings at the level of the heads of government, foreign ministers, and foreign secretaries, has enabled regular contacts and consultations. Even exchanges outside the formal conference venues have been useful and have led to greater trust and confidence at the level of the political elites.

The South Asian Association for Regional Cooperation is basically meant to promote the socioeconomic welfare and cultural development of the people of South Asia. Particular emphasis is given to the concept of *collective self-reliance*: mutual effort and mutual benefit. When SAARC was floated, it was primarily conceived as a vehicle to ameliorate the economic conditions of the people in South Asia. Active collaboration in the social, economic, cultural, technical, and scientific fields is expected to enhance existing bilateral collaboration and open up avenues for cooperation. The focus is to promote interdependent all-around development within the region.

Areas of Cooperation

Areas of cooperation that would be of mutual benefit to all the countries in the region were selected irrespective of their levels of economic development and fiscal status. The agreed areas of cooperation were to be implemented through the Integrated Program of Action (IPA). Originally, five areas of cooperation were identified: agriculture, meteorology, rural development, telecommunications, and health, including population. Subsequently, at the various summit-level meetings and foreign secretary-level talks, more areas of cooperation were included, such as scientific and technological development, postal services, transport, arts, culture and sports, drug trafficking, education, women and development, tourism, and the need to promote greater people-to-people contacts. The organization also envisages the need to strengthen cooperation with other developing countries and regional and international organizations.

South Asian Preferential Trading Agreement (SAPTA)

The South Asian Association for Regional Cooperation has taken some concrete steps to enhance greater intraregional cooperation. The decision to formally establish the South Asian Preferential Trade Agreement (SAPTA) in December 1995 has been a milestone. This is basically a contractual agreement that envisages a framework of rules and modalities for gradual liberalization among the SAARC member states. The fundamental principal that is implicit in SAPTA is "reciprocity and mutuality of advantages." It holds out prospects for greater integration of the regional economies and is expected to intensify trade relations, liberalize intraregional trade, provide concessional tariffs, and provide special and favorable treatment to "least developed" countries in the region. The reality, however, is that SAARC has a long way to go. Trade is one of the core areas of economic cooperation. Intraregional trade as a percentage of total trade of the member states has remained static (between 2.5% and 3.5%) during the period 1980–95. According to one study, Pakistan's exports to India account for a mere 0.93% of its total exports, whereas India's trade with Pakistan is only 0.17% of its overall trade. It has to be conceded that trade has grown at a very slow pace among the SAARC countries. The biggest issue in trade relations between India and Pakistan is the perceived threat to the industries in Pakistan, which may find it difficult to compete with Indian industries in a free-trade environment. Some Pakistani writers term this an *economic offensive.* However, there is no denying the fact that Pakistan also stands to gain in promoting greater intraregional trade with India. Pakistan has by and large enjoyed a surplus balance of trade with India. Liberalized trade will be advantageous to India and Pakistan in the long run, particularly in the present situation in which both the countries are forced to buy the same items from other countries at much higher rates. Liberalization of trade will also help to check the menace of smuggling and the substantial illegal trade that takes place between the borders.

Trade between India and the other countries in the region, particularly Bangladesh and Sri Lanka, has risen considerably. India's share of Sri Lanka's global imports, which stood at 3% in 1989, is today about 10%. Sri Lankan exports to India have also expanded. Similarly, trade between India and Bangladesh has grown such that Bangladesh is today the third largest trade partner of India within SAARC.

Under the SAPTA arrangement, SAARC initially identified 226 items for export at concessional duties. In terms of country-wide distribution, to start with, India offered 106 items for concessional tariffs and Pakistan 35 items, followed by the smaller states in the region. The idea is that the concessional tariffs offered by one country would apply to all the SAARC countries. Although the entire range of concessions offered so far covers only 0.3% of total South Asian trade, it is hoped that the countries in the region will reap the benefits of India's huge market. Since the operationalization of SAPTA, much improvement has taken place in India's relations with her neighbors. These include hydroelectric power accords with Nepal and Bhutan, resolving issues relating to the development of the Mahakali basin with Nepal, and the launching of the Tala project with Bhutan.

The importance of regional economic cooperation should become self evident. South Asia contains one-fifth of humanity. More people are born in South Asia every year (27 million) than the total population of all the Scandinavian countries put together. The World Bank has estimated that South Asia accounts for 40% of the world's poor surviving on less than a dollar a day. The adult literacy rate of 48% in South Asia pales in comparison to 98% in East Asia. Per capita investment in education and health is US$14 in India and $US10 in Pakistan compared to US$150 in Malaysia and US$160 in South Korea (Mahbul ul Haq, 1997: 39–40). Intrasubcontinental trade has to be further enhanced. The present state of bilateral economic and trade ties between India and Pakistan is a far cry from the two countries' economic scale and potential.

The long-term vision of SAARC is to remove internal barriers and allow the free movement of goods, services, capital, and people throughout the region. Trade and commerce are at the core of cooperation. A database at the association's secretariat in Kathmandu, known as SAARCNET, now connects the chambers of commerce of all the member states. This database also includes countrywide export and import regulations and company profiles. The longer term goal is to try to establish a South Asian free-trade area by the year 2000, although the target date can be extended by another five years, if necessary.

The achievements of SAARC, although modest, appear significant. Under the IPA, 12 permanent committees have been set up to plan viable programs of action in vital areas such as agriculture, education, environment, communications, drug trafficking, science and technology, and rural development. On 12 August 1988, the SAARC Food Security Reserve became operational when all the member states earmarked their respective shares. The reserve is to be used by member states in emergencies. Moreover, as the stocks are withdrawn, they are expected to be replaced.

In terms of its larger activities, certain important decisions have been made by SAARC. The year 1989 was declared as the "SAARC Year of Combating Drug Abuse and Drug Trafficking"; 1990 was declared "SAARC Year of the Girl Child" and 1992 the "SAARC Year of the Environment." A "Convention on Narcotics and Psychotropic Substances" came into force in September 1993, and a "Convention on Terrorism" was agreed on. These conventions have to be implemented in both letter and spirit to have an impact on the region.

Observations

There is now a greater realization that the future lies in encouraging economic groupings and interregional groups. Though SAARC has a long way to go, there is an attitudinal change in the approach of the South Asian states to the process of regional cooperation.

Diplomacy at SAARC has been extremely sensitive to the existing state of affairs in relations between India and Pakistan. The state of these relations are central to the success of any South Asian regional organization. The swings in India-Pakistan relations on the one hand and the bilateral equations that the other countries in the region have with India and Pakistan on the other, have consistently cast their shadows on the working of SAARC. In 1989, for example, the association received a major blow when Sri Lanka refused to host the fifth summit on the grounds that Indian peacekeeping forces were still on its soil. The postponement of the summit exposed the vulnerability of the organization, and SAARC has had to overcome this handicap by recommitting itself to its original principals and objectives. Member nations within SAARC have to have greater faith and trust in one another for SAARC to grow.

Since India is at the heart of the association, its role has come in for close scrutiny within SAARC. An active role by India within SAARC has often been perceived with suspicion. Pakistan, for instance, has often been worried about India dominating the region through SAARC, not only politically but also economically. The fear is that by lowering trade barriers, SAARC could lead to India swamping the markets of the other countries with its own products. However, given the type of products identified for such tariff concessions and the present level of intraregional trade, such fears are perhaps unfounded.

Like any other regional organization, SAARC's success depends on the following:

- Its ability to maintain a minimum level of mutual trust and goodwill among member states
- An all-encompassing perspective of the organization in terms of its aims and goals
- Selfless, sustained, and systematic support that has to be extended by all the members.

The attainment of a "security community" along the lines of what ASEAN is trying to develop would go a long way toward nurturing and harmonizing relations among the countries in South Asia. The ultimate success of the organization has to be assessed in terms of the benefits that trickle down to the people in the region. The South Asian Association for Regional Cooperation is also playing the role of "political catalyst" by providing a forum for hitherto hesitant interlocutors in the region. What is required is mutual goodwill backed up by political will in favor of regional cooperation. It is in the interest of India, Pakistan, and the region as a whole that the idea of regional cooperation, both as a concept and as a process, be taken forward to its logical end.

Trade and Security in the Asia-Pacific Region: Asia-Pacific Economic Cooperation (APEC) and Association of Southeast Asian Nations (ASEAN)

Nowhere is the interdependence of trade and security more evident than in the Asia-Pacific region. In just over a single generation, East Asia's share of world output of goods and services has nearly doubled (Larson, in Montaperto, 1993: 70). The Asia-Pacific region has today become the most dynamic area of the globe: we are witnessing the emergence of an Asia-Pacific community, which reflects the need for shared strength, security, prosperity, and commitment. East Asian countries have recorded high levels of economic growth. For example, from 1981–91, total world trade increased by 48%. During the same period, total trade among the core economies of the Common Market (Germany, France, Italy, and the United Kingdom) increased by just under 60%, whereas for all of Asia, it doubled. For Singapore, Taiwan, South Korea, and Hong Kong it is four times larger (Courtis, in Montaperto, 1993: 50–51). East Asia has sustained a steady average growth rate of 6% per annum for the past three decades. In the same period, the American economy and the European community achieved an average growth rate of 3% per year. The role of the Association of Southeast Asian Nations (ASEAN) and the Asia-Pacific Economic Cooperation (APEC) assumes importance in this context. The Association of Southeast Asian Nations is today a group of nine members: besides Brunei, Indonesia, Malaysia, the Philippines, Singapore, and Thailand, three new members–Vietnam, Laos, and Burma–were included later.

Growing economic integration makes APEC an important new post–Cold War institution. The Association of Southeast Asian Nations was launched in 1989 as a loose grouping to promote regional trade and investment. It has grown from an informal dialogue of 12 Pacific Rim countries to a regional organization that is expected to coordinate and facilitate the growing interdependence of the Asia-Pacific region. Today, it links Australia, Brunei Darussalam, Chile, Mexico, Indonesia, Malaysia, the Philippines, Singapore, Taiwan, Thailand, New Zealand, the United States, Canada, Japan, China, Hong Kong, South Korea, and Papua New Guinea. The organization is comprised of weighty nations such as the United States (with a GDP exceeding US$6.7 trillion) and Japan (US$4.8 trillion), and smaller countries such as Papua New Guinea (with a GDP of US$3,850 million). In other words, APEC membership reflects its diversity. Put together, these countries contain nearly half the world's population, control more that half–US$16 trillion–of global income, and represent 50% of the global economy and 45% of global trade. The APEC economies have been drawn closer together through greater exchange of goods, services, capital, and personnel; however, the mechanisms for addressing common concerns have not kept pace. Hence, efforts are being made to identify and remove barriers to greater integration and interaction. To try to maximize this goal, APEC has now established various working groups in the fields of human resource development, telecommunications, transportation, energy, tourism, fisheries, and marine resource conservation (Zoellick, in Montaperto, 1993: 150).

While the end of the Cold War has clearly elected the importance of the economic dimension and added momentum to regional cooperation, a series of potentially destabilizing tensions continue to exist in the Asia-Pacific region. The end of superpower confrontation has resulted in the resurfacing of tensions involving disputes over territories and maritime interests. Many analysts have spoken of the need for APEC to embrace security issues. There is little consensus among the APEC membership on this issue. The Australian foreign minister, for example, believes that "once you start mixing the agenda with political and security issues you run into all sorts of difficulties in keeping the organizational coherence." Former US Defense Secretary William Perry sounded more optimistic when he contended that it is possible to expand APEC to an organization and forum that can seriously discuss security problems, too. Japanese leaders have often preferred the wait-and-see approach.

The ASEAN Regional Forum (ARF) was created by ASEAN in 1993 to discuss international and regional issues that have security implications. The ARF has adopted a threefold strategy vis-à-vis security: confidence building, preventive diplomacy, and conflict resolution. The ARF tries to promote transparency in defense policy, requiring participants to present white papers on defense. The longer-term goal is to maintain a registry of arms. The Association of Southeast Asian Nations is conscious of the potential tensions in the region, particularly those having to do with Sino-US relations. In this context, trade-related issues, human rights, and the status of Taiwan figure prominently.

Technically, APEC is not just a political or military or diplomatic process, although the principal participants in this process are the ministers of foreign affairs. The organization is taking on added responsibilities. It is hoped that common efforts will help APEC achieve common prosperity among the member states and their economies. The Bogor Declaration, which was signed to Indonesia in 1994, has committed APEC to establish a free-trade regime by the year 2010 for the developed economies and 2020 for the developing economies. The ultimate goal is to achieve an APEC free-trade area by the 2010–2020 deadline. Thus far, the path has not been easy, but it is incorrect to assume that the member economies regard security and military questions as unimportant. The APEC forum is gradually heading toward becoming a forum for security related issues, too.

A number of questions and issues remain to be thoroughly addressed. First, there are concerns about the role of the ARF if APEC were to take up security-related issues. Second, the equations between the People's Republic of China and Chinese Taipei within APEC will be closely watched. Third, the role of the European Union and how it could fit into the activities of APEC is still being debated. Finally, it is not clear at this stage whether Russia will be admitted into APEC or not.

There are many flash points in the region that ASEAN and APEC have been concerned about. Some of them deserve to be analyzed in greater depth given the implications they have for security and even trade. The Asia-Pacific region is becoming one of the most important international arms markets. Since 1989, the region has surpassed the Middle East in its imports of arms, in terms of actual deliveries. The issues need to be analyzed against the backdrop of two major considerations: the emerging new configuration of the major powers and their stakes in the region, and the subregional security problems. The Spratly Islands dispute serves as an example of the type of challenges faced by regional organizations in the Asia-Pacific region.

The Spratly Islands Dispute

The Spratly Islands are a group of 500–600 islets, coral reefs, and shoals spread over about 1,000 kilometers, located in the extreme south of the South China Sea. Apart from being a vital sea-lane, the South China Sea is rich in oil resources, which all the countries in the region are keen to exploit. The stretch is also used as the shortest route from the Pacific Ocean to the Indian Ocean via the Malacca Straits by international shipping and the American naval fleet in the region. Hence, territorial disputes that may threaten these sea lanes assume strategic importance, and transcend the interests of individual states in the region. Moreover, the region is also considered to be strategically important for the Southeast Asian and East Asian states. The major claimants include China, Taiwan, the Philippines, Vietnam, Malaysia, Indonesia, and Brunei. This regional dispute came into the limelight in 1995 when China launched an offensive to exercise its rights in what are called the "Mischief Reefs."

The ASEAN members have called for a peaceful resolution of the conflict by asking China to settle the dispute through existing bilateral and multilateral instruments such as the international Convention of the Sea rather than through its own laws. Overall, China's response has been rather ambiguous. On the one hand, Beijing calls for the joint economic development of the Spratly Islands. On the other hand, the Chinese legislature, the National People's Congress, has passed legislation claiming all the dispute lands in the South China Sea. The same legislation also gives China the right to use force, if necessary, to "reacquire" them. The Spratlys are a long-term Chinese objective. Perhaps Beijing will resist a full-fledged assault until such time as the Chinese leaders are convinced that their forces are strong enough to prevail or their adversaries are too weak to resist. Significantly, one of the latest maps of the Chinese maritime borders includes Taiwan, the Paracels, the Spratlys, the sea coast off Vietnam, the Philippine coast, and up to 50 kilometers off the coast of the east Malaysian state of Sarawak.

Within the ASEAN community, there has been a lack of consensus about how to deal with China specifically. Thailand and Malaysia have favored a more gentle, accommodative approach. Vietnam and the Philippines prefer to press China over its claims in the South China Sea. Indonesia has taken a more middle-of-the-road approach: it has tried to galvanize ASEAN on the issue while assuaging Beijing's fears. China's inflexibility on the dispute has been matched by a quiet resolve on the part of the ASEAN nations not to give in on the issue. To quote a senior ASEAN diplomat, "The Chinese can do whatever they want to with Taiwan, but with us it's a different ballgame. They have to take heed of our worries." The Association of Southeast Asian Nations admits that China does have claims in the South China Sea but emphasizes the need to settle the claims in an amicable manner and through negotiations. Any unilateral resolution of the Spratly Islands dispute could affect peace and security in the region.

The region also has a residue of other

unresolved territorial disputes. On the Korean Peninsula, owing to political instabilities in both Seoul and Pyongyong and the frequent skirmishes across the 38th parallel, the potential for a crisis remains real (Pan Zhenqiang, in Montaperto, 1993: 111). The differences between the two Koreas, ideological and otherwise, remain deeply rooted. In a way, the Cold War continues between the two Koreas, and presents the setting for a heavily armed standoff. Another important dispute concerns the Russo-Japanese differences over the Kurile Islands. The region is also faced with the problem of nuclear weapons and the spread of weapons of mass destruction in general.

India and ASEAN Cooperation

Economic reforms in India have laid the basis for deeper interaction within and outside the region. India's ties with Southeast Asia have been recently bolstered. India became a full dialogue partner within ASEAN, placing it on the inner track of decision making. The raising of India's status from a sectoral partner to a full dialogue partner clearly demonstrates India's close association with ASEAN. India is increasingly looking toward the East, and the East is showing greater interest in India.

Perhaps the China factor may have been instrumental in ASEAN's attempt to cultivate India and incorporate it in the ARF as a dialogue partner. The Association of Southeast Asian Nations' move to induct India into the ARF was perhaps to counterbalance China's overwhelming grip over regional security. However, India has a long way to go before it can act as a countervailing power to China. To quote the words of Singapore foreign minister S. Jayakumar, "We recognized India's growing political, economic and strategic importance and the mutual benefits to be gained by India and the ASEAN through closer cooperation." Since India became a full dialogue partner, political ties have deepened and economic links have widened. Apart from tourism, India is also focusing on trade investment, science and technology, infrastructure, human resource development, and greater people-to-people exchanges in its relations with ASEAN. The ASEAN secretariat is presently in the process of completing a study on cooperation in human resource development between India and ASEAN member states.

The liberalization of India's economy and the expansion of ASEAN would afford both sides new opportunities for sharing views, discussing major issues between India and the ASEAN countries, and greater cooperation. This interaction, one hopes, will deepen and expand. The ASEAN countries have been investing heavily in India, particularly in infrastructure and human resource utilization. A thriving ASEAN-India partnership could contribute significantly to the economic growth of the larger Asia-Pacific region.

Concluding Observations

The region's future is full of uncertainties. Washington perceives China as a "critical challenge" that it has to contend with in the Asia-Pacific region. China's growing assertiveness is perceived to run contrary to the familial attitudes underpinning ASEAN's perceptions of how the ARF could contribute to regional security. ASEAN members still find it hard to agree on how to deal with China. As a Malaysian diplomat put it, "the more pressure you put on China, the more allergic it becomes."

It is significant to note that China has been giving high priority to structuring an equilibrium in its relations with the United States, Russia, Japan, and ASEAN, in that order of priority. China's "one nation, two systems" policy will be closely watched, particularly since Hong Kong has now reverted back to the Chinese. On the eve of the Osaka APEC summit in 1995, China released a white paper in which Beijing tried to reassure her neighbors that her defense policy was basically defensive in nature. "Regional conflicts," the paper said, "should be rationally and fairly resolved." The time has come for China and the other regional and global participants to make this a reality.

Thailand has proposed that ASEAN form a military bloc like NATO in order to fill the vacuum left behind following the withdrawal of US forces from some of its bases in the Asia-Pacific region. The purpose is to institutionalize mechanisms to tackle conflicts and situations that may emerge in the future. It is visualized that if this proposal becomes a reality, then a militarily strong ASEAN could perhaps be in a better position to meet the challenge posed by China in the South China Sea. One hopes that the substantial economic clout of the ASEAN region will be topped by growing military cooperation, ensuring that the twenty-first century is the "pacific century" of ASEAN's dreams. To quote the words of Jusuf Wanandi, "In the final analysis, ASEAN countries recognize that their security, both at home and in the region, depends on the pluralism of power. In regional terms, ASEAN needs both great powers (China and the United States) to be present in the region. The Association of Southeast Asian Nations needs the US presence to maintain a balance between the great powers in the region, and ASEAN would also like to have China incorporated in the region in cooperative security arrangements" (Wanandi, 1996: 127).

Security and economic growth should continue to remain the major planks and the main driving force behind ASEAN. The challenge before the regional and extraregional players is to move toward greater cooperation and integration to help achieve common goals. It is important to note that ASEAN is going through a membership explosion. Today, it consists of nine full members. It is feared in certain quarters that since the three new entrants have been isolated from international markets for so long that there are bound to be "clashes of interests" and "clashes of cultures." Yangon's growing military and economic relations with Beijing could be the first test for an enlarged ASEAN.

There could also be difficulties in terms of economic cooperation. Laos, and Burma have been given ten years time from January 1998 to fall in line with the tariff reduction schedule that has been mandated by the ASEAN Free Trade Area (AFTA). Vietnam, which had joined ASEAN in mid-1995, has been given until 2006. The remaining six members have until 2003 to reduce tariffs on 98% of their traded goods to below 5%. Moreover, the lure of emerging markets in Indochina and Burma has also bred competition among ASEAN states. Thailand touts itself as being the gateway to these markets, trying to deny Malaysia and Singapore their due shares. However, Malaysia and Singapore have been able to get some lucrative contracts, though at Thailand's expense.

There is also a proposal to invite Japan, China, and South Korea to join ASEAN in the development of the Mekong River basin. This was proposed by the Malaysian prime minister Mahathir. The main intention of the plan seems to be to prevent Thailand from monopolizing access to the emerging markets of Laos, Cambodia, and Burma. This economic initiative also seems to conceal an important strategic element. The ASEAN community feels that by giving China a stake in the Mekong subregion it can be constrained in security terms.

The above discussion suggests that any effort to understand the role of regional organizations in the context of trade and security would necessarily have to consider the following trends:

- Great power aspirations especially among regional power contenders
- New power dynamics and the shifts in power equations that call for skills to manage transitions and preserve peace
- Balancing the different approaches to peace and security through a consensual approach to security and trade-related issues
- The residue of unresolved territorial and other differences in the Asia-Pacific region. These include unresolved issues between India and Pakistan, the disputes in the South China Sea, and tensions in the Taiwan Straits and the Korean Peninsula
- Regional and subregional dialogues to

promote greater economic interaction and to tackle the common security challenges
- The need to build intraregional linkages in order to overcome suspicion and foster closer cooperation

Although the Cold War has come to an end, peace is still fragile in South Asia and the Asia-Pacific. Both India and Pakistan in South Asia and China and Japan in East Asia have to exercise restraint in the pursuit of their national interests and reconcile their differences through mutual consultation and negotiations. The security environment in East Asia would largely depend on the machinations of the four dominant participants: the United States, Russia, China, and Japan. These countries, in close connection with regional organizations, have to focus on creating a politically stable, economically viable, and militarily secure East Asia. The ARF also has a major role in this context, as it brings together almost all the major powers who have a stake in, or an influence on, shaping or affecting the stability of the region. Its emphasis is very much on the peaceful resolution of conflicts.

How the United States, China, and Japan relate to one another will determine the backdrop for trade and security in the region. Much will also depend on whether China will be smoothly integrated into the international political and economic order. There is even talk of the need for "cooperative security" in the region. The Asia-Pacific region reflects a situation of bilateral relationships interwoven with multilateral arrangements.

What is required is a stage-by-stage consensus approach. The countries in the Asia-Pacific region have to capitalize on all the opportunities to establish long-term conditions for peace and stability. The promotion and implementation of confidence-building measures among the countries in the Asia-Pacific may go a long way in reducing tensions. Preventive diplomacy mechanisms may also help. The trade-security nexus could form the basis for multilateral dialogue and cooperation whereby institutions and mechanisms for dealing with economic matters are broadened to deal with security concerns, too, and vice versa. The precept and practice of "cooperative security" and "cooperative trade" both within and between regional organizations such as SAARC, ASEAN, and APEC are the keys to greater security and greater trade.

Further Reading

Baral, Lok Raj, *The Politics of Balanced Interdependence: Nepal and SAARC,* Kathmandu: Ratna Pustak Bhandar, and New Delhi: Sterling, 1988

This study is country specific and is a thoroughly researched work on the role of Nepal in SAARC. Some of the statistics provided in this book are useful indicators of the role SAARC can play in the economic development of a country.

Bhargava, Pradesh, in *Regional Cooperation and Development in South Asia,* edited by Bhabani Sen Gupta, 2 volumes, New Delhi: South Asian Publishing House, 1986

The focus here is on the working of SAARC following the first SAARC summit held in Dacca. It brings out the challenges and problems of regional cooperation in South Asia.

Chopra, Pran, and Mubashir Hasan, et. al., *The Future of South Asia,* Dhaka: University Press, and New Delhi: Macmillan India Limited, 1986

This volume looks at South Asia from the point of view of the larger issues of development and security facing the region.

Haas, Ernest B., *The Obsolescence of Regional Integration Theory,* Berkeley: Institute of International Studies, University of California, 1975

This is an excellent exposition of the regional integration theory and is considered by many to be one of the best works to date on the subject.

Institute of Regional Studies, *The Future of SAARC,* Islamabad: Institute of Regional Studies, 1992

This study takes stock of the working of SAARC and attempts to look at its prospects.

Institute of Regional Studies, *Regional Cooperation for Regional Security and National Development: A Case Study of SAARC,* Islamabad: Institute of Regional Studies, 1990

This occasional paper identifies the major areas of cooperation in South Asia and analyzes the progress made in each one of them.

Klintworth, Gary, *New Taiwan, New China: Taiwan's Changing Role in the Asia-Pacific Region,* Melbourne: Longman, and New York: St. Martin's Press, 1995

Klintworth is a well-known authority on the Asia-Pacific region. This book traces the history of the Taiwan problem and analyzes the issue in the context of the post–Cold War world.

Mahbul ul Haq, *Human Development in South Asia,* Karachi: Oxford University Press, 1997

This is one of the best sources to focus on the problems of human development in South Asia. The book provides a lot of statistics, which makes it a popular reference book on South Asia.

Melkote, Rama S., *Regional Organisation: A Third World Perspective,* New Delhi: Sterling, 1990

The author tries to come out with a framework to justify the advantages that can accrue to the Third World countries by greater engagement among themselves.

Montaperto, Ronald N., editor, *Cooperative Engagement and Economic Security in the Asia-Pacific Region,* Washington, D.C.: National Defense University Press, 1993

An excellent presentation of how countries in the Asia-Pacific region have stood to gain by having greater cooperation. The various articles in this book are a must for anybody interested in trade and security issues concerning the Asia-Pacific region.

Muni, S. D., *Regional Cooperation in South Asia,* New Delhi: National Publishing House, 1984

This excellent study focuses on the origins and working of SAARC. It is a must for anybody working on regional cooperation is South Asia.

Scalapino, Robert A., editor, *Asian Security Issues: Regional and Global,* Berkeley: Institute of Asian Studies, University of California, 1988

This study makes a useful contribution in terms of the varied issues and themes covered. The focus here is both on the regional and extraregional factors influencing security in the Asia-Pacific.

South Asian Association for Regional Cooperation, *From SARC to SAARC: Milestones in the Evolution of Regional Cooperation in South Asia,* volumes 1 and 2, Kathmandu: SAARC, 1988

This is a document that any researcher working on regional cooperation in South Asia must refer to. The book presents a useful exposure to the evolution and growth of SAARC.

Wignaraja, Ponna, and Akmal Hussain, editors, *The Challenge in South Asia: Development, Democracy, and Regional Cooperation,* Tokyo: United Nations University, and New Delhi: Sage, 1989

This is an excellent study of the problems, challenges, and opportunities for regional cooperation in South Asia.

M. J. Vinod is associate professor of political science at Bangalore University, Bangalore, India. He has published widely, including *The United States Foreign Policy Towards India: A Diagnosis of the American Approach* (Lancers), *Nuclear Proliferation in South Asia: Current and Future Trends* (Center for International and Security Studies, University of Maryland), and *The Assam Imbroglio: Towards a Framework of Analysis* (Wayne State University).

Chapter Seventeen

Competing Asian Giants: Development and State Formation in India and China

Lei Guang and Himadeep Muppidi

China and India have often been conceptualized as "giants" in the international economic system. Their immense populations and their continental-sized economies make them analytically interesting cases for evaluating issues of economic development. Historically, both countries have faced problems of uneven development arising from Western colonialism: both had weak industrial bases and the majority of their populations subsisting on agriculture. Similarly, both were late developing countries with nationalist elites bent on modernizing their economies through inward-looking developmental strategies.

The contrasts between the two countries are no less striking, and both inherited different historical legacies from their revolutionary and nationalist struggles. The resultant political institutions offered different possibilities for domestic transformation. Their varying socioeconomic structures as well as their distinctive positions in the international system presented different challenges to the formation of the two states.

In what follows, we offer a comparative account of development and state formation in China and India. In the first section of the chapter, we argue that the Chinese and Indian states were formed through different historical legacies. These different legacies critically shaped the development strategies that the two states followed in subsequent decades. In the second section of this chapter, we then trace the evolution of these development strategies as they changed in response to changing pressures from domestic and international sources. In the third section, we highlight the economic reform period in both countries as a distinctive phase that differed significantly from their earlier development phases. Finally, we conclude with a brief account of the implications of their different histories of development and state formation for their contemporary economic policies.

Founding the State: Legacies of Revolution and Anticolonialism

India and China were bequeathed different legacies through their nationalist and revolutionary struggles. These legacies shaped the nature of the state, the domestic political arrangements, and the interaction of policy with society. The Chinese revolution radically changed the contours of the new state. Old dominant classes–landowning gentries in the countryside and bureaucratic-comprador capitalists and the old state bureaucrats in the cities–were swept away. With the demise of the older ruling elites, Chinese society offered no strong countervailing sociopolitical forces against the state (Tsou, 1986). The new state, Mao wrote, would be "a state of the people's democratic dictatorship, a state under the leadership of the working class and based on the alliance of workers and peasants" (Mao, 1961).

Jawaharlal Nehru and the Indian National Congress were by no means a revolutionary party such as the Communist Party under Mao in China. The Indian political class that came to power on 15 August 1947 was a nationalist group keen on liberating India from British colonial rule. The contours of Indian nationalism were shaped by a mass struggle against the British that enjoyed the support of existing dominant interests such as the Indian industrialists and rural elite (Chandra, 1965). This meant that the legacy of the nationalist struggle left the postcolonial state with a critical weakness. While enjoying a tremendous legitimacy springing from the freedom struggle, the Indian state's power over existing interests in domestic society was limited in relation to its aims. Given their anticolonial but nonrevolutionary social basis, the state leadership was in no position to act dramatically against the dominant interest groups (Chakravarty, 1987). India's overall developmental strategy thus tended to be more accommodative than revolutionary (Frankel, 1978).

The outcome of the nationalist and revolutionary struggles in terms of their domestic political arrangements differed markedly. The Chinese revolution had, in its long struggle against the Japanese and Nationalist forces, fashioned a variety of institutional arrangements. When the Communist leadership achieved power nationally in 1949, they simply called into being a state that had been in the making since the 1920s, complete with its own ideology, army, bureaucracy, territorial base, and socioeconomic programs. After the revolution, they moved to replace the old regime with this new state (Meisner, 1986).

In terms of the new political arrangements, Mao envisioned a "concentric" model of political power that had been successful in the past (Tsou, 1986). The Communist Party represented the vanguard of the working class in Mao's mind and was to be the core of the circle of power. It was surrounded by the urban proletariat and poor peasants. On the outer circles were the united front comprising the petty and national bourgeoisie and middle-to-rich peasants. Such a model of power reflected the altered political reality brought about by the revolution. The old ruling classes were all but destroyed. Peasants and workers were mobilized and brought from the periphery to the political center. The middle classes in both rural and urban areas were incorporated into the new political order as important peripheral supporters of the new regime.

Unlike the Chinese state, the Indian nationalists did not dismantle an inherited colonial structure of governance. Their accommodative and reformist orientation resulted in a strong reliance on existing institutions such as the army, judiciary, and the bureaucracy, among others. British nationals in these institutions were replaced by Indian nationals. These institutions were given newer missions, but the political arrangements did not change dramatically.

The direct involvement of the Indian masses was relatively limited. It was predominantly mediated through political parties who competed at the polls. Democracy thus provided a useful mechanism for the accommodation of diverse, conflicting interests, but the masses rarely had a direct say in the governing process. Electoral processes were the primary means through which social interests were communicated upward to the state. The bureaucracy was the primary channel for the state to act on society. State-society linkages were thus conducted predominantly through existing institutions such as the bureaucracy and political parties.

This presented a sharp contrast to the Chinese Communist Party's (CCP's) strategy of mass mobilization, which relied on a direct interaction with the masses. Rather than a reliance on competitive political institutions, the Communist Party, as a representative institution, acted as an intermediary between the state and society. The party tried to secure mass political participation through political education and direct campaigns. Owing to its predominant influence and its Leninist organization, the party was fused into one political entity with the state. The combined party/state structure not only possessed great autonomy from society but had a tremendous capacity to bring about social changes in China.

The Developmentalist State: Common Goals and Divergent Paths

Postindependence India was beset by problems brought about by partition and the domestic consolidation of princely kingdoms. Partition resulted in the problem of dividing assets and resources with Pakistan and resettling millions of refugees internally. The consolidation of various small kingdoms into the Indian Union introduced a greater variety of economic diversity. Independent India thus had to reconstitute several economically splintered regions into a coherent economy. Overcoming these initial problems with the help of substantial sterling balances, the Indian state set out to promote rapid industrialization within an historically backward economy (Kumar, 1983).

By and large, China in 1950 faced similar problems of national consolidation and reconstitution of a nationally integrated economy. The new state moved to achieve territorial integrity by suppressing opposing forces in various parts of the country. It was also confronted with a war-ravaged country that was divided economically. With crucial backing from the Soviet Union, China embarked on an ambitious modernization drive similar to the efforts of the Indian state.

Indian nationalism was strongly anticolonial but also extremely promodern (Chatterjee, 1992; Chakravarty, 1987). The nationalist vision of a free India was heavily shaped by the need to catch up with Western levels of economic and social development. This resulted in the idea of a modern India that, like the West, was democratic, secular, and economically developed. The experience of British colonialism meant, however, that the nationalist elite was also wary of cultivating linkages that could make them dependent again on the West. They thus came to view development as involving the twin projects of modernizing India while protecting it from colonial relationships in the international system.

The Indian state under Nehru arrogated to itself the primary mission of "developing" India. Given the economic weakness of the domestic industrialists and the poverty of a majority of the people, the state saw itself as the only actor in an economic position to undertake the project of rapid industrialization. Given the extreme diversity and the sectarian attitudes of various social groups, the state also assumed that such groups could not take a truly "national" view of important problems. It saw itself, therefore, as the only body powerful and "national" enough to transcend sectarian societal interests and modernize India.

The main elements of Indian industrialization strategy in the 1950s included a central role for the public sector, an emphasis on heavy industries, and a focus on self-reliant development based on "export pessimism" (Ahluwalia, 1994). The state acquired control over the "commanding heights" of the Indian economy. The role of the private sector–domestic and foreign–was essentially subsidiary to the state's own efforts. Rapid capital accumulation was to be balanced by an emphasis on the reduction of social and regional economic disparities. Democratic processes, such as the holding of regular elections, would give the state institutional legitimacy in carrying out these projects.

The Soviet experience of rapid industrialization through the promotion of heavy industry clearly had an important effect on Indian planners. Following the Soviet experience in some ways, the Indian state developed and implemented its long-term vision for economic growth (Chakravarty, 1987). The important influence of the Soviet model on Indian planning notwithstanding, India did not adopt an antagonistic attitude toward the West. It sought to benefit by getting aid and economic assistance from both camps. In a highly controversial decision, India also retained its membership in the Commonwealth as a republic. The Commonwealth connection, Nehru argued, would be useful to India in terms of maintaining its Western economic linkages in the international economic system. At the same time, India actively resisted the efforts of the superpowers to set the global agenda on Cold War terms. Nonalignment allowed India to focus on economic development rather than defense. Similar considerations as well as a genuine sense of Asian solidarity went into Nehru's promotion of friendly ties with China (Gopal, 1975).

The Communists' vision of a postrevolutionary China was an equally modernizing one animated by the imperative of achieving "wealth and power." The Communists were highly confident that a socialist route, as had been demonstrated by the Soviet example, was eminently viable for achieving that goal. So development for the Chinese Communists became a dual project of modernizing China and transforming it along socialist lines at the same time.

In the first three years after the revolution, the Chinese state nationalized the urban industries owned by the bureaucratic-comprador capitalists but left the national and petty bourgeoisie alone. Under situations of scarce capital and even fewer entrepreneurs, the state initially looked to the latter for crucial resources to rebuild the economy. Private-sector firms and employment actually grew until 1952 (Meisner, 1986). In the countryside, the party pushed through a radical land reform and created a new peasantry in the process. But even when the new Chinese state tolerated the private sector and a new peasant class, it was moving to eliminate both. At the end of 1952, the regime crafted its First Five-Year Plan, modeled after the Soviet Union's plan of 1928–32. Like the Soviet version, it was an ambitious blueprint for rapid industrialization propelled by the development of heavy, capital-intensive industries.

The pace of nationalization increased under the plan as the private sector was reduced to a minuscule scale by 1956. Investment capital came under the state's monopoly control. The state also blocked the free movement of labor by bringing it under "unified allocation" schemes. To mobilize resources for the industrialization drive, it tightened up the control of grain procurement and further monopolized grain purchase and sale by 1953. It then pushed for the reorganization of agriculture through a series of gradual collectivization programs that culminated in the commune system in 1958 (Riskin, 1987; Selden, 1988).

Parallel to the domestic development, China pursued an even closer political alliance with the Soviet Union during this period. Soviet aid and technology were crucial in shaping the plan and its implementation in the 1950s. The new state also had to tackle the issues of the Western colonial occupations of Hong Kong and Macau and the United States support to the nationalist regime in Taiwan. The Chinese leadership took a pragmatic attitude toward confronting the Western powers over their colonial possessions, preferring to see them as potential windows to the outside world. During the same period, however, China's relationship with the West was further strained by the Korean war and the containment policies adopted by the latter.

By the end of the 1950s, China had a greatly expanded and strengthened party/state apparatus. The membership in the Communist Party and the number of state cadres increased greatly. A multitude of economic ministries, including the State Planning Commission and the powerful State Council, were established. Coercive arms of the state, such as the Ministry of Public Security, reached ever deeper into Chinese society, especially after the establishment of "substations" at the local level in 1954. With the reorganization of industry and the collectivization of agriculture, the state also extended its administrative reach to the enterprise level in the cities and the villages in the countryside. The urban "work units" and rural communes were creations of the new state, and as such, became the lowest echelons of the state structure that mobilized local resources for the statist development projects. Mass mobilization by the state would not have been so effective without these institutional changes.

The Indian state, in contrast, was not as strong or effective in its impact on society.

The dominance of the ruling Congress Party was premised on its ability to accommodate the interests of various local elites. It acted more as an umbrella organization: rather than the party, various bureaucratic organizations were endowed with the developmental mission. The Planning Commission and the five-year plans were important institutional manifestations of the state's systematic effort to transform the society and economy. Such efforts, however, had to contend with opposing societal pressures articulating different goals.

One of the biggest societal challenges to the state was the demand for a linguistic reorganization of states in the 1950s. To a developmental state reorganizing what it saw as a premodern society along modern lines, this was a backward-looking, sectarian demand that detracted attention from more pressing economic and national concerns. However, with the agitations threatening to get out of hand and given the nature of political pressure in a democracy, Nehru conceded the various regional demands for separate states on linguistic lines (Gopal, 1975). What we see then is an interesting institutional contrast in the state's relative abilities to transform society in India and China.

The developmentalist state in China took a populist turn in the 1960s and 1970s. The change was visible in the late 1950s with the start of the Great Leap Forward (GLF) program. In the next two decades, the Chinese government oscillated between bureaucratic control and more populist approaches to development and social transformation as it tried to deal with the social and political consequences brought about by each.

The ultimate goal of the GLF was still to generate the funds necessary for rapid industrialization in China, but the route taken was distinctly Maoist in that it emphasized populism and "voluntarism" (White, 1993). As a development program, it called for a strategy of "simultaneous development" that shifted more resources to labor-intensive industries and agriculture. Institutionally, the GLF validated "mass movement" as an effective instrument of resource mobilization. The government, through "mass campaigns," short-circuited cumbersome bureaucracies and blurred the state/society distinction. Mass movements had helped in harnessing popular enthusiasm for the project of state socialism, but in order to stay on a stable and constructive course, such movements required of cadres, "rebels," and other participants delicate political skills and ethics. A classic scenario of "political decay" as described by Huntington emerged in 1958–60 as mass movements spiraled out of control. The Cultural Revolution (CR) several years later reenacted mobilizational politics and development on an even grander scale. It elevated "class struggle" and ideological reform over and above the order of "objective" economic performance. Frequently degenerating into fratricidal struggles at local levels, the CR plunged China into a state of constant political infighting and violence for many years (White, 1989).

Both the GLF and the CR were followed by periods of adjustment that sought to impose bureaucratic control of the economy, but at least in rhetoric, the Maoist developmental principles carried the day since the GLF. They emphasized local self-reliance, decentralization of economic policies, egalitarian distribution of income, and balanced development of industry and agriculture. There was a huge gap, however, between rhetorical commitments and practice. Self-reliance made peasants and workers more dependent on the cadres while decentralization failed to empower ordinary citizens. Social stratification, though not so much property based, increased as social groups acquired unequal powers in the political hierarchy enforced by the new regime. State plans continued to prioritize heavy industries and squeeze agriculture in the process, and the livelihood of peasants deteriorated in spite of their gains in the areas of health care and education. What added to the peasants' plight were the continuing efforts by the state to stamp out any sign of market-based transactions and thus their sense of freedom and efficacy in the marketplace. Mass participation notwithstanding, the organizational innovations during the period distinctly lacked an element of democratic accountability. Scholars have ultimately attributed the post–GLF famine in China to the lack of democracy in the country (Drèze and Sen, 1995). It is plausible that the lack of democratic institutions and popular control in China were largely responsible for the disastrous outcomes of the Cultural Revolution.

When China was embroiled in the Cultural Revolution, it went through a long period of isolation from international society. The public rift with the Soviet Union made China more belligerent in its foreign relations in the 1960s. After China's rapprochement with Japan and the United States in the 1970s, it seriously began to look to the international arena for advanced technology and investments.

On the other side of the border, a series of events unhinged various aspects of India's developmental strategy. India's defeat in 1962 in its war with China and the ensuing Sino-Pak accommodation forced India to divert greater resources to its defense. India's nonalignment came under strain as the government relied on the United States for arms, economic assistance, and food imports. The death of Nehru, a war with Pakistan, a cutoff of aid from Western countries, and a series of droughts within the Indian economy soon put additional pressure on Indian resources. Nehru's death was followed by internal convulsions in the ruling Congress Party and a succession of economic and political crises. A brief attempt at liberalization was an economic and political disaster and soon heralded a much more populist turn under Indira Gandhi. Initially installed by a "Syndicate" of powerful regional leaders, Indira Gandhi progressively outmaneuvered these regional leaders by adopting a radical and populist economic policy. Under Indira Gandhi, the state significantly increased its direct administrative control over the national economy. It nationalized major banks, introduced antimonopoly legislation, abolished the privy purses of princes, changed patent laws, and instituted new foreign exchange regulations.

The United States sought to use its greatly increased leverage over the Indian state in this period to bring about a transformation in Indian economic policy. The Indian state responded to such external pressures through the pursuit of an aggressive economic self-reliance policies, minimizing the country's dependence on external resources. This inward direction in development policy was matched by an active and realpolitik foreign policy. India utilized the internal crisis in Pakistan and the diplomatic support of the Soviet Union to reduce American and Chinese influence within South Asia. The Simla Agreement committed Pakistan and India to a bilateral resolution of their problems. This was also the period that saw India conduct a "peaceful nuclear explosion." Nonalignment acquired a new North/South economic dimension that argued for a more equitable distribution of global resources. India also effectively managed the macroeconomic consequences of two oil price hikes and avoided the debt crisis that affected a lot of developing countries (Mansingh, 1984; Joshi and Little, 1994).

Successes in specific sectors such as agriculture notwithstanding, the economy–especially the industrial sector–continued to stagnate (Ahluwalia, 1985). As a way of gaining greater independence in intraparty struggles, Indira Gandhi progressively "deinstitutionalized" the Congress Party. She bypassed elected Congress Party leaders and appealed directly to the people through various programs that promised a direct attack on poverty. While initially electorally successful, this did not stem the increasing political opposition to her rule. Faced with rising domestic political dissent and troubles, Indira Gandhi's government came down heavily on Indian society, declaring an internal emergency in 1975 and suspending the fundamental rights of citizens. This was perhaps the zenith of state control over Indian society.

Parallel to the developments in India, Mao centralized his personal control of the state but undermined the state's institutional power in many ways. Mao inspired a popular assault on the bureaucracy that turned the country increasingly toward "charismatic" politics. Mass leaders wielded great influence, frequently outside of the state machine. Emphasis on self-sufficiency and local initiatives created a "honeycomb structure" that limited the "reach of the state" (Shue, 1988). Patronage networks characterized much of the politics at the basic levels of government, in urban work units, and in rural villages (Walder, 1986; Oi, 1989). The state itself became a site of intense ideological struggles that reduced its effectiveness in achieving its development goals.

The 1960s and 1970s thus marked a break away from the Soviet-style development pattern in both India and China. As the two states looked for ways to renew their development efforts, they turned increasingly toward populism under the charismatic leaderships of Indira Gandhi and Mao. The personal powers of these leaders grew at the expense of their states' institutional capacities to govern. Economic growth slowed down during this period and did not change the conditions of the poor significantly in either country.

The Chinese and Indian States in the Reform Era

The beginnings of economic reform in both countries can be traced to the late 1970s and early 1980s. Mao's death in 1976 was followed by an internal power struggle and an intense debate within the Chinese Communist Party on China's development options. After a two-year interregnum under Hua Guofeng, the regime came under the pragmatic leadership of Deng Xiaoping, who reoriented the state to economic development and articulated a gradual reform strategy. This strategy downplayed ideological struggles and refocused attention on modernizing the economy. It stressed the inviolability of "objective economic laws" and the need for introducing market mechanisms into the Chinese economy (Sun, 1995).

There were two main aspects to the reform program. The first was an open-door policy that was oriented toward a greater participation in the international economy. Its purpose was to attract foreign capital and utilize advanced technologies available on the international market (Pearson, 1991). It also paved the way for the large communities of Chinese expatriates to invest in China's development. China's foreign policy during this period shifted to securing a peaceful international environment conducive to domestic economic development. The state even tried to bring Hong Kong, Macau, and Taiwan back into the fold of a unified China by its one-country-two-systems formula.

The second aspect of the reform program was domestic economic restructuring aimed at reducing administrative control and increasing the role of the market in the economy. It started with the implementation of the "household responsibility" system in the countryside in 1979 and continued to urban sectors and beyond. The pressure for decollectivization in the countryside came mostly from the peasants themselves (Kelliher, 1992). For that reason, most peasants embraced economic reform enthusiastically in the beginning. Grain output and agricultural productivity improved significantly after 1979. Increasing numbers of peasants found jobs in rural industries, which expanded rapidly in the 1980s and 1990s (Byrd and Lin, 1990). Restrictions on the peasants' movement to the cities were relaxed. At the same time, however, state investment in agriculture declined. Infrastructural projects such as irrigation, road, and agricultural extension service suffered. The social welfare of peasants also eroded as the collectives were dismantled (Meisner, 1996). On balance, Chinese peasants gained a lot of freedom and autonomy from the state through economic reform but also lost many of their entitlements.

The ensuing urban reform struck at the core of China's socialist values and practices, shaking the foundations of the party/state rule in fundamental ways. Whereas the planning institutions and the state/collective enterprises formed the bedrock of Chinese state socialism, the reformers pared down the planning machine and pushed individual firms onto the market, especially after Deng Xiaoping's southern tour in 1992. Under economic reform, the state/collective enterprises acquired considerable autonomy in production, employment, and marketing. In terms of macroeconomic structure, reform greatly enlarged the service sector and brought a new emphasis on light and consumer-goods industries. Reform also spawned a diversified ownership structure as a variety of ownership patterns–private businesses, individual proprietorships, joint-stock companies, foreign enterprises, joint ventures–emerged on the scene. The nonstate sector, including the collective sector, saw the fastest growth since the beginning of the reform. This was especially the case in the coastal regions where the special economic zones (SEZs) were located. The regional impact of the reform was very uneven, with the coastal and southern provinces reaping most of the benefits of the various reform programs. Growing regional disparity compounded problems of increasing inequality among the Chinese.

Economic reform has significantly altered the role and capacity of the Chinese state. On the surface, the postreform Chinese state appears to have lost power in many areas. Its unitary political structure is increasingly challenged by strong provincial assertions of power. The party/state unity has come unhinged at various levels of the government. The direct administrative control of the state over public enterprises has given way to market mechanisms. At a deeper level, however, the Chinese state has actually managed to increase its power. For directing the economy, the fiscal and monetary instruments in the hands of the state are more effective than before. The integration of the economy and the formation of a unitary national market have made the central government indispensable in China's development. Compared to the Maoist era, the postreform Chinese state may have lost some formal authority, but it has gained power in many respects (Lieberthal, 1995; Migdal, Kohli, and Shue, 1994)

A stronger Chinese state does not necessarily mean, however, that it is more autonomous than before. Economic reform has fundamentally reorganized the relationship between society and the state and in the process has created new social constituents whose fate is bound tightly with the new state. The social basis of the state has shifted to the entrepreneurial community and the growing "middle class." The implicit Maoist social contract between the Chinese state and the urban workers and peasants has been dissolved (Meisner, 1996). As the state moves further away from the Maoist path, it has also lost its mobilization potential with the masses and become increasingly embedded in an inegalitarian society.

What the Chinese reform has singularly not done is bring about a democratization of the political structure. In spite of the repeated attempts by the Chinese people to push toward a democratic political reform from 1978 on, the party/state has not yielded political ground in any significant way (Friedman, 1994). China's handling of the democracy movement in 1989 shows its willingness to absorb the political costs of enforcing "internal stability." International pressures on China have in recent years led to a form of nationalism that threatens to make meaningful political reform difficult. Such reactive nationalism can only strengthen the power of the state over society.

Unlike economic reforms in China, Indian economic reforms had a more tentative beginning. Buffeted by a series of economic and political crises and under pressure from international institutions and the United States, India had gone in for its first experiment in economic liberalization as early as the 1960s. India declared a plan holiday from 1966–69, devalued the rupee, and liberalized its import-export regime. Economically and politically, however, this experiment was a disaster. It did not improve the domestic economic situation but only made the ruling Congress Party vulnerable to charges of caving in to foreign pressures (Bhagwati and Desai, 1970). It also resulted in significant internal political struggles within the dominant Congress Party, forcing Indira Gandhi to take a populist turn.

During Indira Gandhi's second term, India renewed its unsteady effort at economic reforms. It negotiated one of the largest loans ever from the International Monetary Fund (IMF) in 1981 in order to undertake "structural adjustments" and to begin a serious reevaluation of state controls over the economy. This period also witnessed the broadening of the social base of the Indian middle class. This broadened base was reflected politically in the rise of sectarian parties–regional, linguistic, religious–and the growing assertion of "other backward classes" and Dalits (oppressed) in the north. The Congress Party tried initially to manipulate these movements but could not stop their growing power. The dominance of the Congress Party was increasingly challenged electorally and through violent separatist agitations.

The Indian state saw these new movements and their rapid politicization as arising at least partly from the slow growth rates of the economy. A greater stress on rapid economic growth was needed to meet these rising internal conflicts over limited resources. With these factors in mind, the Indian state started introducing various economic reforms on a piecemeal basis. However, the legacy of decades of socialist rhetoric and the historical fears of Western colonialism made any drastic change in economic polices politically difficult. Rajiv Gandhi tried but failed to translate his massive electoral mandate into an effective program of economic liberalization. Under his governance, the state targeted specific sectors, such as electronics and the information-technology industry, to lead economic growth. Growth rates improved, but intraparty dissension and political controversies over corruption slowed down reforms and eventually led to the Congress's downfall (Kohli, 1990). The fall of the Congress ushered in a period of volatility in Indian politics that led to a growing economic crisis.

The economic crisis came to a head in 1991, with India coming close to defaulting on its rising international payments obligations (Nayyar, 1996). It propelled a minority government under P. V. Narasimha Rao to undertake economic reforms on a dramatic scale, a scale calculated to reinspire international confidence in the Indian economy. The state, through a series of policies, liberalized its license-control raj and ceded space within the economy to domestic and foreign private concerns. It also moved in the direction of a greater reliance on indirect fiscal rather than direct administrative controls (Bhagwati, 1993). This direction was maintained in the rest of the Congress rule under Rao's political tenure.

The United Front government that succeeded the Congress paid greater attention to social welfare and the decentralization of power to the states, but it also continued the policy of economic liberalization. One indication of this was the continuance of crucial political and bureaucratic personnel within various ministries handling reform. With the Communists and the Congress parties both supporting the United Front government, the economic process appeared increasingly insulated from the rhetorical claims of political parties.

Externally, India made an effort to portray itself as an attractive, "emerging market" for foreign capital. Visits to the World Economic Forum at Davos by the prime minister and his finance and commerce ministers were now routine. The state also took the initiative in prioritizing the economic agenda of regional associations such as SAARC (South Asian Association for Regional Cooperation). The foreign exchange reserves of India increased to record proportions, giving it greater leeway in managing its external transactions. Politically, India's refusal to sign the Nuclear Nonproliferation Treaty and the Comprehensive Test Ban Treaty brought it into increasing conflict with the United States, but the growing convergence of economic interests also resulted in a muting of such disagreements.

For the Indian state, liberalization has replaced planning as the buzzword of economic policy. The state has actively used the idea of a 200 million strong middle class to portray India as an attractive market for foreign capital. While the emigration of skilled personnel was at one time seen as a "brain drain," it is now seen as a "brain bank" that can be utilized to draw on external sources of capital and technology. The state no longer sees itself as the predominant agent for promoting economic development. Private enterprises are encouraged to compete and accumulate capital to generate growth while the state increasingly concentrates on running a smaller but more efficient apparatus. In spite of opposition from social groups such as organized labor and environmental activists, political support for economic reform extends across the spectrum of political parties, including significant sections of the left. Arguments related to reform–such as greater decentralization of state power to regional and local levels and more transparency in government–also enjoy a broad political consensus. Democracy has meant a slower institutionalization of economic reforms, but it has also ensured a greater continuity in the development policies of the state.

Conclusion

Since the late 1940s, the Chinese and Indian states have presided over major political and economic changes in two of the most populous countries in the world. Both attempted to modernize their economies through a state-sponsored process of rapid industrialization. Both were initially influenced in significant ways by the Soviet model of economic development, though they took different lessons from it. But the specific development strategies adopted by these two states depended critically on the distinctive nature of state-society relations and their varying positions within the international system. As we have shown, state formation took a different trajectory in the two countries based on a combination of internal and external pressures.

A strong party/state combination in China transformed its socioeconomic structures significantly by relying on mass mobilization and periodic organizational changes. Externally, the state relied in the first phase of its development on Soviet assistance. During the Cultural Revolution, it largely disengaged itself from the international economy. Reform policies from the 1970s restored China's participation in the global economy as a crucial aspect of its development strategy. The Indian state, by contrast, was more constrained by a democratic framework in the transformation of its socioeconomic structures. Externally, a policy of nonalignment allowed it to benefit from the economic assistance of Soviet and Western countries. Within the context of the Cold War clash of ideologies, the Indian concern with development and democracy lent itself to a sympathetic hearing in Western policy-making circles. For an important group of influential academics and policy makers within the West, the Indian state's experiment with democracy and development posed a good counterexample within Asia to the Chinese model of development and was an important way of curbing the influence of communism. While India did try to minimize its external linkages in the late 1960s and early 1970s, economic reform initiatives have reintegrated the state into the global economy from the 1980s on.

While superficially similar, economic reforms have manifested themselves differently within the two countries. In China, the state has, among other things, presided over the introduction of a capitalist sector and its attendant legal and economic institutions into a predominantly state socialist economy. The resulting contradictions–two economic systems under one state–will increasingly shape state/society interactions in the future. In India, an active capitalist sector was already present, though under the overall control of the state. The Indian state has primarily focused on the reduction of state control and the introduction of greater competition into the economy. While this has not affected its basic agenda of modernization without colonization, globalization has changed the international context within which the Indian state can pursue these twin objectives.

Further Reading

Ahluwalia, Isher Judge, *Industrial Growth in India: Stagnation since the Mid-Sixties,* New Delhi: Oxford University Press, 1985

A minor classic in the literature on Indian economic development. Ahluwalia takes up the issue of why industrial growth in India slowed down after the mid-1960s and argues that the causes lie largely in the adoption of an "industrial policy framework" that discouraged an "efficient" use of factors of production.

——, "The Role of Trade Policy in Indian Industrialization," in *Trade Policy and Industrialization in Turbulent Times,* edited by G. K. Helleiner, London: Routledge, 1994

A review of the role of trade policy in India's industrial strategy that focuses particularly on the changes in the 1980s. This chapter is part of a broader study reviewing the relationship between trade policy and industrialization in 14 developing countries.

Bhagwati, Jagdish N., *India in Transition: Freeing the Economy,* Oxford and New York: Oxford University Press, 1993

An analytical review of India's development strategy by an economist who has been a consistent advocate of economic liberalization since the late 1960s. Based on the author's Radhakrishnan Lectures at Oxford University, the book offers a quick account of what went wrong with the Indian model and why economic reforms were long overdue.

Bhagwati, Jagdish N., and Padma Desai, *India, Planning for Industrialization: Industrialization and Trade Policies since 1951,* London and New York: Oxford University Press, 1970

An in-depth study of the economic policy framework regulating Indian industrialization from 1951 to the late 1960s. Significant for being one of the earliest systematic criticisms of the "inefficiencies" promoted by this policy regime.

Byrd, William A., and Qinsong Lin, editors, *China's Rural Industry: Structure, Development, and Reform,* New York: Oxford University Press, 1990

This book grew out of a collaborative project between the World Bank and the Chinese Academy of Social Sciences. It offers a comprehensive empirical survey of China's rural industries, including township, village, and private enterprises.

Chakravarty, Sukhamoy, *Development Planning: The Indian Experience,* Oxford and New York: Oxford University Press, 1987

An overview of Indian development planning by a prominent economist cum policy maker. The book deals with the economic themes underlying India's development strategy as well as the changes that came about as India faced different problems over the years.

Chandra, Bipan, *The Rise and Growth of Economic Nationalism in India: Economic Policies of Indian National Leadership, 1880–1905,* New Delhi: People's Publishing House, 1966

An exhaustive account of the economic understandings that shaped the nationalist struggle against the British colonial rule and also of the nationalist program for the development of independent India.

Chatterjee, Partha, *The Nation and Its Fragments: Colonial and Post-Colonial Histories,* Princeton, N.J.: Princeton University Press, 1993

A theoretically nuanced argument about the distinctively anticolonial nature of Indian nationalism and the ways in which it influenced the development of the postcolonial Indian state and nation.

Drèze, Jean, and Amartya Sen, *India: Economic Development, and Social Opportunities,* Delhi: Oxford University Press, 1995

A broad perspective on and evaluation of India's development in terms of both economic and social opportunities for the Indian population. Includes insightful comparisons with other countries, including China.

Frankel, Francine R., *India's Political Economy, 1947–1977: The Gradual Revolution,* Princeton, N.J.: Princeton University Press, 1978

A detailed analysis of the economic and political development strategies adopted by the Indian leadership from 1947 to 1977. While dated in some respects, the book is insightful in tracing the central tensions structuring the accomplishment of India's economic and political goals.

Friedman, Edward, *The Politics of Democratization: Generalizing East Asian Experiences,* Boulder, Colo.: Westview Press, 1994

This edited volume is a collection of essays on democratic transition in East Asia. It argues against "East Asian exceptionalism" with respect to democratic transition even while it explores the historical contexts, coalitional dynamics, institutional arrangements and transnational factors associated with it in East Asian countries.

Gopal, Sarvepalli, *Jawaharlal Nehru: A Biography,* volume 2, Delhi: Oxford University Press, 1975

This is part of a three volume biography of India's first prime minister. Gopal's use of Nehru's private papers makes this a valuable resource for understanding the factors structuring independent India's economic and political policies.

Huntington, Samuel, *Political Order in Changing Societies,* New Haven, Conn.: Yale University Press, 1968

Huntington offers a conservative statement on the relationship between societal changes and political development or decay in developing countries.

Joshi, Vijay, and I. M. D. Little, *India: Macroeconomics and Political Economy 1964–1991,* Washington, D.C.: World Bank, 1994

Forms part of a World Bank study on the responses of 18 countries to external and internal domestic economic crises. It is useful primarily for its detailed outlining of the economic policy interventions adopted by India between 1964 and 1991.

Kelliher, Daniel, *Peasant Power in China: The Era of Rural Reform, 1979–1989,* New Haven, Conn.: Yale University Press, 1992

Based on his fieldwork in China, the author provides a bottom-up view of the rural reform process during its first decade. He argues that peasants themselves, not the central government under Deng, instigated the most radical changes of the reform era.

Kohli, Atul, *Democracy and Discontent: India's Growing Crisis of Governability,* Cambridge and New York: Cambridge University Press, 1990

Kohli argues that India, while continuing to be democratic, became increasingly difficult to govern from the late 1960s onward. His study locates the causes of this problem in India's weakening political institutions more than in its socioeconomic structures.

Kumar, Dharma, editor, *Cambridge Economic History of India,* volume 2, Cambridge and New York: Cambridge University Press, 1983

A survey of the economic history of India with individual chapters focusing on specific aspects of Indian political economy from 1757 to 1970. The volume has invited critical comment for its downplaying of the impact of colonialism on India's economic structure. It remains useful, however, as a broad account of some historically crucial processes affecting Indian development and state formation.

Lieberthal, Kenneth, *Governing China from Revolution through Reform,* New York: Norton, 1995

Explores the problem of governance in China by situating it in the broad context of its imperial traditions, the Republican experience, and the Communist revolutionary legacy.

Mansingh, Surjit, *India's Search for Power: Indira Gandhi's Foreign Policy 1966–1982,* New Delhi and Beverly Hills: Sage, 1984

An analytically well-grounded account of Indian foreign policy under Indira Gandhi. Focusing on the crucial phase in Indian economic development, the book is particularly strong on the state's external relations as they affected India's political and economic interests.

Mao Zedong, *Selected Works of Mao Zedong,* vols. 1–4, Beijing: Foreign Languages Press, 1961

A collection of Mao's writing before 1949.

Meisner, M., *The Deng Xiaoping Era: An Inquiry into the Fate of Chinese Socialism, 1978–1994,* New York: Hill and Wang, 1996

The author offers a judicious and scathing critique of China's reform under Deng. He argues that reform in China has led to the rise of "bureaucratic capitalism" in spite of the rhetoric of "socialism with Chinese characteristics."

—, *Mao's China and After,* New York: Free Press, 1986

This is an expanded and updated version of the author's classic historical study of the Chinese revolution and the socioeconomic transformations in China under Mao and his successors up to the mid-1980s.

Migdal, Joel S., Atul Kohli, and Vivienne Shue, editors, *State Power and Social Forces: Domination and Transformation in the Third World,* New York and Cambridge: Cambridge University Press, 1994

A collection of essays that offers an alternative perspective–"state-in-society" approaches–on the question of state-society relations in developing countries.

Nayyar, Deepak, *Economic Liberalization in India: Analytics, Experience, and Lessons,* London: Sangam, 1996

An analytical assessment of economic liberalization in India structured around the Indian government's management of the 1991 economic crisis. The author's theoretical critique is ably supplemented by his practical experience as an economic adviser to the government.

Oi, Jean, *State and Peasant in Contemporary China: The Political Economy of Village Government,* Berkeley: University of California Press, 1989

Oi provides a detailed analysis of the local politics of collective agriculture during the Maoist period by studying the state-peasant relationships, which she characterizes as "patron-client" relationships, in the area of grain procurement.

Pearson, Margaret, *Joint Ventures in the PRC: The Control of Foreign Direct Investment under Socialism,* Princeton, N.J.: Princeton University Press, 1991

Examines China's experience in simultaneously absorbing and controlling foreign direct investments from 1979 to 1988.

Riskin, Carl, *China's Political Economy: The Quest for Development since 1949,* New York: Oxford University Press, 1987

A comprehensive historical account of China's development course from 1949 to 1984 and the controversies around various development alternatives.

Selden, Mark, *The Political Economy of Chinese Socialism,* Armonk, N.Y.: M. E. Sharpe, 1988

In this collection of essays surveying four decades of China's rural development, the author critiques the Maoist model of socialist development and examines implications of the post-Mao reform in China's quest for socialism.

Shue, Vivienne, *The Reach of the State,* Stanford, Calif.: Stanford University Press, 1988

Four critical essays examining the state-society relationships in contemporary China. The author argues that, contrary to the totalitarian mode of Chinese politics, the "reach of the state" was limited by the "honeycomb structure" of Chinese society during the Maoist era.

Sun, Yan, *The Chinese Reassessment of Socialism: 1976–1992,* Princeton, N.J.: Princeton University Press, 1995

Provides a well-documented analysis of the Chinese reassessment of the Marxist-Leninist ideology and its relevance to changing Chinese society. She shows how ideological reassessments inform policy making and factional struggles in the reform era.

Tang, Tsou, *The Cultural Revolution and Post-Mao Reforms: A Historical Perspective,* Chicago: University of Chicago Press, 1986

A collection of Tsou Tang's essays offering astute analysis of the Cultural Revolution and its legacies for the reform period.

Walder, Andrew G., *Communist Neo-Traditionalism,* Berkeley: University of California Press, 1986

Examines the nature of authority and politics at the level of the workplace in China. Argues for a "clientelist" perspective on workplace politics.

White, Gordon, *Riding the Tiger: The Politics of Economic Reform in Post-Mao China,* Stanford, Calif.: Stanford University Press, 1993

The author traces changes in China since the late 1970s and focuses on the political aspect of China's post-Mao market reform.

White, Lynn K. III, *Politics of Chaos: The Organizational Causes of Violence in China's Cultural Revolution,* Princeton, N.J.: Princeton University Press, 1989

This book uses Shanghai as a case study for understanding what policies were responsible for the violence and ostracism during the Cultural Revolution.

Whyte, Martin K., and William L. Parish, *Urban Life in Contemporary China,* Chicago: University of Chicago Press, 1984

Offers a comprehensive analysis of urban life under communist rule and documents the extension of the state into the lives of urbanites and the transformation of urban social structure before the 1980s.

Lei Guang and Himadeep Muppidi are doctoral candidates in the Department of Political Science, University of Minnesota.

Chapter Eighteen

International Monetary Fund and World Bank Involvement in India's Economic Reforms

Farida C. Khan and Roby Rajan

Early in 1991, India came close to defaulting on its international payment obligations. Foreign exchange reserves stood at US$1.2 billion, barely sufficient to cover two weeks of imports. With falling industrial production and inflation rising to 17%, confidence in the Indian economy had hit rock bottom, and international sources of funds had all but dried up. It was against this backdrop of a possible default on its external debt that India negotiated a special drawing rights (SDR, a special currency unit created by the International Monetary Fund to expand global reserves) 1.7 billion standby arrangement with the International Monetary Fund (IMF) and a US$500 million structural adjustment loan from the World Bank. The intervention of the IMF was needed to restore confidence in India's ability to repay and thereby reopen access to borrowing from individual and commercial sources in capital markets. The government that took office in June 1991 immediately set in motion the twin processes of macroeconomic stabilization as well as microeconomic structural reform as required by the IMF and the World Bank.

The first half of this chapter examines India's long and evolving relationships with these institutions, relationships that culminated in the IMF and the World Bank playing a significant role in restructuring India's economy. The second half of the chapter examines the particular reforms undertaken, their effects to date, and potential problem areas, and it concludes with the possible changes in the nature and scope of multilateral institutional involvement under the new economic structures.

The Early Years

The International Bank for Reconstruction and Development (IBRD), commonly referred to as the World Bank, and the International Monetary Fund (IMF) were established in 1944 during the Bretton Woods agreement to rebuild and stabilize the industrial world and its peripheries. The two sister institutions had the same fundamental agenda but carried them out with their separate instruments and over different time horizons.

The IMF's role was to lend short-term funds to member countries to help tide over balance of payments difficulties that would be encountered in the new international monetary system–the dollar standard. The dollar had become the main international reserve currency, although the importance of the pound sterling and gold continued. The IMF was to provide liquidity to its member countries by lending them reserves when needed. It was to give repurchases to members by buying their currency with internationally liquid currencies, thereby providing loans for two to three years.

The IBRD's mandate was more amorphous. Although it was concerned foremost with reconstruction (the IBRD's first loan was to France in 1947 for US$250 million; in inflation adjusted amounts, this is largest loan that the World Bank has ever made), it would also be involved in enabling long-term development for poorer countries. However, the IBRD's resources paled before the Marshall Plan, which disbursed US$13 billion to Europe from 1948 onward. The IBRD subsequently diverted its efforts from Europe and made its first loan to a developing country (Chile) in 1948. Loans soon followed to Mexico, Brazil, Yugoslavia, and India. It still continued to lend mostly to middle-income countries, such as Australia and Japan, and by the end of the 1950s, it became clear that it was simply competing with private funds. At this time, India's and Pakistan's difficulties with financing even the small amount of debt they had accumulated made the World Bank reconsider its direction.

India's foreign exchange crisis of 1958 marked a turning point for the World Bank. The United States and other countries formed an Aid India Consortium, which was coordinated by the World Bank. The hope was that concessional finance could help poorer countries in the same way that the Marshall Plan had allowed Europe and Japan to recover after the Second World War. In 1960, the soft-lending arm of the Bank, the International Development Association (IDA), was created. It was primarily from the IDA that India borrowed in the late 1960s and throughout the 1970s. India and China have been the largest recipients of IDA loans, with India's Green Revolution financed primarily through such loans. By the end of the 1970s, and particularly after the debt crisis in the early 1980s, the World Bank changed its lending philosophy away from a development orientation to that of structural adjustment. Rather than lending primarily for state-assisted expansion of agriculture and industry, it began to provide non-project-specific loans that were to be used to carry out policy reforms in developing countries so that they could develop along a market-oriented path. The success of market orientation and export-led growth in East Asian economies, such as Korea and Taiwan, provided the rationale for these policies. Structural adjustment loans or sectoral adjustment loans have allowed the bank to follow the IMF's agenda more closely, so that in recent years, they have worked in tandem, the IMF insisting on macroeconomic reforms and the World Bank urging microeconomic reforms to move countries quickly to a liberalized market regime.

The World Bank has two other smaller agencies–the International Finance Corporation (IFC) and the Multilateral Investment Guarantee Agency (MIGA). The IFC was established in 1958 to cofinance private sector investment and advise countries on improving their local capital markets. The MIGA began to operate in 1988 and acts as a guarantor of private sector loans to developing countries.

Sources of Financing and Lending Policies

Countries join the IMF by paying a subscription fee or "quota." The quota is composed of the country's national currency (75%) and hard currencies such as the dollar or the SDR (an international accounting "currency" created by the IMF itself) and is based on the size of the country's gross domestic product (GDP), foreign exchange reserves, and international trade. The size of the quota determines the number of votes a country has in the IMF's decision-making process. The United States has 18% of the voting power, which is well above the 15% needed for a veto. The IMF's quotas make up most of what it lends to other countries. The process of lending is done through a repurchase so that the borrowing country gives the IMF its own currency and receives the "hard currency" it is seeking. When repaying the loan, the borrowing country reverses this transaction. (A borrowing country, when it borrows, is said to have a "positive reserve position" at the IMF.) Since only about half the listed amount of the quotas is in hard currencies–the other half being billions of "soft" Nigerian naira, Indian rupees, and the like–the

IMF lends far less than the total of its quotas. The total available funds for lending was increased in 1962 through the General Agreement to Borrow (GAB), which is a pool of funds created by a group of industrialized countries. Small amounts were borrowed from the Bank of International Settlements (BIS) to enhance lending capability.

A small part of the World Bank's loans is raised from its capital contributions from member countries. Like the IMF, voting rights at the World Bank are also allocated according to capital contributed. The board of directors at the World Bank never reject the recommendation of the staff members. The staff members are largely located in Washington, D.C., with only a small fraction working in resident offices of member countries. Most of its lending is funded by selling bonds in international capital markets in which the Bank is the single largest borrower. The Bank's bonds have a triple A rating or are considered sound investment. The IDA obtains its funds from grants given by developed countries. Because donor countries must replenish IDA funds every three years, legislative approval in these countries has become a concern. The United States has delayed funding several times, virtually shutting down the IDA and reducing its size in the 1980s and 1990s. In the last replenishment of the IDA in 1994, the US Congress agreed to replenish for only two years, with the third year's replenishment being subject to further US scrutiny of the IDA's decisions. The IFC borrows from the IBRD and also floats its own bonds in world capital markets. The MIGA began with a small capital of US$1 billion with member governments paying 10% of the capital, giving promissory notes for another 10%, and the remaining 80% being callable capital.

The IMF usually lends a country 25% of its quota, called the *reserve tranche,* without any interest charged or any conditionality. Thereafter, each *credit tranche* is lent with monitoring of the country's economic policy. The first credit tranche requires "reasonable effort" on the part of the debtor to correct its foreign exchange problem, and subsequent credit tranches require "substantial and viable programs" to be followed. Credit tranches must be paid back in three to five years at a charge of 6.5%. Once 100% of the country's quota has been borrowed, it can apply for further loans from one or more of IMF's special facilities. These facilities were mostly created during the debt crisis in the 1980s and consist of the Extended Fund Facility, the Enhanced Structural Adjustment Facility, the Compensatory and Contingency Financing Facility, the Buffer Stock Facility, and others. Most of these are lent with the explicit purpose of bringing about structural adjustment or averting unusual crises, such as commodity price fluctuations.

The IBRD lends only to less developed countries and graduates its borrowers when they reach a per capita income of US$4,080. Interest on its loans is set at 1% above the rate at which it borrows in capital markets. When it decides to lend for a project, the rate of return on the project must be above 10%. Almost 80% of the projects have an internal rate of return of more than 14%. Most of the projects are cofinanced with member-country governments and other countries or regional development banks, such as the Asian Development Bank (ADB).

International Development Association loans have long maturities–40 years or more. They are interest free (except for a 0.75% service fee) and have a grace period of ten years before repayment of principal begins. To be eligible for IDA loans, a country must have per capita GDP of US$805 or less. Project performance for IDA loans must be similar to IBRD loans, but more IDA loans have been made for agriculture, and there have been more failures in these loans, with about 10% of them having a negative rate of return. The IFC also cofinances with governments and entrepreneurs in host countries so that the actual value of investment is three or four times the amount of IFC loans.

Early Involvement in India

The Import Substitution Phase

At independence, India inherited some of the basic preconditions for industrial development that were not present in many newly independent countries. These included transport infrastructure, an administrative and legal system, and a manufacturing sector consisting of steel and engineering, ordnance, cement, bicycles, and motorcycles. The Nehru government favored public sector industrialization and put into place a strategy that (1) focused on increasing investment, (2) favored industry over other sectors of the economy, and (3) favored those subsectors in industry that had strong linkage effects, such as steel. With the First Five-Year Plan, a tightly regulated system was established with controls put on foreign and domestic investment and foreign trade, particularly on imports. The Second and Third Five-Year Plans increased the emphasis on heavy industry, particularly capital goods production. The Soviet-style planning that was adopted sought to promote self-sufficiency through industrial development. In the years of the first three five-year plans, India borrowed abroad to finance defense expenditure, purchase raw materials, and purchase capital goods. Beginning in 1960, the IDA became a primary source of borrowing to balance the trade deficit. India received 26% of all IDA credit and 60% of all IDA commitments to the South Asian region. When there were bad crop years, food imports needed to be financed. The governments of prime ministers Nehru and then Shastri relied heavily on Public Law 480 to meet grain shortfall. The two big aid donors at this time were the United States and the World Bank, with US bilateral aid far exceeding money received from the World Bank.

The Green Revolution

Agriculture figured quite prominently in the First Five-Year Plan (1951–56), and crops were good most of those years. By the time the second plan was prepared, the country's commitment to industrialization was established, and the extraction of resources from agriculture to industry was seen as the appropriate development strategy. By 1958, the Rockefeller Foundation and the Ford Foundation began to introduce high yield plant breeding and supported a joint US-India "food crisis team," which in 1959 called for increased food production and agricultural reform. These efforts were also supported by the World Bank. During 1965–67, India experienced a severe food crisis. Two successive droughts during this period increased food imports to over 10 million tons annually. The Woods-Mehta agreement was signed in 1966 and was intended to create a program of incentives, technologies, and institutional changes to make for a more productive agricultural sector. World Bank lending for agriculture increased from US$124 million (6% of all lending to India) in 1968 to US$687 million (31%) by 1974. The resulting Green Revolution was made possible by increases in fertilizer imports and subsidies, supplies of high yield variety seeds, agricultural credit, and expansion of irrigation. A public procurement system was widely used. The IDA played a critical role in financing and sustaining the agricultural reforms.

The Onset of Market Reforms

At the beginning of the 1970s, the basic thrust of India's development policy continued to be state-led industrialization and agricultural expansion through the Green Revolution. Government licensing of new industrial capacity remained in force, and in 1969, the nationalization of all domestically owned banks increased state control. However, the two oil shocks introduced new elements in India's policies and its relationships with donors.

The first oil shock hit when economic performance had been weak for some years. Agricultural production had been slow, so that with the rise in international crude oil prices, the current account deficit rose to 1.4% of GDP in 1974–75, representing a sizable deterioration from the 0.4% average in the previous three years. The crisis was met by withdrawing from the IMF gold tranche, first credit tranche, and the Oil Facility. Along with short-term official financing, loans from multilateral institutions and other external borrowing rose substantially, and the trade deficit turned into a surplus within two years. Export growth played an important part in the adjustment while imports were moderated. A restrictive macroeconomic policy was adopted at this time, with the government taking a series of steps to freeze wages, limit dividend distributions, and increase taxes. A conscious effort to liberalize imports was taken after 1976 and was reflected in the 1978–79 import policy order.

With the second oil shock, oil prices doubled in 1979, raising India's import bill and the current account deficit to 1.6% of GDP. The adjustment to the second oil shock differed greatly from the first. The current account deficits continued to increase, and India had to look to external borrowing. However, the availability of long-term concessional loans had dried up. The IDA had run into difficulties and did not receive replenishments easily from donor countries. Also, the entry of China as a borrower meant less funds available for India. Although commitments from the IDA were offset by higher IBRD funding, this meant less favorable terms of borrowing. Aid commitments from bilateral sources were also falling.

In 1981, the IMF committed 5 billion SDRs to India to be paid over three years, because India's foreign debt position was considered to be "healthy," with a debt service to export ratio of only 9.4%. This was the single largest loan the IMF had ever made, representing 30% of IMF resources. The loan amounted to 290% of India's quota. An interesting element to the loan was its intended use: it was to increase investment in vital sectors, such as coal, oil, rail transportation, and agriculture. The IMF had never before funded oil exploration and production. The United States registered a protest abstention in the vote to approve the loan. Donald Regan, then US Treasury Secretary, criticized the plan, saying that such loans should come from private banks, given that India was very credit worthy. The IMF began to work closely with the World Bank at this time, and the two institutions agreed that the problem in the Indian economy was structural. The IMF and India differed, however, on the timing and specific policies to be used for adjustment. One significant area of disagreement was over the length of the planning period: the IMF wanted the planning process to be broken down into annual plans, whereas the Indian government continued to insist on a five-year planning horizon. To demonstrate the sincerity of its intent to reform, the Indian government did, however, undertake further liberalization in 1981 by removing subsidies on many basic goods and allowing prices to rise to near market levels. In addition, the loan committed the Indian government to increase export incentives, increase the ceiling on domestically administered interest rates, deregulate foreign investment in oil, banking, and industry, and expand foreign borrowing. Unlike most IMF programs, however, these objectives were stated in broad terms, without explicit targets.

Throughout the 1980s, public expenditure increased far in excess of resource mobilization, culminating in a fiscal deficit of 8.5% of GDP for the fiscal year 1990–91; much of this expenditure went toward subsidies, government consumption, and defense. Internal debt that the government owed to its citizens rose from 36% of GDP at the end of 1980–81 to 54% at the end of 1990–91, and the external debt nearly quadrupled, from US$20 billion to US$70 billion in the same period. Interest paid by the government on its debt increased from 10% to 19% of central government expenditure. With roughly one-fifth of government expenditure consumed by interest payments, the difference between income and expenditure of the government had to be financed by borrowing, for which the government had to raise the interest rates to make lending attractive to creditors.

The current account deficit, which stood at US$10 billion (3.5% of GDP) in 1991, was accentuated, due to the Gulf War, by a drop in remittances by Indians working in the Middle East, an increase in the oil import bill, expensive imports of weapons systems financed by foreign borrowing, and a lackluster export performance. During this time, India's receipts of concessional borrowing had fallen to unprecedented levels. Aid utilization as a percentage of imports had also decreased to an average of 5% for the period 1980 through 1985. Simultaneously, there was a run on deposits held by nonresident Indians, and it became difficult even to roll over the existing short-term debt as international creditors held back from lending to India. Commercial bank borrowing increased rapidly during this period, with total long-term claims of commercial banks increasing from US$1.8 billion in 1982 to US$8.3 billion by 1992–almost all of it to US banks. As a percentage of export earnings, the debt service burden (interest payments and amortization) rose from 15% in 1980–81 to 30% in 1990–91. For the first time in its history, India was teetering on the brink of defaulting on its payments. The government was left with no option except to turn to the IMF and World Bank to avoid default and restore its credibility with international creditors. In January 1991, the use of the first credit tranche and the Compensatory and Contingency Financing Facility at the IMF raised US$1.8 billion.

Stabilization and Structural Adjustment

International Monetary Fund programs of macroeconomic stabilization have as their twin objectives the reduction of the current account deficit and the reduction of the fiscal deficit, which gives rise to monetary expansion and consequently inflation. The underlying theory holds that an increase in the fiscal deficit raises aggregate demand, giving rise to inflation and therefore higher prices for goods produced in India. As a result, foreign goods become cheaper relative to Indian goods, leading to a larger trade deficit that increases until it equals the fiscal deficit. A reduction in the fiscal deficit is therefore thought to reduce inflation, bringing down the price of Indian goods and making them more attractive in world markets, thereby leading to larger exports and smaller trade deficits.

The principal instruments used in the pursuit of these objectives are the government's fiscal policy (cuts in government expenditure) and the central bank's monetary policy (increased interest rates) combined with a devaluation of the currency to boost export earnings. This package of policies aimed at compressing demand to curb inflation was also implemented in India. Simultaneously, a World Bank program of microeconomic structural adjustment, intended to shift resources from the government sector and import-competing activities to the private sector and export activities, and of improved resource utilization, by means of increased competition, was also adopted. These measures, required by the IMF and the World Bank, have had their effects felt in the areas of industrial policy reform, trade policy reform, public sector reform, financial sector reform, and tax reform.

Industrial Policy Reform

The area of industrial policy reform has witnessed the most drastic change by means of the virtual dismantling of the system of industrial licensing that had characterized earlier policy. License requirements for entry of new firms and for the growth of existing firms have been abolished, except for a few strategic and hazardous industries and some industries reserved for the small-scale sector. The law regulating monopolies has been greatly amended to facilitate the growth of firms through mergers and acquisitions.

Portfolio investment (shares, debentures, and other financial assets in the domestic market) has now been opened to foreign institutional investors. The earlier policy of encouraging selective foreign investment in high technology areas has undergone a major change, with many consumer goods now thrown open to foreign direct investment. The limit of foreign equity holding has been raised from 40% to 51%, with higher levels permitted on a case-by-case basis. Under the earlier policy, the Foreign Exchange Regulation Act (FERA) prevented companies with more than 40% foreign equity (FERA companies) from borrowing or raising deposits in India, from taking over any business interest from an Indian resident, from acquiring or disposing of any physical assets (including land within India), and from appointing managers or using trade marks without central bank clearance. The 1993 amendment to FERA eliminated the differential treatment of FERA companies and, except for restrictions in the agriculture and plantation sectors, FERA companies are now treated on par with domestic investors and permitted to remit dividends abroad without limitations or restrictions. Remaining restrictions in FERA have to do with controls of capital movements, reflecting the fact that India's capital account is still closed; most capital account transactions involving acquisition or repatriation of assets continue to require prior government approval.

Trade Policy Reform

Prior to reform, India's trade regime was characterized by a complex licensing system with pervasive quantitative restrictions on imports and high import tariffs. As part of the reforms, import restrictions have been dismantled (except for some consumer goods), allowing free import of machinery and other inputs. Tariff levels, which were among the highest in the world, have been reduced significantly, and remaining licensing restrictions on imports of consumer goods are also being removed. Given the high dependence of central government tax revenue on import duties, the tariff reductions are being complemented by significant tax reform.

The rupee has been devalued significantly to make exports more competitive and imports more expensive, and the exchange rate has been floated to a market-determined rate. All permitted current and capital foreign exchange transactions now take place at the inter-bank market rate, and all surrendering requirements have been abolished. Foreign exchange dealers are free to maintain balances in convertible currencies, both domestically and abroad, and deal in spot and forward markets in all major currencies. Except for some services, such as purchase of insurance abroad, travel, and imports of some consumer goods, all current account transactions are permitted without government approval.

Public Sector Reform

Unlike other countries undergoing similar reforms, India has not undertaken a massive privatization of public sector enterprises. Government ownership is being diluted through sale of government equity (with government share of equity being kept to a minimum of 51%) as well as through fresh issue of capital to the public. The induction of private shareholders and the trading of stock in the stock markets is intended to make these enterprises more profitable and increase managerial accountability. Budgetary support to loss-making public enterprises is being phased out, and areas formerly reserved for the public sector, such as civil aviation, petroleum exploration and refining, and parts of telecommunication, have been opened for the private sector to compete with the public sector. For the long term, public sector involvement is being reoriented toward strategic and infrastructural sectors. The sale of government equity in the public sector has also resulted in considerable capital receipts, which have been put toward reducing the borrowing needs of the government and narrowing the fiscal deficit. An exit policy, allowing the closure and restructuring of unviable firms, is being put into place, and a World Bank-assisted National Renewal Fund, to provide compensation and retraining for displaced public sector workers, has been established.

Financial Sector Reform

Reform in the financial sector has been aimed at strengthening banks and at deregulating capital markets. India's banking system was dominated by several large public sector banks and characterized by heavy mandatory reserve requirements designed to support government borrowing at low administered interest rates. Both the statutory liquidity ratio (the minimum percentage of deposits that banks must hold in government securities) and the cash reserve ratio (the minimum proportion of deposits that banks must hold in cash) have been reduced to ensure that resources available in the form of bank deposits are released for the private sector and not preempted by the government. The complex structure of differential interest rates charged and paid by commercial banks has been simplified and rationalized to make them more profitable. Regulated interest rates have been replaced by maximum deposit rates and minimum lending rates. Interest rates on long-term government securities have also been raised close to market levels to reduce the burden on commercial banks imposed by preemptive government borrowing. Accounting practices and prudential norms now conform to the international standards of the Basle Accord. New private sector banks are also being licensed to inject competition into the banking sector; to improve performance and accountability in public sector banks, government ownership is being diluted (subject to a minimum government share of 51%) in banks that raise fresh capital by inducting new private shareholders.

Apart from the banking system, financial sector reform also encompasses the reform of capital markets aimed at financing investment in the private sector and at attracting foreign portfolio capital. Foreign institutional investors, such as mutual and pension funds, are now allowed to invest in the capital market. A second route for foreign portfolio investment has been the issue of shares abroad by Indian companies, and several Indian corporations have mobilized significant amounts of capital in this way. Interest rates in the domestic capital market have been deregulated, and the need for prior government approval of the size and price of equity issues in the primary capital market has been dispensed with. The deregulated capital market is to be overseen by the Securities and Exchange Board of India, an independent statutory regulatory body that has introduced a framework of rules and regulations to govern trading practices, standards of disclosure, speed of settlement, and transparency of transactions. This appears to have helped the market recover from a major securities scandal in which bank funds were illegally siphoned off into the stock market. A new National Stock Exchange with computerized screen trading has commenced operation, and other stock exchanges are being computerized and modernized.

Tax Reform

Far reaching tax reforms involving a broadening of the tax base and a reduction in exemptions have been introduced to compensate for revenue loss through tariff reductions. Important features of tax reform include reduction in the maximum marginal rate of personal income tax, exemption of all financial assets (including shares and other corporate securities) from the wealth tax, reduction of corporate income taxes, reduction of customs duties to bring them in line with other developing countries, and shifting the bulk of taxes to an ad valorem basis. The ad valorem tax replaces the specific excise duties on many domestic manufactured goods that were charged at varying rates on different goods. The long-term objective is to move to a unified value-added tax, but this is not a prospect in the foreseeable future, given the constitutional provision for taxation on production by the central government and taxation on sales by state governments. Further simplification and rationalization of domestic indirect taxes while extending the tax to services are, however, being considered. The tax reforms taken together are intended to avoid microeconomic distortions and to attain macroeconomic buoyancy of tax revenues.

Effects of the Reforms

The results of the reforms have been generally encouraging when viewed from the perspective of macroeconomic indicators: inflation has dropped below 7%, exports are growing at an annual rate of 20%, gross savings have risen to 26% of GDP and gross domestic investment to 27% of GDP, foreign direct investment has risen from US$155 million in 1991 to US$2.1 billion in

1995–96, foreign portfolio investment crossed the US$3 billion mark in 1996, the rupee has stabilized at about 35 to the US dollar, and the fiscal deficit has fallen to 5% of GDP. When various other aspects of the reforms are taken into account, however, the results appear to be mixed, and the long-run effects still indeterminate.

The theoretical justification for the mix of policies advocated by the IMF and the World Bank derives from the neoclassical school of economics that had been in the ascendant in these institutions since the seventies and that has now achieved a position of dominance within them. In India, however, the Keynesian tradition remains strong, and one of the remarkable features of the IMF and World Bank-sponsored reforms is their implementation in an intellectual milieu that has never been favorably disposed to neoclassical theory. Predicting whether or not the reforms will have their intended effects in the long run is of course a theoretical question, and it is therefore worth recalling briefly some traditional Keynesian objections to the neoclassically derived stabilization and structural adjustment policies.

The fundamental issue is whether a reduced fiscal deficit necessarily translates into a smaller current account deficit and lower inflation. Neoclassical theory holds that by compressing aggregate demand, a smaller fiscal deficit decreases the demand for imported goods and reduces the current account deficit. Here, the traditional Keynesian objection turns on the exact mechanism through which the economy will adjust. Rather than prices falling, it is theoretically quite conceivable that output would fall, leading to higher unemployment. It is also possible that the accompanying devaluation of the currency increases rather than decreases inflation by raising the cost of imported inputs in domestic production and consumption in the short term. If the inflation rate exceeds the rate of increase in money wages, real incomes may fall, leading to further demand contraction and a fall in output.

Neoclassical theory tends to view fiscal deficits as intrinsically inflationary, whereas Keynesian theory tends to take a more sanguine view of fiscal deficits. In the latter theory, the effect of the fiscal deficit depends on the rate of return on government investment expenditure relative to the cost of borrowing. If the productivity of such investment is higher than the real rate of interest and so long as the income flow exceeds the burden of servicing the debt, fiscal deficits can lead to growth. Deficits become a problem when they are incurred to support consumption rather than productive investment.

In Keynesian theory, deficit reduction through reduced government expenditure could well lead to decreased purchasing power in the economy and–because of the "multiplier effect"–the magnitude of this decrease could be larger than the original reduction in the deficit, with consequent loss of income and increased unemployment. The trade deficit would still fall, because of reduced imports, until it equals the fiscal deficit, but this would happen through falling employment and output rather than falling prices. A restrictive monetary policy could further aggravate this by squeezing credit and dampening investment.

Ever since independence, one position that the majority of Indian economists hold is their consistent hostility to foreign capital. The reforms in the financial sector have now brought these issues to the fore because they have led to large capital inflows for portfolio investment by foreign institutional investors. The share of productive investment has been relatively low, and this has given rise to fears that the reforms have only led to existing financial assets changing ownership with little by way of additional long-term productive capacity. Decisions on investment in physical capital assets are largely made by the managements of multinational corporations, whereas portfolio investment decisions are made by mutual and pension fund managers. Attracting portfolio investment has been much easier, because these investments are easily repatriable at short notice whereas productive investments involve much longer time commitments. The high interest rate required to attract financial investments and maintain a strong rupee has invited the charge that India is being turned into a "rentier economy." It is also feared that the large inflow of capital to take advantage of high interest rates could lead to an increase in the money supply and consequently inflation–contrary to the original intent of decreasing inflation.

The distributive effects of the reforms have also attracted criticism. Tariff reductions, for example, were sought to be made revenue-neutral by increases in indirect taxes that fall disproportionately on the poor through increases in the prices of essential commodities. Import liberalization may similarly switch domestic demand away from domestically produced goods to foreign substitutes, leading to increased unemployment. There has also been concern about widening regional disparities, with states such as Maharashtra, Gujarat, Karnataka, and Tamil Nadu attracting the lion's share of investment while states such as Bihar and Orissa lag far behind.

The crux of the difference between the neoclassical and Keynesian approaches lies in whether the management of public finance should entail balanced budgets, as the neoclassicals maintain, or whether under conditions when private investment is not forthcoming, public deficit spending is permissible, as the Keynesians maintain. The setting out of IMF–World Bank criteria–such as a 5% upper limit on fiscal deficit as a percentage of GDP or elimination of the monetized deficit caused by borrowing from the central bank–stems from the underlying neoclassical belief that government spending must be subject to a balanced budget discipline. This is perhaps the most fundamental bone of contention between the two major schools of economics, and the only safe prediction to make here is that the matter is unlikely to be decisively resolved anytime in the foreseeable future; what both schools *do* agree on, however, is that with greater global integration, the power of national governments to incur large deficits or set low interest rates to encourage domestic investment is largely taken away by the fear of capital flight to more profitable destinations.

Future Reforms and the Prospects for Growth

Considerable discretionary power over the pace of future reforms still rests with the central government and, in a federal democracy such as India's, the future of liberalization is likely to be dictated at least as much by the exigencies of regional and national politics as by considerations of economic efficiency. Subsidies to the politically powerful farming sector are unlikely to be cut by any political party, and it is also improbable that there will be any large-scale privatization or drastic change in labor laws. Even as income and corporate taxes were being slashed in the 1997 budget, food, fertilizer, and rural credit subsidies were increased. Government consumption expenditure also increased: the defense budget was up, and the allocation to government salaries swelled to US$1.1 billion on the recommendations of the fifth Pay Commission. Tight constraints have been put on the government's ability to monetize the deficit through the practice of drawing "ad-hoc treasury bills" on the central bank at administered interest rates. The hope is that tax cuts will rebound in high revenue realization through high growth as well as through a broadening of the tax base and increased compliance. Divestment in some state-owned companies is also expected to raise about US$1.4 billion.

A major determinant of India's growth rate has always been the monsoon rains, which have been good for many successive summers, contributing to annual growth rates averaging 7% in the last three years; yet the monsoons remain notoriously unpredictable, and a failure of the monsoon season can have serious repercussions throughout the economy. The other significant impediment to growth is the infrastructural bottleneck. Private investments in power have been constrained by the low tariffs set by state governments, and power output for 1995–96 was up just 3.4% against a 10% rise in demand. Domestic crude oil production also fell by 10%. Serious problems lie ahead if power generation and energy resources continue to lag disproportionately behind galloping demand.

With the assistance of the World Bank, several states are undertaking large-scale reforms in their power sector. The eastern state of Orissa has disbanded its State Electricity Board and created three separate agencies for the generation of coal-fired power, hydroelectric power, and distribution. Orissa Power Generation Corporation, which owns two thermal units of 210 megawatts each, is to sell 25% equity to private parties, with the Orissa Hydro Power Corporation and the Orissa Gridco Corporation expected to follow suit. The World Bank has made a loan of US$350 million to the government of Orissa for the restructuring of its power sector, and this is now seen as a model for other states. The Asian Development Bank has approved a loan of US$350 million to the western state of Gujarat to bolster financial restructuring and infrastructure privatization. Talks are now under way between the World Bank and the government of the state of Andhra Pradesh to finalize what is likely to be the World Bank's first package of loans for state-level structural adjustment negotiated directly with the state government. These loans are expected to focus on moves to enhance government revenues while cutting subsidies; a major condition of the loans is likely to be the reorganization of the state's electricity board into separate units for generation and distribution on the Orissa model accompanied by increases in power tariffs and cuts in staffing. Similar plans are being prepared in the northern states of Haryana and Rajasthan. All this is part of a shift by multilateral lenders toward specific state-level lending packages in India; because of the limited borrowing and fiscal resources of the states, it is likely that this trend will gain added momentum in the near future.

Despite these measures, overall progress in adding capacity in the power sector has been slow. Of the eight "fast track" power projects approved since delicensing in 1992, only one unit has been commissioned; two more projects are under construction, and the rest are still under negotiation at various levels of the central and state governments. India's state electricity boards incurred a combined loss of US$1.8 billion in 1994–95, and despite the reforms initiated in some states, most state governments are hesitant to introduce any changes in the electricity subsidies for farm use that contribute to the large operating losses.

Other significant changes urged by the World Bank in its 1996 country study for India are

1. Further lowering of the fiscal deficit to bring forth a strong private investment recovery and to contain food subsidies. The study recommends more effective management of the record levels of food stocks in the warehouses of the Food Corporation of India and suggests that food storage activities be allowed by private traders.
2. Continued tax reform and simplification so as to increase tax revenues
3. An acceleration of public enterprise reform, including privatization of chronically loss-making public sector enterprises that serve to crowd out private investment. The study says that public enterprise managers have not been given the authority to carry out large-scale retrenchment, reorganization, closure, and selling of units, or to undertake joint ventures. It points out that progress with respect to liquidation of loss-making enterprises and shedding excess employment has been very slow.
4. Strengthening the administration of existing poverty programs. The study is critical of additional programs, as they could "strain the delivery system" and are funded by the banking sector.
5. Further liberalization of the import regime, especially the removal of restrictions on imports of consumer goods and the trade of agricultural commodities–two important obstacles to the fuller integration of India into the world economy.
6. The development of a more competitive financial system with a wider range of financial products and services and

greater electronic capability to support intercity payments and a rising volume of transactions.

The most politically sensitive areas are food and agricultural subsidies, labor laws, privatization of insurance and other public enterprises, and intellectual property rights. The 1997 Indian Infrastructure Report had urged significant insurance reform to help create long-term funds to finance an infrastructure bill estimated at US$130 billion in the next five years for road, power, telecommunications, and urban projects. The latest budget, however, only allows for minority foreign participation in private health insurance joint ventures. Fuel prices also remain a sensitive issue, although with the subsidy bill approaching US$4.4 billion, the government is likely to have no option soon but to raise the administered prices of petroleum products. Complete deregulation of oil prices, however, does not appear to be politically feasible in the near future.

The Dunkel draft of the General Agreement on Tariffs and Trade (GATT) regarding intellectual property rights protection had generated intense debate in India, especially regarding the effects on domestic pharmaceutical companies, farmers, and traditional healers by the international patenting of drugs, seeds, and medicinal plants. The prospect of retaliation by other countries and the consequent loss of export markets eventually prompted the government to adopt a pragmatic approach to intellectual property rights protection. Although this means that India would have to pay more for pharmaceutical patents, it can expect to gain through patent protections on its software, video, and film exports. The case of intellectual property rights illustrates the confluence of cultural, political, and economic factors that eventually determine the outcome of particular reform initiatives.

In the vast ethnic, religious, and political mosaic of India's democracy, it is hazardous to venture any guess regarding the future course of specific reform proposals. Nevertheless, as the fiercely independent Indian press is fond of saying, there now appears to be "broad consensus" on the necessity for reforms across the political spectrum, and differences center mostly on the pace of the reforms and their particulars. The old "license-permit raj" has definitely been consigned to history; something new is taking its place, even if it has not come fully into view. It will affect the fates of a fifth of all humanity–and the other four-fifths will be watching closely in the years to come.

Further Reading

Ahluwalia, Montek Singh, "Balance-of-Payments Adjustment in India 1970–71 to 1983–84," in *The International Monetary System and Its Reform: Papers Prepared for the Group of Twenty-Four by a United Nations Project Directed by Sidney Dell 1979–1986,* edited by Sidney Dell, Amsterdam: North Holland, 1990

Alhuwalia discusses how India coped with the two oil shocks and other changes in the international lending environment. He presents a model that simulates the balance of payments adjustment.

Bhaduri, Amit, and Deepak Nayyar, *The Intelligent Person's Guide to Liberalization,* New Delhi: Penguin Books, 1996

The recent reforms in India are discussed in very accessible language without sacrificing detail and scope.

Bhagwati, Jagdish N., *India in Transition: Freeing the Economy,* Oxford and New York: Oxford University Press, 1993

This book is the text of Bhagwati's Radhakrishnan lectures at Oxford. It has an overview of India's early "inward looking" economic policies and how they led up to the economic reforms.

Cline, William R., *International Debt Reexamined,* Washington, D.C.: Institute for International Economics, 1995

Cline updates the debt problem with comprehensive and detailed statistics on borrowers and creditors. India is listed as a moderate debtor nation.

Faruqi, Shakil, editor, *Financial Sector Reforms, Economic Growth, and Stability: Experiences in Selected Asian and Latin American Countries,*

EDI Seminar Series, Washington, D.C.: The World Bank, 1994

This is a collection of papers presented at a seminar in the Economic Development Institute at the World Bank. One chapter is devoted to India.

Hogendorn, Jan S., *Economic Development,* 3rd edition, New York: HarperCollins College Publishers, 1996

This economic development textbook has a good section on the role of the IMF and World Bank in global lending.

Lateef, Sarwar K., *The Evolving Role of the World Bank: Helping Meet the Challenge of Development,* Washington, D.C.: World Bank, 1995

This publication evaluates the role of the World Bank for the last fifty years in rather positive terms. There is a chapter on the success of World Bank assisted food and agricultural policies.

Lewis, John Prior, *India's Political Economy: Governance and Reform,* Delhi: Oxford University Press, 1995

Lewis, who was with the US government in India for many years, evaluates India's economic policy and foreign assistance receipts.

Lipton, Michael, and John Toye, *Does Aid Work in India? A Country Study of the Impact of Official Development Assistance,* London: Routledge, 1989

This is a revision and updating of a previous work called *Does Aid Work?* Lipton and Toye are skeptical of what development assistance does for the Indian economy.

Muir, Russell, and Joseph P. Saba, *Improving State Enterprise Performance: The Role of Internal and External Incentives,* World Bank Technical Paper Number 306, Washington, D.C.: World Bank, 1995

This World Bank study looks at public sector reform and privatization efforts in various countries including Indonesia, France, Italy, and Poland. India's case study is about Hindustan Machine Tools.

Oliver, W. Robert, *George Woods and the World Bank,* Boulder, Colo., and London: Lynne Rienner, 1995

This biography of a past president of the World Bank is a journalistic rendition of the development of the IDA and the involvement of various personalities in that process.

Payer, Cheryl, "The IMF and India," in *Seminar Proceedings No. 18: The IMF and the World Bank in Africa. Conditionality, Impact, and Alternatives,* edited by Kjell J. Havnervik, Uppsala: Scandinavian Institute of African Studies, 1987

Payer is a well-known critic of IMF and World Bank lending policies. In this chapter, she describes how India "fell into the IMF debt trap."

Srinivasan, T. N., "'Economic Liberalization and Economic Development: India," in *Journal of Asian Economics* 7, no. 2 (1996)

Srinivasan has an approving account of India's economic reforms and the improvements they are likely to bring about.

Stiles, Kendall W., *Negotiating Debt: The IMF Lending Process,* Boulder, Colo.: Westview Press, 1991

This book has excellent institutional details of the IMF lending process. The loan to India in 1981 is included as a case study.

World Bank, *India: Five Years of Stabilization and Reform and the Challenges Ahead,* A World Bank Country Study, Washington, D.C.: World Bank, 1996

This study applauds the reforms carried out so far in India and suggests directions for further reform in trade and the public and financial sectors.

World Bank, *World Bank Support for Industrialization in Korea, India, and Indonesia,* Washington, D.C.: World Bank, 1992

Roby Rajan is an associate professor in the business department at the University of Wisconsin, Parkside. His work has appeared in numerous journals including the *International Economic Review,* and his current work focuses on the predicament of traditional cultures within the global economic system

Farida C. Khan is associate professor of economics at the University of Wisconsin, Parkside. Prof. Kahn's research has focused on international trade policy in small developing countries, and he is currently examining trade relations between India and Bangladesh as well as the empirical modeling of household production in rural economies. He has also been a consultant to the World Bank.

Future

Chapter Nineteen

India's Development at 50 Years: The Center, the States, and the National and World Economy

Subbiah Kannappan

India: The Multinational State

There are three arenas in which India's development has been influenced since its independence half a century ago. Although they have shaped the country's political economy, they are not well studied. One reason is that people do not consider India in the context of its details, or they are fazed by its complexity. Economists and decision makers also tend to focus on the country as the unit of analysis at the level of decision making (Guhan, in Cassen and Joshi, 1995: 73). My concern is with the governmental and developmental legacy that enveloped the country and defined its political economy, the equation of power between the center and the states, the inevitable changes in this equation, and the incipient, as yet inchoate challenge to the underlying political economy today. My concern is not with the usual center-state breakdowns but with how democratic development affected the multinational state and vice versa as well as the portents for the future. This is, therefore, only an exploratory essay, raising issues that are important to studying India's future development. To ignore the center-states power relations and the parameters that govern them is to miss a basic ingredient of India's political economy and the significance of what is taking place today. In addition, the tendency to treat the country's class, caste, and religious breakdowns as an India-wide phenomena is to bypass the importance of the states, which are diverse in these respects, as the basic building blocks. The interactions of the external economy with the multinational state and its component states have also escaped analysis (with the exception, of course, of orthodox Marxists, who view the state as an agent of international capitalism). The first requirement is thus to get a grasp of the singular position of the country in terms of both its size and diversity.

Much attention has been given to the economic significance of countries that are comparable to India in terms of population and output, such as the European Economic Community. Among the giants, India is like Europe with its many nations, but unlike Europe, it is under the authority of an established central government. It is also unlike China, the United States, and Russia, which have a central government but are not multinational. The sui generis position of India reflects its legacy as the "crown jewel" of the British Empire and the long history before.

When India gained independence, it lost Pakistan, but it incorporated the rest of British India. British India in turn consisted of earlier empires, kingdoms, and nationalities that had been incorporated under an effective central political authority only under Britain in the country's long history–actually, not even then and not under the Mughal Empire. However, Britain's rule lasted for over a century. The new India acquired most of princely India, a collection of major kingdoms and fiefdoms (some as large as France or Italy), which had not been under Britain's direct colonial rule. Significant for our purposes is that the new Government of India (GOI), or the center, as it is often called, inherited intact the imperial structure of governance. This included a vast governmental hierarchy, led by the viceroy, the governor general, and the governors of the major presidencies. They were complemented by well-knit, all-India services, led by the famed Indian Civil Service (ICS, labeled the "steel frame" of the Raj), which provided the heads of the GOI departments, district collectors, revenue and railway officials, magistrates, and lesser officials in charge of the towns and rural subdivisions. These were backed by the security services, including the Indian army and police forces. The chain of command covered all the major jurisdictions and exercised civil, managerial, judicial, fiscal, and monetary authority. This apparatus was inherited intact by the GOI and adapted and extended to cover the wide range of new responsibilities of independent India. At a political level, the Indian National Congress, the party of freedom, assumed authority over all of India. Endowed with a legitimacy that the departing British government never had, New Delhi, the capital, emerged as far more significant in the nation's affairs and with much greater authority. It was not just the center, but an imperial center. The political structure and popularity of the ruling party, the preferred ideology of development, and the charismatic leadership of those united in the struggle for independence contributed to this result.

New Delhi: The Imperial Center

New Delhi's emergence as an imperial center is seen best when compared with the features of India as a multinational state. The developmental significance of such a large and diverse country is lost in analyses that treat it as a composite entity and its heterogeneity as comparable to other countries that lack India's size, diversity, or both. The Union of India is today a political entity of about a billion people, about a sixth of the world's population. It is larger than continental Europe (including Russia), North and South America put together, or all of Africa. It comprises some 25 major nationalities in established territorial boundaries, each with a dominant written language, some even with separate scripts, and distinctive literary and cultural legacies. These should not be confused with minorities, for no one nationality is dominant in the entire country. Every nationality, some as large as Russia or Germany, and others comparable to a medium-sized European nation, reigns supreme in its territorial domain. Andhra Pradesh, home of Telugu, and Tamil Nadu, home of Tamil, in 1991 housed only 8% and 6.6% respectively of India's total population, but their populations of 67 and 60 million each equaled or exceeded the populations of the Philippines, Thailand, Poland, Iran, the United Kingdom, Italy, and Ethiopia. They also regularly organize worldwide convocations in their home states, the United States, or Europe to salute their global diaspora. Even the polyglot, cosmopolitan megacities, such as Bangalore, Bombay (now Mumbai), Calcutta, Greater Delhi, Madras (now Chennai), and many others, and their metropolitan labor markets are dominated by the language of the immediate nationality. The autonomy of the constituent states is not likely to disappear with economic growth or interstate commerce, as happened to the 13 founding "sovereign" states of the United States, if for no other reason than the enduring difference of language and their diverse historical heritages. The organizational and political processes, based on universal adult franchise, underwrites the autonomy and equality of each state. English rather than any Indian language has served as the official language, the language of the Constitution and the associated *travaux preparatoires,* the common language between the different states and of international transactions, modern education, business, and communications. Finally, the hallowed national anthem is the great poet Rabindranath Tagore's expression of the Union of India and names each major nationality in a sacrosanct acknowledgment of a union.

The Role of Development Ideology and Competitive Coexistence

A number of factors led to and sustained the imperial center. From its founding in 1885, the Indian National Congress had equipped itself for self-government over several decades. Unlike other developing countries, it had worked out tentative plans well ahead of independence in 1947. This envisaged a dominant governmental role for realizing economic development, industrialization, and social justice. The Congress High Command, as it was called, consisted of the senior leaders of the party who were representative of the different nationalities, and they continued at the helm. Several of them also made the transition to high office in the new setup. At the head of this was the charismatic Prime Minister Jawaharlal Nehru. Esteemed nationally and worldwide, he made rational economic planning within a secular, democratic framework, the creed of the new India.

Global developments greatly added to the center's dominance within the country and outside. Security considerations, already important because of India's history of colonial subjugation, emerged as an even greater priority because of the Cold War and its destabilizing alliances. The "British Revolution," Moynihan's term for the liquidation of the European empires and emergence of new nations, underscored the ideological significance of the Indian approach in a world split between Moscow, Washington, and the nations shaking loose from colonial rule. It was widely hoped that India would be an appealing alternative to communism, authoritarianism, or instability. The Cold War and China's turn to communism further underscored India's ideological and strategic significance and gave rise to the phase of competitive coexistence. At the time of independence, India already had a monopoly of foreign reserves on government account, including substantial nonconvertible sterling balances, to be rationed out on a priority basis. The inflow of foreign assistance further reinforced the center's allocative role and dominance.

The prevailing ideology of the development state emphasized the pivotal role of governments. Indeed, most discussions do not distinguish between the state and government. In India, there was also the widespread belief that the British Indian government had held back Indian development and that a positive, activist role by the GOI would by itself turn the country around. Much has been written about the socialist, left, right, pro-Western, and pro-Soviet biases in Indian planning and priorities. Most of this is beside the point and misleading, for the real emphasis was one of building an autonomous indigenous capability and reducing national vulnerability. This resulted in a high priority for defense, scientific and technological capability, heavy and basic industries ("machines to build the machines which would build the machines"), large-scale infrastructural development, and, since the food crisis of the 1960s, self-sufficiency in food. These were mostly capital intensive, with a long gestation in terms of returns. They represented trade-offs between deferred and collective "nationalist" goals and present distributive and consumptionist objectives. It is pertinent to note that there was no major challenge to these objectives within the country. Although high levels of nonconsumption, to borrow Gershenkron's terminology, were characteristic of communist countries, India was an exception among the low-income democracies in its emphasis on stepping up investment. It was the paean to nation building that enabled the prime minister to emphasize sacrifice in his rhetoric ("This generation is condemned to hard labour") with, of course, the expectation of benefits for a later generation.

The charisma was not merely the reflection of one man but of the prevailing ideology and of the supporting cast he had mobilized, indeed galvanized, for the tasks of democratic development. The nation's top brains, and there were some world-class people among them, went to work on the outlines of the five-year plans and pressed into service the available economic studies and data. A series of major documents, all home spun, covered all major economic issues and sectors, including industry and technology development. Since the thrust was on control and direction of the economy, especially of its "commanding heights," the inherited bureaucracy had an even larger role to play than before. An imperial, all-India vision that transcended provincial ties and was insulated for public pressures gave rise to self-confident, even smug assertions, such as "the writ of the GOI runs throughout the length and breadth of the country," or "we act without fear or fervor" in discharging our responsibilities. The term *civilian* was supposed to indicate an elite category of officials privileged to act thus. Since a directed, controlled economy required discretionary decision making, this power was also centralized at the highest levels.

The technical expertise needed to formulate and work out the details of economic policy for the long haul and for the whole economy also reinforced centralization. The formulations were visionary, abstract, aggregative, and mathematized. It dealt with sectoral priorities and interactions over time, which is why I. M. D. Little called the exercises "magnificent dynamics." The investment strategy went beyond the spontaneous *tatonnements* of the market and what might result toward planned targets and coordinated actions aimed at convergent and simultaneous outcomes on many economic fronts. The dialectics of India's political economy are thus that interplay of the prime minister and his close associates, the sophisticated bureaucracy and technocracy that were responsible for the multiperiod, rational economic formulations, and the relatively earthbound political leadership at large. The leadership was even more earthbound and dependent at the state levels, with their more parochial concerns and shorter time horizons. The expertise was available within the country to elaborate on the "magnificent dynamics" in the form of intersectoral, multistage planning models. They were elusive even to economists. The required empirical underpinnings posed novel challenges. The chosen coefficients reflected the normative drives of the elitist planners and leaders, although every effort was made to bring everyone into the planning process. These directions of policy and relationships changed as economic and political developments led to, even forced, changes in them.

A major point to remember is that the Indian development approach commanded influential and broad-based support among the world's donors and economists. There was widespread support for Indian plan targets to step up saving and investment. There was also implicit endorsement of the crucial role of the government in directing investment. Prevailing theories had defined development as moving from a 5% to a 15% or 20% saver and talked in terms of a "Big Push," "critical minimum effort," "low-income equilibrium trap," "market failure," "investment criteria," capitalistic transformation drawing on "unlimited labor supplies," etc. Donor agencies lauded India's effort as a model of economic planning. An influential "two-gap" theory stressed the importance of sustaining the investment strategy with inflows of foreign savings to redress resulting balance of payment deficits. There was also the global vision of India (a proxy for developing nations) and the United States (standing for the West) pooling their abundance of labor and capital in a concerted effort to overcome poverty and advance democracy. These ideas were reflected in the doctrines of the Alliance for Progress and the leapfrogging envisaged in Walt Rostrow's accelerated "take-off" into self-sustaining growth. There were dissenting voices, of course, on the size of the leap, on sectoral priorities involving agriculture or foreign trade, and, by a minority of conservative economists, on the overweening role of government. But these were set aside even in the West. The promise of an economic axis between rich and poor was undoubtedly an important factor.

The US-USSR rivalry and the foreign aid regime of "competitive coexistence" also strengthened the center's dominance as the conduit to the country of foreign capital and technology.

These confident assumptions of a global partnership were to founder for their political naïveté and the complications of national and world politics (Rosen, 1985). Economic policy judgments are also political statements. India's technocratic and decision-making elite, which after all prided itself on being able to chart its own course, balked at too close an involvement with foreign agencies on policy choices. In turn, the Indian technocratic and bureaucratic expertise that combined forces at the center had to reckon with powerful internal factors. The Government of India Planning Commission, which appeared at one stage as a potential super cabinet, an Indian version of the Gosplan authority that would supersede ordinary political processes, faced major political challenges at home. Plans need to be dovetailed with appropriate macropolicies and expenditure controls. While the finance ministry and the Reserve Bank of India (RBI) were reinforcements on the conservative and budget constraint side, the priorities and procedures of the "spending" or "operating" ministries were at variance and reflected the distributional and capital-building goals expected of them. Far more important were the states, the constituents and interest groups they served, and the populist and nationalist pressures they faced from within their own states. Their leadership showed more commitment to spending than raising resources. Their priorities were different. They also had political clout.

Center-States Economic Regimes: Dependency and Bargaining Relationships

A council of chief ministers emerged alongside the cabinet to meet directly with the prime minister to formally approve the plans. As the political processes asserted themselves, the Planning Commission evolved over time toward the status of a staff authority, and its proposals became working documents rather than diktats. This role diminished further when the Planning Commission became mainly a review body to check proposals against plan priorities. The influence of the states was felt in allocational decisions, especially in plan projects and the location of public-sector enterprises. Pricing of key inputs, such as steel and rail tariffs, sought to reduce or eliminate differential costs of unequal access because of location. Under the regime of monetary repression and in the absence of capital markets, the states pressed the center for needed funds. These were handled by successive Indian Finance Commissions, which set periodically revised guidelines for the distribution of central revenues. The RBI, acting under the authority of the finance ministry, dealt on a more continuous basis with the states' requirement of overdrafts. Foreign exchange was a particularly critical focus of convergent pressures, for it was allocated at a very low rupee price, far below its scarcity value for the economy. The original rationale was for stringent and rational direction, but the directive role of the center was diluted and distorted over time.

The state economic regimes replicated the center but with important differences. The state planning mechanisms were less sophisticated, being reactive to and improvisations of operating departments without the independent technical strength. The states also had their own priority projects for development by government agencies or parastatal enterprises such as electricity boards, transport corporations, housing authorities, industrial finance corporations, etc. The state governments adopted the center's practice of discretionary allocation of resources and favors in their control. The states were also subject to pressures from influential landed classes and populist pressures (there being no general pattern for all of India), and they responded with measures befitting the politics of the states concerned.

It must be noted that the state authorities were closer to the people and were more vulnerable to political pressures and opportunistic as well. Inevitably, ad hocism, nepotism, and corruption gained ground. Examples include appropriations of sanctions for cultural or educational purposes, bread-and-butter goals, preferences for backward areas and communities, tax relief to protesting farmers, concessions to attract industrial ventures (electricity tariffs [Vijayamohanan N. Pillai, "Seasonal Time-of-Day Pricing of Electricity under Uncertainty," Ph.D. diss., University of Madras, 1993], excise taxes, land allotments), lobbying on behalf of the state industrialists at the center, photo-opportunity sessions for chief ministers (and others who could not be left out), highlight gifts for the poor and needy, and discretionary decisions in routine administration of programs for low-income housing, admissions to schools and colleges, etc. These were aimed at building a political base where and when it counted, and one can understand them only in terms of the dynamics and politics of each state. At the risk of generalization, however, it is possible to emphasize some common aspects.

Although the state governments reflected diverse objectives and pressures, they echoed the center's secular, socialist, and industrial goals. They challenged only the size of the cuts of the pie the center was dishing out and clamored for more. The states' primary interest was in their political base, and this meant priorities that furthered their autonomy, culture, language, and literature. Populism inevitably dominated because of universal adult franchise and a political process in which the mother tongue dominated, which meant neither English nor Hindi (in the non-Hindi states). The states were also more interested in pursuing what the development literature has termed "basic" needs. Examples include food or consumption subsidies, midday meals for school children, rural electrification, shelter, basic health services, guaranteed employment, land reform, schooling, etc. However, the power structure in the individual states varied substantially, as did the commitment behind the drive to eliminate poverty or help the backward castes. The impact of this structure could be regressive or negative in terms of growth because of competing aims or preferences of the dominant castes, business, and landed interests. The state leaderships also pressed for center projects and allocations of expenditures, revenues, scarce resources, or credit on concessional items. The center's "failures" were, of course, publicized.

The bargaining relationship with the center emphasized the center's pipeline as the important resource to tap rather than other sources. The system was not designed to encourage the states to be self-reliant or to raise funds from private markets. Initiatives such as state industrial finance corporations or housing development authorities were stillborn, stifled by a preemptive bureaucracy. The states were also not keen on sharing their surpluses or relative abundance with the directed national economy. The food surplus states balked at the center's efforts to commandeer grain surpluses at procurement prices. The states with the highest contribution to foreign exchange earnings pressed hard for more foreign exchange and plan allocations. State enterprises have generally posted huge losses. The states had a limited tax base but lagged in broadening or exploiting it, especially in terms of agricultural incomes, whose taxation is in their sphere in the Constitution. They have also voiced complaints about their share of official development assistance. In the course of time, the states became more or less equal partners in negotiations on loans (for instance, with the World Bank) for major infrastructural projects. The RBI was continuously importuned for overdraft accommodation at concessional terms. However, these were not strategies aimed at competitive access to the money and capital markets but instead at increasing their share of the national grants economy. Some states did make efforts to attract private investment from within the country; these were on negotiated terms involving concessions on land, local taxes, electricity tariffs, labor policy, etc. The states were not in the forefront, however, in seeking private foreign investment. The few who ventured did not meet with encouragement, either. For instance, the state of Rajasthan teamed up with Tatas to bring in PepsiCo but ran into stonewalling delays at the center as well as opposition from Indian business interests.

The bargaining relationships between the center and the states were for the most part tussles within one ruling party and were part of a centralized economic regime in which they had a joint stake. The leadership at the center and states were thus collaborators maximizing their respective, sometimes opposed interests. This was a unique phase of Indian history, with the center in almost exclusive command of the economy and the states as among the major clients.

The Dirigiste Economy, the Center, and the States: Senior, Junior, and More or Less Equal Partners

A triad of forces led to and held together this dirigiste regime: a dominant ideology, the governing and organizational framework, and the resource base.

Development policy was a product of an appealing economic viewpoint and parallel political purpose. A pan-Indian and visionary ideology of secularism, economic progress, and social justice, inspired by a charismatic leader and the euphoria associated with an independent government at the helm, proclaimed a strategy that was the very opposite of the "minimalist state" approach to development. This was backed by the Congress Party, the party of independence, whose national standing had been forged over three quarters of a century in opposition to Britain's rule. The technical elites, the professional classes, and the opinion makers wrote and functioned in English, and their vision was more expansive than the parochial visions of grassroots leaderships, especially the burgeoning nationalisms of the states. An example is Tamil Nadu's cognitive elite, which included talented Brahmins who served with distinction in the center and everywhere else but were limited in their political clout because of the state's anti-Brahmin sentiments.

As for the organizational framework, the center held command through its framework of appointed governors who worked alongside elected chief ministers and an all-India framework of civil, judicial, security, and professional cadres, who also manned the senior ranks of the state governments. The center had also the reserve authority to suspend elected state governments as well as to intervene to maintain law and order. The political structure of the Indian National Congress was also centralized, and the leaderships were acting within an established all-India framework of ideology, old loyalties, and party patronage. Getting a Congress "ticket" was essential for electoral success in the state, and there were any number of able politicians vying for the nod, there being no viable countrywide alternative.

The Indian National Congress was the only party with an all-India standing. Neither the confessional Hindu parties nor the secular communist parties had a pan-Indian appeal that could carry them beyond their enclaves in specific states in the Hindi belt, Kerala, or West Bengal. Generally, when the Indian National Congress lost in the states, it was not to any all-India alternatives but to parties restricted to the language frontier of the specific state and that had no standing beyond. In cases where the Congress experienced a total rout, its leaders were without a power base in the state, except for their access to the center's favors. This process had begun even under Prime Minister Nehru. He had a unique appeal throughout the country, faced no challenge within the party, and developed an active, primus inter pares working relationship with the chief ministers, whose political standing was in many cases even greater than that of his cabinet appointees. After his death, the center-states relationships embarked on a more uncertain phase.

The resource base is somewhat more involved. The monopoly of money supply and foreign exchange, including the accumulated sterling balances and inflows of foreign assistance, naturally enhanced the center's leverage. Other factors that were accommodative of a dirigiste economy included the dominance of direct taxes in the Indian fiscal system and a thin but national system of commercial branch banking and finance, organized industry and plantations, and the infrastructure of railroads, harbors, and transport. The center's control was further formalized in the Constitution and legislative instruments and distinguished from the areas of exclusive state jurisdictions (especially agricultural income taxes) and overlapping jurisdictions (some commodity and direct taxes). An excellent overview is provided by S. Guhan.

> The wide-ranging responsibilities of the Centre . . . extend far beyond the core functions of any central . . . government. . . . They include control over the financial sector . . . [including commercial banks and insurance, both nationalized] . . . conduct of elections and audit at both the central and State levels . . . [and] . . . recruitment of all-India civil and police services. It is concerned with research, standards, basic labour laws, and institutions of national importance. It holds a monopoly over radio and television. Legislation for promotion and regulation of industries is a central responsibility and the Centre exercises control over mines and oil development. The Centre is also responsible . . . for . . . umpiring in matters in so far as they concern more than one State. Furthermore, . . . the Centre . . . [is empowered] to extend itself to important areas in which the States have primary responsibilities. . . . Most importantly, the catch-all entry . . . "economic and social planning" has been very widely interpreted. (Guhan, in Cassen and Joshi, 1995: 74)

The overall setup thus represented a unique turn in history, but it could not be expected to last, as the conditions that gave birth to it were bound to change. The dramatic and sweeping economic reforms of 1991 were thus prompted by an external payments crisis rather than a major political shift. The evolution was not only gradual but endogenous to the political economy established at independence. Initially, the polity extended and deepened the features of the controlled economy. This consolidation took place in both the center and the states, but the economy's growth also set in motion forces that weakened and eventually rendered obsolete major features of the directed economy.

Self-reliance was the watchword, but it was soon subverted by the goal of self-sufficiency as each interest group pressed its claim for recognition and special favors for their actual or potential contribution to the national economic interest. There were of course real accomplishments. The import-substitution-industrialization (ISI) strategy aimed at national control of the "commanding heights" of the economy, the development of heavy industry, and technological autonomy. The comprehensiveness of the effort was such that the country seemed poised to produce virtually everything within its shores. A vast infrastructure of transport, shipping, harbors, irrigation, and statistical, economic, and scientific institutions was put in place for the planned economy. Self-sufficiency, conserving foreign exchange, technological independence, and capacity production became the mantras of policy and of discretionary decisions. Claims of equity in distribution to the states and others were also built into the process (for instance, subsidies to equalize steel delivery prices in the different states. However, it was an expensive approach, and it became unwieldy as well. It led inevitably to a lusty, rent-seeking, pork-barrel economy in which everyone participated: private and public enterprises, non-profit institutions and groups, public authorities, and state governments.

There is a case in theory, and a strong case at that, for India's approach, which aimed at self-reliance. Henry Bruton, Sanjaya Lall, and others have ably argued the point. Also, despite the neoclassical swing toward markets and government failure, it is by no means assured that India would have developed its economic, institutional, entrepreneurial, and human capital base without the initiatives of the Nehru period. Nevertheless, the excesses of the planned economy and the opportunity costs in terms of lost growth and built-in inefficiencies (such as the high domestic resource costs for foreign exchange saved, the chaotic structure of effective rates of protection, etc.) were also real. These have been extensively documented in the literature, and there is no need to recapitulate them here. Crucial for our purpose is the viability or lack of it of the resulting political economy. The claims on the economic system by what the Rudolphs (1987) term *demand forces* encouraged gains beyond the resources of the directed economy.

The interventionist economic regime blanketed the entire economy, established solid roots, and won repeated endorsement in free elections. There was of course the inevitable griping, but its overall legitimacy–of objectives as well as means–was regularly renewed in open general elections. One reason was a widespread and continuing faith in the power of the *Sicar.* The role of a benign sovereign has it roots deep in Indian tradition and belief, and faith in the ability of a committed national government did not disappear. Criticism was vociferous and plentiful, for failures were obvious and could not be hidden in a democracy, but it focused on lapses in execution rather than the need for alternative policies. The new political economy was sustained further by a vast and live system of dispersion of power and patronage, common to most democracies. It went beyond the transience and euphoria of a new nation.

Independent India's political economy evolved over its entire existence and saw the growth of a network of supporting institutions. It was institutionalized in laws, administration, judicial decisions, practice and precedent, and legitimacy of expectations that are the heart of a democracy. It was Nehru's rare distinction to be a revolutionary who turned into a nation and institutions builder. He put in place a stable, democratic, secular, political structure to implement his goals of planning and socialism. He was prime minister for 17 years, until his death in 1964, and was reelected regularly in free elections. Further, the democratic system has governed every change of government since, including the general elections in the late 1970s that ended Indira Gandhi's short-lived emergency and the subsequent election that restored her to power. In sum, the sweeping economic policy changes introduced in 1991 were prompted by action from the top to deal with an economic crisis and not by radical political shifts. Democracy has not been an instrument of liberalization. Instead, the dirigiste philosophy and practice spawned many vested interests in both public and private enterprise as well as in the rest of the economy and permeated the political process and ideology.

One may now look at parallel developments in the states, at the political and economic structure that evolved over the years. The developments were state specific, as each was self-governing with varying internal features. Linguistic nationalism transferred power to the state rather than to parties within the state. This led to one-party governments and fiefdoms. Linguistic and local nationalist themes were important, especially when the center seemed too partisan in favor of Hindi, the Hindi states, or, in the case of the Hindi states, not sufficiently partisan. Industrial and infrastructural development were obvious priorities in terms of claims on the center's allocations but not necessarily in terms of actual outlays. Even agriculture, a state subject, suffered in the absence of an independent resource base (Jalan, 1992). The states emerged as independent sources of patronage and rent-seeking in the concessions offered to private businesses, the write-offs of loans offered to agriculturists, the distribution of water and irrigation benefits, the awards of public works projects, the negotiated tariffs for power, and in the administrations of laws and regulations (such as housing for the poor or worksites for the low-caste occupations). Although observers have noted countrywide platforms to combat poverty and backward status, the commitment has varied from state to state, ranging from mere photo opportunities (a distribution of clothes and petty cash by a minister, token allotment to backward or tribal areas) to full-scale caste wars (either between two or more dominant caste groups or dominant and subordinate caste coalitions). Some were more to the left or right than the others. Some identified as pro-Hindu or anti-Muslim turned out on closer scrutiny to reflect local Hindu high-low caste divisions. Whether it is policies toward growth or equity, the political structure of the individual state, including its caste and social structure, has been a powerful influence in shaping policy. It is difficult to discern critical factors that cut across the states or to label them as "regional" developments, and it is premature to place them as part of an emerging all-India party system.

The evolution of the states and of India's political economy can now be placed in context. The states have emerged as rival centers of power but replicate (by and large) the center's economic regime. They are also among the major vested interests spawned by the political economy and resistant to being cut loose. One should add to the Rudolph's depiction of the relationship between business and state as "dependent capitalism" the term *dependent nationalism* or *dependent sovereignty* to describe the center-states relationship. It has been an enervating and less than optimal partnership.

In 1989, there was the usual spectacle of an all-India strike of government employees with support from non-Congress state governments. As for the economic reforms of 1991, the most comprehensive in over 40 years, the actual progress on liberalization was much less than the expectations that were aroused. It also lost steam in a few years. Established vested interests among public and private enterprises, trade unions, and others, including the central bureaucracy, certainly contributed to the halting advances. One must also include the network of state-based research institutions sponsored and funded both by the center and the states, many of whom were skeptical of the shift toward markets and the terms on which privatization would proceed. A major factor in the halting advance has, however, been the center-states nexus. The Congress Party in the center faced non-Congress governments in key states. India's center did not collapse or lose its legitimacy overnight as Moscow did in the Soviet Union, nor could it swing the economic structure around as China did, where the Communist Party retained a tight rein on power while permitting an economic free-for-all in key regions. India's dirigiste framework never did break up and could by no means be dislodged easily.

In 1996, again by democratic elections, the Congress was displaced at the center by a coalition that had at its base various state-based parties. There is no basis for any alarmist interpretation of the country's stability. The new government at the center has reaffirmed at the highest level the resolve to continue on the path of economic liberalization and associated reforms, and independent observers confirm that "there is no going back." There is doubt, however, about the effectiveness of the political economy, with its deeper roots.

It is now possible to sum up the economic impact of the Indian centrist economic regime without making excuses for the poor performance of the states or taking sides in the center-states debate. If the states were a drag on the Indian economy, the reverse is also true. There is a long litany of grievances the states have, pointing to actions or inactions by the center that have adversely affected their budgets or performance: inflationary pressures arising from the center's fiscal and monetary policies or center pay scales that impinge on state budgets, the inadequacy of the finance commission process, failure to pay agreed-on shares of central revenue, all-India labor standards, biases in the administration of controls, export and import controls, and so on and so forth (Guhan, 1995; Mukherji, 1991; Nair, 1991). The center and the World Bank have an equally compelling list. Regardless, changes in the Indian scene and the world economy indicate that India would be much better off with liberalized measures and less detailed management of the economy.

Center-State Relations and the Economic Challenge

The economic regimes of India–at the center, the states, and in the center-state relations–are dominated by political transactions involving organized governments and increasingly autonomous nationalities. The political market place is wanting in providing a shared commitment and machinery for economic efficiency and productivity growth. Democratic politics has actually entrenched the rent-seeking economy in both the center and the states, which is manifested in chronic excess demand, rationing, inefficiencies, delays, and misallocations. The governmental structure has become more fractious and will require much more skilled political juggling (Chalam and Mishra, 1997), but no amount of it can deal with the increasingly complex economic situation. The issue is not whether India's slow economic growth can sustain its political economy but whether this outdated framework can respond to the country's continuing need for economic growth and welfare. With the weakening of directives from the center, cumbersome coordination will have to take over, but neither will be adequate to reverse the growth of the rent-seeking economy or handle in a discriminating manner the diverse shortfalls and promises of multinational India.

Indian democratic politics nurtured rent-seeking regimes in the center and the states. It is possible to view the states as challenging not the rent-seeking economy but only the center's control of it. There are two aspects to this: first, the rent-seeking economy of the country, where all came to feed, including the states, and its replication in the states, and, second, the fallout of this arrangement for the entire economy and its direction. Even the aggregate share of the economy under direct control of the state governments is quite significant, as are the deviations and aberrations from economy-wide objectives (such as growth or equity). The state governments, being closer to the local electorate, have naturally reflected and been vulnerable to the local power structure, its priorities, and its trade-offs. Universal adult franchise has brought to the forefront very popular figures (for example, movie stars, literary figures, writers, singers) who could effectively wield the language of the state and reach mass audiences. Some state leaders became cult figures. Grandiose concessions to important groups (such as farmers), negotiated terms to industrialists, photo-opportunity sessions featuring gifts to the needy, etc., have taken the place of disciplined budgetary management and leadership. The problem is likely to grow as each state reflects its particular dynamics.

A parliamentary system, rather than a US type of system with checks on executive authority, even with active legislatures and an independent press, has not been of much help, particularly in the absence of well-developed party or program alternatives. The ICS and the Indian Administrative Service (IAS), the steel frame, has begun to erode, too, as the states have increasingly supplanted the center's authority over them. The corruption, patronage, and abuse of authority would have been more difficult on the national stage, but even this is not inviolate now. The center is also under increasing pressure to accommodate nonpriority pressures. For a weakened center, the problem is one of securing and implementing a political consensus on growth and equity. For the country and economy as a whole, it is the more important one of encouraging the diverse potential of the states as well as others and internalizing the opportunity costs of different trade-offs.

Tamil Nadu, with a population of 60 million, is an illustrative example. In 1962, the Dravidian Party's crown prince and Tamil orator, C. N. Annadurai, ousted the National Congress. However, as chief minister, he developed a working relationship with the center, which could only be termed a New Delhi-Madras axis, especially after Nehru's death in 1964. Annadurai's successor, M. G. Ramachandran, became renowned for his populist propensities. His consort and successor, Jayalalitha, carried the tradition farther and earned the status of an unchallenged political boss. She refused to take calls from the governor of the state, had installed giant cutouts of herself that dominated Madras (now Chennai), the capital, and hosted a lavish wedding dinner for some 100,000 guests for her son's friend. Ironically, the political deal struck by her with India's prime minister on the eve of India's general elections in 1996 cost the prime minister his job and contributed to his party's debacle and Jayalalitha's defeat. The Tamil Nadu government is a key partner in the United Front. A coalition of states rather than of parties, it formed the Government of India at the center. A key figure of the Tamil coalition and a prominent business leader, P. Chidambaram, emerged as India's finance minister and vowed to continue the liberal economic policies initiated in 1991.

West Bengal, with a population of about 75 million, is the only other state I will have space to discuss. Quite early after independence, its affairs were directed by an elected Marxist government. It faced and overcame severe challenges to law an order from radical Naxalites in rural areas, *gheraos* (a coercive tactic in collective bargaining in which aggrieved workers imprison managers in their offices or residences) in industry, and urban violence. It developed a governance system for the huge metropolis, Calcutta, and negotiated, with support from the Government of India, long-term loans from the World Bank for its development. Its spokesmen have been articulate and at odds with the center for its alleged niggardliness toward the state. In 1985, Prime Minister Rajiv Gandhi went to Calcutta and met with West Bengal's chief minister, Jyoti Basu, to present his case in negotiating sessions that had all the formality of a get together of heads of state. Basu outfoxed him, just on the eve of the pourparlers, by releasing to the public in his state a major white paper (in a Bengali version as well) setting forth West Bengal's case. Difficult border issues and pressures from separatist groups have also clouded the center-West Bengal relations. The communist government has strengthened itself by the acclaim it has received from India's press for its integrity and respect for press freedom. West Bengal has also taken important steps to attract private investment into the state. Its overall economic record has received praise from ranking national and world scholars. Jyoti Basu is also a legend, but unlike Jayalalitha, he represents a program and party and has won very favorable notice. In fact, its success has been so well publicized that carefully documented "debunking" has also come in that stresses uniquely state-specific, middle-class leadership (including antilabor policies) as important ingredients. However that may be, in 1996, West Bengal was part of the United Front that ousted Congress at the center and Marxist Chief Minister Jyoti Basu was the first choice of the Front to be India's prime minister.

The cementing factor in the United Front coalition is states rights in opposition to a center that is considered too powerful. Although the Congress won the largest percentage of the popular vote, it could not form a government lacking an absolute majority in the Lok Sabha. The openly "Hindu" Bharatiya Janata Party got the nod to form a government, but it lasted exactly 11 days. The Lok Sabha members from the United Front states were able to form a dominant coalition. West Bengal, a dominant state in the United Front coalition, could not dominate it with its ideology, which had limited reach beyond the state; nor could Tamil Nadu, with its probusiness finance minister. There is no basis on which to predict capitalism, socialism, communism, Hinduism, Hindu-Muslim holocausts, or even caste wars as the wave of the future (Alam, 1966). Constitutional democracy will continue if for no other reason than that no single opposed state, ideology, or trend is dominant throughout the country. The question is how effective the political economy will be.

Assessing the Future

The Received Legacy: Pluses and Minuses

Observers have reacted with gloom to the recent turn of political events as indicative of a leaderless, rudderless India. Much of what happened had to happen and is arguably desirable. The British had to go, and on the whole, it was an orderly change, compared to the liquidation of other, much smaller colonial legacies. The imperial center that followed was also destined to disappear. It was born of opposition to Britain and held in place by a cohesive national leadership that could talk in terms of countrywide priorities and implementation. With India's political and economic development, this could not last in a country that is at least as diverse as all of Europe or Africa. The Rudolphs, in their significant *In Pursuit of Lakshmi,* document the unraveling over nearly four decades of this powerful all-India structure in its various dimensions. The focus and critical turning point has always been what would emerge when the states successfully challenged the dominance of the center when diverse indigenous forces asserted themselves vis-à-vis the vision and direction of the modernizing, anticolonial elite, when what I call the second revolution came of age (the first being the major national initiatives when Britain quit and an imperial center took over).

Independent India would not have taken shape without the first revolution, and its evolution would have been aborted without the second. Both were necessary. Selig Harrison sounded the alarm bell in 1960, raising the specter of a breakup of India, and humorous writing has envisaged the emergence of a United States of India, with 20 to 25 seats in the United Nations. What has emerged is a commonwealth of the states of India within the framework of democratic change and the Indian Constitution rather than the pell-mell succession or collapse as in the Soviet Union or the butchery and civil conflict as in the former Yugoslavia.

What of the future, then? Not a million mutinies, as a well-known title says, but one would hope steady, perhaps even accelerated progress building on the legacy of 50 years of democratic self-government. This point is of more than ordinary importance. Democracy entrenches vested interests. It also provides continuity and the opportunity to learn from experience. A careful look at India's legacy and emerging trends indicates the basis for optimism.

Along with the Indian Union, what else has survived is important. Secularism still stands as a fundamental tenet. There is no basic disagreement in the country about a foreign policy of independence and national defense capability. One should note also the emergence of a constitutional framework for reconciling the autonomy of the states and the authority of the center in place of the capricious practices of the past. The educational and scientific base is the most advanced outside the most developed countries and meets all the country's essential requirements. A stable framework of legal, administrative, and economic institutions has been developed and remains in force. Of particular importance is the solid macroeconomic structure: the RBI and its control over money supply, the monetary system, and financial institutions, and continuing machinery for regulating center-states fiscal relations.

In a more complex sense, there is a broad consensus on the country's goals of economic progress and justice and the importance of government leadership. True, there is not the clarity there was in the halcyon days of Nehru, and untold new problems and issues have surfaced. It must be emphasized, however, that current political struggles aim at capturing power within the system rather than overthrowing it, that the secret ballot is the overriding political weapon, and that the protests and revolts are aimed at making the system work. The IMF and World Bank, despite the economic clout of these institutions, will (and should) have only an indirect role in this process, for success in policy requires commensurate political sense and accountability. The center's think tanks and scholars who see Indian problems and solutions in all-India terms also need to be wary. India's economic success will depend on the dialectics of change within India and among the different parts of India, as well as on the adaptability of India's political economy to changing economic realities at home and abroad. The future is going to depend a good deal more on India's skills and disposition toward effective economic management to complement the superior, nation-building political leadership that staved off chaos or drift in the last half-century and achieved a measure of consolidation. The monumental scale of this accomplishment is best described in the words of an astute observer who placed at the head of India's problems the country's cultural and linguistic diversity. "India is not really a nation; it is an ontology, a set of beliefs and perceptions which barely hold together a people who would seem to have as little in common as a single nation formed out of Ireland, Hungary, Mexico, and the Philippines. One senses that India has had to devote enormous amounts of energy to holding herself together" (Nick Eberstadt, "Letter from Madras," in *Rockefeller Foundation Illustrated* [December 1976]).

The pluses cited above also constitute drags, given the entrenched vested interests and related ideology and practice that has grown over five decades. The political economy will need to move away from the crippling constraints of received legacy. The basis for optimism rests on a more assertive public opinion, a high caliber leadership that has placed sound economic management as a top priority, the democratic give and take that continues to guide change, and the emergence as well as promise of attractive options in the global economy.

Pressures for More Effective Government

The directive central authority and the fervent linguistic nationalisms have both had their say; they are increasingly recognized as inadequate to the tasks ahead in terms of the country's economic and social goals. In the states, they brought to the forefront chieftains who had little to offer and banked mainly on adulatory masses and bargaining chips with the center rather than on developing sound economic policies. Nor could they effectively lead or represent the state without developing sound forms of democracy. Although each state has a dominant language, the average state is more populous than most single nations in the world and has its source of tensions reflecting breakdowns of class, ideology, caste, religion, and other divisions (including those of language and dialect). In the worst cases, they have aggravated the tensions. More generally, in the country at large and in the center, the leadership that emerged has been skilled at, and has devoted its major energy to, playing the game of political balance rather than shaping an efficient, flexible economic system. It is of course a credit to India's accommodative politics that passions and fears over language have become less central, but it is not good enough now. India's progress as a democracy also means that neither brutal repression nor virtual collapse are the only alternatives. In spite of the erosions, the belief in an ideology of democratic development and core institutions and practices that are independent of the vagaries of the political process endures. The emerging expectation was effectively voiced by the demand that elected politicians should serve as responsible "officials" rather than parade as "leaders" (Guhan, *Financial Express,* 6 October 1986).

Loosening Up and Facing the World Economy

The economic changes of the past 50 years have also opened up many new avenues. There is emerging a wide range of economic opportunities that are not under the control of the political and administrative structure and lie beyond the states, the center, and even India. Many are outside the spheres that have engaged the attention of economists and planners, such as employment or formal sector job creation. These are particularly relevant for the skilled and venturesome, of which India has many. A supporting family and kinship structure has also made it possible for many to adapt to the risks and take advantage of emerging opportunities. As an indicator of such opportunities, more and more people living within confining jurisdictions are voting with their feet, so to speak, and relocating to areas that are more favorable for employment. Rural-urban migration is a good example of such opportunity-induced shifts. The efforts to regulate and license migration to the gulf areas were ineffective even in the case of semiskilled labor and despite the high costs of labor permits. A subcontinental identity has built up among millions whose life and economic base is outside their state, in the major metropolises of India and other emerging centers of economic importance. The changing world economy is at the backdoor of India's capitals, cities, towns, and homes.

Wherever one goes outside India, one finds substantial resident populations of expatriate Indians who are relatively well off. These go beyond the significant number in the scientific and professional categories who got the most attention in earlier discussions of the "brain drain." Many in these expatriate populations now are independent entrepreneurs who retain contracts with the home state (an important example is Gujarat) and raise capital at home and abroad for productive ventures. New York alone is reputed to have nearly half a million Indians, with possibly the highest average income of ethnic groups in the United States; the Silicon Valley, in California, boasts of more than 400 millionaires who are of Indian origin.

In the United States more generally, professionals, with their command of English and appropriate credentials, have reached the highest ranks of industry, finance, publishing, writing, journalism, other media, academia, medical and scientific fields and research, and government and corporate leadership. They have established positions in the new locations, with good schooling and housing and a record of public service in their new communities (Cheney and Cheney, 1997). The larger settlements display political sophistication and participation in both general and ethnically bound issues, and many have graduated to political prominence. Correspondingly, foreign enterprises and subcontracting operations are providing opportunities in India that were formerly scarce. More and more foreign corporations and units in India are led by Indians. A migration of operations from rich countries is taking place in search of India's technical manpower, which is well trained and cheap compared to the wage and salary costs in developed countries. A specialized category of Indian skills in demand abroad is in the arts, dance, music, and various spiritual and healing services: it is not clear how much is done by emigrants, itinerants abroad, and India-based exports. All this is witness to the vibrancy and appeal of the expanding world economy, global labor markets (particularly for the skilled), and open democratic societies.

The world economic order has changed substantially. The United States is still the dominant economic power, but it has to share its leadership role with others in Europe, Japan, and increasingly the successful Southeast Asian economies. The Cold War has ended, and with it has ended the significance of the bootstrap economies of the communist world and the grants economies of the developing world, with only the poorer African nations getting a reluctant recognition. The concessional regimes of the World Bank and IMF are less important as compared to the competitive capital and money markets of Europe, Japan, and the United States. Even powerful countries such as the United States or Japan cannot toe an independent line without risking adverse consequences (for example, a rise or fall in the external value of their currencies). Further, new and old global institutions have emerged with expanded jurisdictions that undermine the appeal and authority of less-developed countries' protectionist regimes. Examples include the World Trade Organization, issues of labor or environmental standards, regional organizations (such as NAFTA), the Law of the High Seas, copyright conventions, etc. The new world economic order is simply not hospitable or indulgent to incompetent, or even hesitant, sovereignties that fail to adapt.

India's liberalization policies initiated in 1991 have a way to go yet, but there has been no turning back despite the general elections since 1991, the changes in governments, and the continuing shift in power from the center to the states. One reason is the gradualism of change within an established democratic order. Perhaps more important is the lack of any real alternative, the manifest weaknesses of old policies, and the perception and promise of change. In the important sphere of center-states relations, the creaky and cumbersome old machinery has been a major influence on Indian economic reforms.

The continuity of able leadership in key positions such as finance and the RBI attest to the country's steadfast course. Credit and monetary tightness, and the realization of the importance of the world economy for India's development goals, have already shifted the focus from discretionary decision making at the center or the states to decisions that will be governed more by market imperatives. Nothing is more important than the issue of rupee convertibility. The slow progress toward it since it was announced as a major objective by former Minister of Finance Manmohan Singh indicates the resistance and high stakes involved. Singh's able successor, P. Chidambaram, has kept up the pressure for convertibility. This has been India's goal in recent years, and there is no doubt it will be realized. The retention of able economic statesmen in the top positions in finance and the RBI (C. Rangarajan) attest to the continuing commitment. A top priority for India is to set a defensible value for the rupee, to let everyone adjust to its scarcity price, and to remove it from dubious political fora.

India is lucky in the sense that it has already won the political battle posed by conflicting sovereignties (in contrast to Europe, which is just edging toward the Euro and a European Central Bank). Enhanced participation in the world economy is clearly producing favorable results (for popular, euphoric accounts, see Clifton, 1997; Strasser and Mazumdar, 1997; and Zakaria, 1997). It will further boost decentralization (John Echeverri-Gent, "Governance in a Globalizing World," unpublished paper, 1997). It should also lessen the dependence of the third revolution on the layers of India's oligarchies. A more open economy will loosen the constraints of a directed economy, such as interest groups in rural areas despite the success they have had at political mobilization (Varshney, 1995). India's local authorities, unlike China's, show little economic enterprise or dynamism. A liberal economic foundation will advance India's democracy and economy as well as the cultural and political goals of its diverse constituencies. The positive developments in the relations between India and its South Asian neighbors, especially Pakistan, also signal promise: the South Asian Association for Regional Cooperation (SAARC), a product entirely of South Asian leadership and not a regional treaty in response to an external threat, is taking the first steps toward economic cooperation ("SAARC Leaders for SAFTA by 2001," and "CII Paper on SAARC Targets Multifold Increase in Trade," in *India News* [15 May–1 June 1997]).

Further Reading

Alam, Javeed, "Behind the Verdict: What Kind of a Nation Are We?" in *Economic and Political Weekly* 31, no. 25 (1996)

Bhagwati, Jagdish N., *India in Transition: Freeing the Economy,* Oxford and New York: Oxford University Press, 1993

Three lectures critical of India's planning legacy, providing a highly intelligible survey of developments leading up to the liberalization, 1991.

Bhagwati, Jagdish N., and T. N. Srinivasan, *India's Economic Reform,* New Delhi: Associated Chambers of Commerce and Industry, 1993

A short work from two leading economists that provides a brief appraisal of the reform process in India. This study was undertaken on behalf of the Finance Ministry of the Government of India.

Cassen, Robert, and Vijay Joshi, editors, *India: the Future of Economic Reform,* Delhi and New York: Oxford University Press, 1995

Chalam, K. S. R. V. S., and Rajiv Mishra, "Streamlining Norms: A Renewed Approach for Finance Commission," in *Economic and Political Weekly* 32, no. 25 (1997)

Cheney, Susan A., and Charles C. Cheney, "Adaptation and Homebuying Approaches of Latin American and Indian Immigrants in Montgomery County, Maryland," in *Cityscape* 3, no. 1 (1997)

Clifton, Tony, "Making Sense of India," in *Newsweek,* 4 August 1997

Drèze, Jean, and Amartya Sen, *India, Economic Development, and Social Opportunities,* Delhi: Cambridge University Press, 1995

A major review of the broader social consequences of India's economic development. It links poverty, population, health, and education with issues of social and gender inequality.

Frankel, Francine R., and M. S. A. Rao, editors, *Dominance and State Power in Modern India: Decline of a Social Order,* 2 volumes, Delhi and Oxford: Oxford University Press, 1989, 1990

A comprehensive survey of state politics in all the major states, focusing on the rise of backward castes and middle peasants.

Harrison, Selig S., *India: The Most Dangerous Decades,* Princeton, N.J.: Princeton University Press, 1960

Jalan, Bimal, editor, *The Indian Economy: Problems and Prospects,* New Delhi and New York: Viking, 1992

A very good collection of essays on current economic issues by an adviser on economic policy.

Joshi, Vijay, and I. M. D. Little, *India: Macroeconomics and Political Economy, 1964–1991,* Washington, D.C.: World Bank, 1994

Traces the roots of the 1991 reforms through the macroeconomic policies of Indira and Rajiv Gandhi.

—, *India's Economic Reforms, 1991–2001,* Oxford: Oxford University Press, 1996

An excellent overall review of the issues of economic reform in India in the 1990s.

Kannappan, Subbiah, "India, the Imperial, Welfare, or Pork-Barrel State?" in *South Asia Journal* 2 (1989)

—, "The Economics of Development: The Procrustean Bed of Mainstream Economics," in *Economic Development and Cultural Change* 43, no. 4 (1995)

Khurso, Ali Mohammed et al., editors, *Industrial Policy: A Panel Discussion,* New Delhi: Vikas, 1990

Kohli, Atul, *Democracy and Discontent: India's Growing Crisis of Governability,* Cambridge and New York: Cambridge University Press, 1990

An influential work that maintains that the Congress Party has suffered institutional decline since the 1960s.

—, *The State and Poverty in India: the Politics of Reform,* Cambridge: Cambridge University Press, 1987

A landmark comparative study of the relationship between party organization and effectiveness at implementing redistributive policies in Karnataka, West Bengal, and Uttar Pradesh. Kohli presents a critical analysis of the Indian government's failure to significantly reduce poverty after 40 years of independence.

Mallick, Ross, *Development of a Policy of a Communist Government: West Bengal since 1977,* Cambridge and New York: Cambridge University Press, 1993

Nicholson, Mark, "India: Power of the States Is Increasing," in *Financial Times,* 17 November 1995

Parikh, Kirit S., editor, *India Development Report, 1997,* Delhi and New York: Oxford University Press, 1997

Rosen, George, *Democracy and Economic Change in India,* Bombay, Vora, and Berkeley: University of California Press, 1967

Excellent review of India's economic development through the early 1960s, continuing an evaluation of rural and urban winners and losers, and the political implications.

—, *Western Economists and Eastern Societies: Agents of Change in South Asia, 1950–1970,* Baltimore, Md.: Johns Hopkins University Press, 1985

Excellent analysis of the history of Western economic ideas in Asia and the people who formulted them.

Rudolph, Lloyd I., and Susanne Hoeber, *In Pursuit of Lakshmi: The Political Economy of the Indian State,* Chicago: University of Chicago Press, 1987

One of the best analyses of the political economy of India's development since independence, marked by empathy and critical reflection. The authors focus on the bureaucratic institutions and interest groups that dominate political life.

Sen, Amartya, "Indian State Cuts Population without Coercion," in *New York Times,* 4 January 1994

Singh, Inderjit, *The Great Ascent: The Rural Poor in*

South Asia, Baltimore: World Bank and Johns Hopkins University Press, 1990

Strasser, Steven, and Sudip Mazumdar, "A New Tiger," in *Newsweek,* 4 August 1997

Thakur, Ramesh, "India in the World," in *Foreign Affairs* (1997)

Varshney, Asutosh, *Beyond Urban Bias,* London: Frank Cass, 1993

Zakaria, Fared, "From the Old to the New," in *Newsweek,* 4 August 1997

——, *Democracy, Development, and the Countryside: Urban-Rural Struggles in India,* Cambridge and New York: Cambridge University Press, 1995

World Bank, *India: Recent Economic Developments and Prospects,* Report no. 12940-IN, Washington, D.C.: World Bank, 1994

Subbiah Kannappan is a professor of economics at Michigan State University, East Lansing, Michigan.

Appendices

Michael Newbill

Appendix 1

Chronology

1947	**February**	The British Government of India announces that it plans to grant independence to India no later than June 1948. Lord Mountbatten is appointed Viceroy to oversee the transfer of power.
	June	Nationalist leaders and Lord Mountbatten, failing to come to an agreement over the nature of an undivided India, agree to partition British India into two independent states, one primarily Muslim in population. Mountbatten announces 15 August 1947 as the date for the formal handing over of power.
	June–August	Mountbatten and nationalist leaders scramble to work out arrangements for dividing the country, including the allocation of the Indian army, the Indian Civil Service, financial assets, guidelines for protecting minority rights, trade agreements, and the integration of the semiautonomous Native States (princely rulers) into the two new nations. In regions that were to form the new border areas, communal riots erupted between Hindus, Muslims, and Sikhs.
		British India is partitioned into the independent states of India and Pakistan. Jawaharlal Nehru becomes prime minister of India, Muhammad Ali Jinnah the governor general of Pakistan. Several hundred thousand die or are murdered in violence associated with the mass transfer of population over the next few months, roughly 4.5 million Hindu and Sikhs moving to India, 6 million Muslims traveling to Pakistan. Refugees pour into cities on both sides.
	October	Tribesmen from northern Pakistan invade the princely state of Kashmir, which until then had not acceded to either India or Pakistan. India rushes troops to Srinagar, pushing the tribesmen back to the current Line of Control. This line essentially divides Kashmir into two sections, one-third controlled by Pakistan, two-thirds by India. Though India's portion is integrated into the Indian Union, Nehru resolves to hold a plebiscite to determine the future status of Kashmir.
	December	Most foodgrains and other commodities are decontrolled. India comes to an agreement with Pakistan over financial and fiscal issues, including outstanding trade balances and foreign reserves.
1948	**January**	Gandhi is assassinated by Hindu nationalists in Delhi (January 30). India appeals to the United Nations (UN) to mediate the conflict in Kashmir. The UN appoints a commission to investigate.
	April	The government issues the Industrial Policy Resolution. The United Nations Security Council asks India and Pakistan to withdraw troops.
	September	The Reserve Bank of India is nationalized under the Reserve Bank (Transfer of Public Ownership) Act.

		Indian troops enter the princely state of Hyderabad, whose ruler had resolved to remain independent. Hyderabad territory comes under Indian administration.
1949	**January**	Undeclared war with Pakistan ends in a United Nations arranged cease-fire.
	March	The Banking Companies Act, which invests supervision and control over banks in the Reserve Bank of India, comes into effect.
	September	India announces devaluation of the rupee at the suggestion of the International Monetary Fund.
1950	**January**	The Republic of India is declared, with the promulgation of the Constitution (26 January). Dr. Rajendra Prasad is elected by parliament as the first president.
	February	The Indian Planning Commission is formed with Nehru as its chairman.
	May	At a conference of Commonwealth finance ministers, members recommend the formation of the Colombo Plan, a consortium of Commonwealth nations for economic and agricultural development in South Asia and South East Asia.
	November	Industrial Development Committee is organized to encouraged the development of efficient industries as well as to make recommendations for production levels and planning.
	December	Indo-US Technical Aid Pact signed, making US aid available.
1951	**May**	United States authorizes US$190 million loans to India to enable it to buy 2 million tons of wheat. China and the Soviet Union agree to supply rice and wheat to India.
	July	Colombo Plan comes into operation. Draft of First Five-Year Plan (1951–56) published.
1952	**January**	Financial assistance agreement signed between India and the United States. The United States promises US$50 million in aid for development projects. India contributes the same amount in rupees.
	May	A new parliament is elected under a Congress government. Nehru is reelected prime minister.
	August	National Development Council established. Its purpose is to evaluate and make recommendations about the national planning process, taking into account social and economic development at the state level.
	October	The Indian government nationalizes air transport companies.
	December	First Indo-Soviet trade pact signed. Final draft of the First Five-Year Plan published.
1953	**February**	The Bhakra Dam, one of the largest in the world, opened by Nehru.
	October	After months of demonstrations, the first reorganization of Indian states occurs with the inauguration of Andhra Pradesh.
	December	Hindustan Steel Ltd. established as a private Indo-German venture.

1954	**March**	Nehru rejects military aid from US President Eisenhower.
	June	Chinese Prime Minister Zhou Enlai visits India. He and Nehru work out trade and diplomatic agreements to encourage closer relations.
	November	French possessions in India, including Pondicherry, succeed to the Indian Union.
1955	**April**	Bill introduced to parliament that would nationalize and amalgamate the Imperial Bank of India, along with ten other banks associated with state governments, into the State Bank of India.
		Nehru and the heads of other "nonaligned" nations meet in Bandung, Indonesia for the first global conference of the movement.
	November	Soviet Prime Minister Bulganin and Communist Party Secretary Nikita Khruschev visit India.
1956	**January**	Government issues ordinance nationalizing all life insurance companies; becomes law in July.
	May	Nehru presents Second Five-Year Plan to Parliament.
	December	Zhou Enlai returns to Delhi to resume talks with Nehru.
1957	**January**	Nehru inaugurates India's first atomic reactor near Bombay.
	May	Second general elections held; Congress maintains power.
1958	**January**	The United States approves a US$250 million development loan to India, the first of several in 1958. Further grants come from the United Kingdom, Canada, West Germany, and the Soviet Union.
	August	Indian and Pakistani soldiers clash on the northern border.
1959	**March**	Fleeing Chinese-controlled Tibet, the Dalai Lama seeks asylum in India.
	May	Nehru announces formation of Indian Oil Company to handle distribution and refining of oil products countrywide.
	August	Chinese troops cross border into India in eastern Kashmir and in the northeast provinces.
	October	Chinese forces attack policemen outside Ladakh, in eastern Kashmir.
	December	US President Eisenhower visits India.
1960	**February**	Now prime minister of the Soviet Union, Khruschev visits New Delhi.
	May	Bombay state is divided into the new states of Gujarat and Maharashtra.
	June	Indo-Soviet agreement on joint oil and gas exploration signed.
1961	**December**	Indian Defense Forces enter the Portuguese colony of Goa, on India's west coast. Goa succeeds to the Indian Union.
	August	Third Five-Year Plan is presented to parliament.

1962	**July**	The International Monetary Fund sanctions a US$100 million credit to raise foreign reserves to buy needed imports.
	October	Chinese troops invade sectors of the far north and northeast India, initiating a border war.
	November	China declares cease-fire from a position well inside Indian territory.
1963	**July**	The International Monetary Fund authorizes over US$100 million in loans to India over the next 12 months.
1964	**February**	India rejects United Nations plan for a plebiscite in Kashmir.
	May	Prime Minister Jawaharlal Nehru dies. Lal Bahadur Shastri sworn in as prime minister.
1965	**September**	After continuing border skirmishes, Pakistan declares war on India (September 6). Shastri warns China not to interfere in Indo-Pak conflict. With the intervention of the United Nations, India and Pakistan agree to a cease-fire (September 22). The Soviet Union offers to hold peace talks in Tashkent.
1966	**January**	Five-year trade agreement signed with Soviet Union. Lal Bahadur Shastri passes away in Tashkent hours after signing peace agreement with Pakistan. Indira Gandhi, daughter of Jawaharlal Nehru, sworn in as prime minister.
	February–May	Poor harvests lead to the threat of famine. India receives food donations and monetary aid to avert crisis.
	June	Rupee is devalued 36%.
	November	Newly reorganized states of Punjab, Haryana, and the Union Territory of Chandigarh come into being.
1967	**March**	Congress again wins national elections. Indira Gandhi remains prime minister.
1968	**April**	Construction begins on Bokaro Steel Plant.
1969	**August**	US President Richard Nixon arrives in New Delhi for talks. Bank Nationalization Bill passed by parliament.
	November	Government vows to reserve some government positions for Scheduled Castes and Scheduled Tribes.
1970	**July**	The New Industrial Policy, setting new standards for production and diversification, is announced.
	October	First Indian-made MIG fighter plane handed over to Air Force. India objects to resumption of military aid to Pakistan from the United States.
1971	**March**	Fifth general elections held. Indira Gandhi remains prime minister.
	April	Refugees from East Pakistan fleeing repression by West Pakistan, flood the Indian border state of West Bengal, sparking a humanitarian and political crisis. India expresses its solidarity with the people of East Pakistan.

	August	India and Soviet Union sign 20-year Treaty of Peace, Friendship, and Cooperation.
	December	Pakistan declares war on India (December 3). India informs United Nations of Pakistani aggression and recognizes the People's Republic of Bangladesh. India and Bangladesh agree to unified military command (December 10). Pakistan surrenders (December 17).
1972	**February**	Indian government discontinues subsidies to former Indian princes.
	May	India nationalizes all general insurance companies.
	July	Indira Gandhi and Pakistani President Z. A. Bhutto sign agreement to abjure the use of force in future conflicts.
1973	**December**	First draft of the Fifth Five-Year Plan published.
1974	**January**	In response to decisions by the oil companies ESSO and Shell to double oil prices, the Indian government acquires a majority share in ESSO.
	May	India successfully carries out a nuclear weapons test in the Rajasthan desert.
	December	Nine-year-old trade embargo between Pakistan and India ends.
1975	**June**	High Court rules against Indira Gandhi on irregularities stemming from her 1971 election campaign. Gandhi files an appeal with the Supreme Court, which grants her a stay of operation until a final ruling. Opposition parties launch a campaign for her resignation. In response, Gandhi declares a state of emergency, arresting opposition leaders and rescinding some civil liberties.
	July	Gandhi announces her 20-point Economic Program, emphasizing measures to aid the poor.
	December	Regulations enforcing press censorship come into effect.
1976	**May**	To encourage development of the telecommunications industry, the World Bank provides India a loan of US$80 million.
1977	**January**	Parliament is dissolved and new elections are scheduled. Press censorship is relaxed.
	March	Indira Gandhi and the Congress Party are defeated and a coalition–the Janata Party–comes to power with Morarji Desai as prime minister.
	May	India and China resume trade after a 15-year hiatus.
1978	**January**	US President Jimmy Carter visits India.
	December	Indira Gandhi is arrested on corruption charges, then released.
1979	**July**	Morarji Desai submits his resignation. Charan Singh leads a new coalition as prime minister.
	December	Draft of Sixth Five-Year Plan released.

1980	**January**	Seventh general elections held. Indira Gandhi returns to parliament and is elected prime minister at the head of a Congress government.
1981	**January**	Cabinet recommends the establishment of the Export-Import Bank of India.
1982	**April**	Some Foreign Exchange Regulations Act rules relaxed to encourage foreign investment in key industries.
	October	Maruti Udyog, operated by the Indian government, and Suzuki Motors of Japan sign a joint venture agreement to produce cars and minivans.
1983	**February**	The ongoing conflict between anti-immigration extremists, native Boro tribespeople, and Muslim immigrants in the northwest state of Assam erupts in violence during elections. Over 1,200 people are killed.
	August	Foreign ministers meet for the inaugural session of the South Asian Association for Regional Cooperation in New Delhi.
	December	First Maruti cars roll off the assembly line.
1984	**June**	After a long standoff between Sikh separatists lodged in the Golden Temple and the Indian Security Forces in Punjab state, the Army storms the temple, killing over 1,000 of the separatists and their leader. This action, known as "Operation Bluestar," marks the beginning of a long period of unrest in the Punjab.
	October–November	In retaliation for her support of "Operation Bluestar," Indira Gandhi is assassi nated by her Sikh bodyguards (October 31). Her son, Rajiv Gandhi, is elevated to prime minister. Innocent Sikhs are attacked and murdered in northern Indian cities in the days following the assassination.
	December	Toxic gas leak at the US Union Carbide plant in Bhopal kills over 2,500 people and injures thousands more. Thousands die of residual effects over the next decade. Rajiv Gandhi and the Congress Party win overwhelmingly in the December 1984 general elections.
1985	**April**	First budget of the Gandhi government introduces modest liberalization policies, lowering taxes and encouraging the private sector.
	December	Creating controversy, Minister of Finance V. P. Singh embarks on a vigilant anti-tax fraud policy. He raids and arrests prominent businessmen and industrialists. South Asian Association for Regional Cooperation heads of state meet for the first time.
1986	**April–June**	Increased terrorist attacks by Sikh separatists force the migration of Hindus from the Punjab.
1987	**April**	The Bofors scandal–in which government officials are implicated in receiving bribes from the Swedish defense contractor Bofors–breaks, further discrediting the Gandhi government, which is plagued by accusations of mismanagement, corruption, and disunity.
	May	Responding to increasing instability and violence, the government dissolves the Punjab Assembly and places it under president's rule.
	August	Gandhi and Sri Lankan Prime Minister J. R. Jayawardene sign Indo-Sri Lan-

		kan accord that commits an Indian Peace-Keeping Force to northern Sri Lanka to monitor the cessation of hostilities between Tamil separatist groups and the Sri Lankan government.
	October	Caught among fighting be'ween Tamil separatist groups and the Sri Lankan government, Indian troops go on the initiative, taking the northern city of Jaffna despite heavy casualties.
1988	**June**	Government reduces licensing obstacles on smaller industries as part of liberalization.
1989	**July–August**	Gandhi agrees to begin the withdrawal of the Indian Peace-Keeping Force following mass anti-Indian sentiment and domestic political violence in Sri Lanka.
	December	In the general elections, the Janata Dal–a coalition of non-Communist left and regional parties–defeats Rajiv Gandhi and the Congress Party. V. P. Singh, the former Congress finance minister, is installed as prime minister.
1990	**January**	Indian Security Forces establish a significant military presence in the Kashmir Valley following increasing violence by pro-Pakistani and separatist militants.
	August	The decision by Prime Minister Singh to implement the recommendations of the Mandal Commission–a study calling for increasing the reservation of government positions for Scheduled Castes and other disadvantaged groups to 60%–draws criticism and mass protest. Violence and riots associated with this decision continue for months.
	November	Amid mounting protests over the Mandal Commission decision and internal power struggles within the Janata Dal, V. P. Singh resigns as prime minister. Rajiv Gandhi vows to support the minority government led by former coalition member Chandra Shekhar.
	December	Agitation by Hindu religious and political organizations over the rebuilding of a Hindu temple on the site of the Babri Masjid mosque in Ayodhya sparks communal violence across India.
1991	**January–March**	Rising oil prices related to the Gulf War compound inflation and deplete India's foreign reserves.
	March	The Congress Party withdraws support for the Chandra Shekhar government and parliament is dissolved. New elections are scheduled for May.
	May	While campaigning in Tamil Nadu, Rajiv Gandhi is assassinated by a member of a Sri Lankan Tamil separatist group (May 21).
	June	Congress returns to parliament as the largest single party and is asked to form a government. P. V. Narasimha Rao becomes prime minister.
1992	**March**	New budget promoting economic liberalization unveiled by Manmohan Singh.
	December	In defiance of police and government orders, right-wing Hindu organizations, led by the Bharatiya Janata Party, demolish the Babri Masjid mosque at Ayodhya in order to build a Hindu temple at the disputed site. Communal riots break out across India. Thousands, especially Muslims in Bombay and other large cities, are murdered or injured.

1993	**January**	Communal violence continues in the wake of the mosque demolition and right-wing Hindu agitation. Over 700, mostly Muslims, are murdered in Bombay before the army is called in.
	March	A series of 13 bombs explode in Bombay, taking more than 600 lives in a matter of hours. The bombs go off in strategic targets, such as the Bombay Stock Exchange, the Air India building, hotels, and movie theaters.
	July	The finance minister devalues the rupee by 20% against the dollar.
	September	A devastating earthquake hits central India, killing over 15,000 (September 30).
1994	**June**	Rao travels to meet US President Bill Clinton with a trade delegation in order to increase economic and political ties and to reassure US investors about India's commitment to liberalization.
1995	**August**	Citing corruption and financial irregularities in the bidding process, the ruling Bharatiya Janata Party and Shiv Sena combine in the state of Maharashtra and decide to cancel the multibillion dollar contract with US-based Enron to build a power plant.
1996	**January–March**	A number of Congress and opposition members of parliament resign after being brought up on charges of corruption and bribery. Prime Minister Rao is also linked to the case but is not charged.
	May	In the general elections, no party is returned with a majority in parliament. As the largest single party, the Bharatiya Janata Party is asked to form the government but loses a vote of confidence 13 days later. H. D. Deve Gowda of the National Front coalition becomes prime minister, as a compromise candidate supported from outside by the Congress.
1997	**February**	New budget unveiled by Minister of Finance P. Chirambaram addresses inflationary trends and lagging economic growth by attempting to stimulate growth and investment through tax cuts.
	April	Congress Party withdraws support from National Front government; Gowda resigns. After weeks of negotiations, respected Minister for External Affairs in the National Front cabinet, Inder Kumar Gujral, is accepted as a consensus prime minister.

Appendix 2

Glossary

All-India Congress Committee (AICC): Made up of elected delegates from state Congress Committees, the AICC is the secondary decision-making body of the Indian National Congress. It must meet annually and can establish rules that govern lower party committees. Members of the primary decision-making body, the Congress Working Committee are elected from the AICC.

Association of Southeast Asian Nations (ASEAN): An organization founded in 1967 of Southeast Asian nations to promote economic growth, regional stability, and cultural exchange. Brunei, Indonesia, Philippines, Singapore, Thailand, and Vietnam are member nations, while India, the United States, Japan, and other Western and East Asian nations are "dialogue partners."

Ayodhya: Town in north India (Uttar Pradesh) that is the alleged birthplace of the Hindu mythological hero Ram. On 5 December 1993, Hindu nationalist groups tore down the Muslim mosque Babri Masjid, which was built in the sixteenth century over a Hindu temple marking the putative birthplace of Ram. The historical details of the site have engendered considerable dispute, but the 5 December riot and destruction of the mosque sparked the deadliest communal violence in the nation's history and made Ayodhya a symbol of communal conflict.

Bharatiya Janata Party (BJP–"Indian People's Party"): Formed in 1984 as the renamed Jana Sangh, the BJP is the latest incarnation of a series of Hindu nationalist parties and currently the largest single party in parliament. It attracts support from higher castes and the urban middle-to-lower-middle classes. Led by L. K. Advani and Atal Behari Vajpayee, the BJP advocates the privatization of public sector industries, a more aggressive foreign policy, and the removal of special protection or advantages for minorities, particularly Muslims. The party regularly uses Hindu symbols to advocate nationalist platforms; it was the prime organizer behind the destruction of the Babri Masjid at Ayodhya in 1993, which sparked a series of deadly riots around the country. Asked to form a government following the May 1996 elections, Prime Minister Atal Vajpayee could not garner a parliamentary majority, and the government fell after 13 days.

Bombay Plan: A proposal put forth by Indian industrialists supporting an approach to India's economic development after independence. This plan proposed state action in planning equitable growth, protecting national industries against foreign competition, and concentrating on developing heavy industries. Though leading industrialists, the supporters of the Bombay Plan recognized the dangers of a free-market approach in India's development and sought to work in a partnership with the state.

Caste System: *Caste* is a term used to describe the hierarchically ranked endogamous kinship groups, known as *jatis*, that are the basis for social structure in South Asia. *Jatis* often have a regional base and center around the performance of traditional occupations (such as leatherworkers, priests, merchants, or tailors) in an interdependent relationship with other *jatis*. The caste system also sanctions social and economic inequality by associating certain *jatis* with physical and spiritual impurity. Lower-class groups, such as *Dalits* and scheduled and "backward castes" (who were once known as "Untouchables"), make up the majority of India's population. Upper-caste groups, such as *Brahmins* and merchant and ruling groups, have traditionally discriminated against lower-caste groups, but the ranking of upper- and lower-caste groups has varied by region and through time. Recent efforts by the Indian government to rectify these gross inequalities has brought marginalized groups into the mainstream (particularly through government employment), though not without resistance from dominant castes.

Commonwealth: A loose and informal organization of independent nation-states that formerly made up the British Empire. Some of these countries chose to recognize the British Crown as their titular head, while most, including India, did not. The Commonwealth promotes modest economic, cultural, and educational exchanges, and in recent years, it has met only biannually.

Communist Party of India (CPI) and Communist Party of India–Marxist (CPM): Founded in 1928, the CPI attracted workers and peasants as the most prominent far left alternative to the Congress in the pre- and postindependence eras. Historically plagued by struggles between ideological factions, the party split in 1964 over the issue of its approach to the dominant Congress Party–the CPI favored working with the progressive elements within the Congress while the CPM sought to oppose it entirely. More radical groups employing revolutionary tactics have split off from the communist parties, such as the Naxalites in West Bengal and the People's War Party in Andhra Pradesh, but from their inception, the communist parties favored participation and reform through the parliamentary system. They stand as the only democratically elected communist parties in the world, first in Kerala in 1957 (CPI) and in West Bengal in 1977 (CPM), where it remains in office. The communist parties favor state ownership and stimulation of industry, strong unions, and equitable land distribution. In recent years, however, they have sought to attract foreign investment and have promoted private sector growth. The Left Front of the CPM and the CPI are partners in the current National Front government; the CPI returned to head the Kerala government in the 1996 state elections.

Community Development Programme: Launched in 1952 to coincide with the agricultural reforms included in the First Five-Year Plan, the Community Development Programme sought to implement a number of localized development projects (Community Development Projects). These projects were intended to initiate social change at a village level by encouraging cooperative activities and economic self-sufficiency. Designed to be spread thinly and evenly throughout rural areas, projects focused on more intensive agricultural development through crop rotation, farm management, and the use of fertilizers. Other projects concerned with rural issues such as health, education, and local self-government were also introduced.

Congress Working Committee (CWC): Made up of the Congress Party president, the party leader in parliament, and 19 other members, the CWC is the Congress body with the highest authority. It is responsible for carrying out party policies and programs.

East India Company: One of the first European joint-stock ventures, the East India Company was granted a charter in 1600 to initiate monopoly trade in spices, foodstuffs, silk, and luxury products in India and Southeast Asia. Company merchants established trading posts on the Indian coasts and vied for trading privileges with other European powers. By the mid-eighteenth century, the company had pushed out most of the other European trading companies and, taking advantage of the decentralization of Mughal power to provincial rulers, became a factor in subcontinent politics. It soon found itself as kingmaker, military power, and territorial ruler controlled by stockholders back in London. In parliament, interest groups resentful of company power, tired of its financial troubles, or eager for economic opportunities combined to gradually strip the company of its trading privileges. This chain of events turned the company into an administrative entity by the 1830s. In 1858, the East India Company was dissolved, and India became part of the British Empire, ruled directly by the English government.

Emergency: This term refers to the two-and-a-half-year period (June 1975 to December 1977) during which Prime Minister Indira Gandhi suspended democratic and civil liberties in response to mounting economic grievances, "indiscipline" in the workplace, and mass protests against her increasingly authoritarian rule by the opposition. Granted wide powers by the Emergency Regulations, Indira Gandhi arrested and detained without trial over 110,000 people (mostly from the opposition), severely censored the press, banned labor action, overrode the prerogatives of state assemblies and courts to contest acts of parliament or defend civil liberties, and embarked on highly controversial social programs (the most notorious of which was her son Sanjay's sterilization campaign, in which there were numerous reports of forced vasectomies). Her resounding defeat in the 1977 general election took her by surprise and revealed the enormous public backlash against her authoritarian measures.

Fabian Socialism: A popular strand of socialism developed by intellectuals and artists in England in the 1880s, Fabian socialism advocated the parliamentary road to socialism and was influential in the formation of the Labour Party. It was gradualist in design, scorning revolutionary methods, yet did favor public ownership of industries, class cooperation, and elitist bureaucratic leadership. This philosophy made an impact on Jawaharlal Nehru during his student years in England and subsequently influenced his economic policy decisions as prime minister.

Fifth Five-Year Plan (1974–79): Drafted during the emergency, this plan sought to reduce poverty and to increase self-reliance in small and large industries. Again, targets were agriculture and critical industries. The continuing support of these two areas was hoped to reduce rural poverty and unemployment. Additionally, the energy-production sector was given special attention.

First Five-Year Plan (1951–56): The first attempt at planning economic growth in five-year blocks following the Soviet precedent, the First Five-Year Plan emphasized balanced industrial and agricultural development. As the first planning program, it was a largely a confirmation of economic policies already in practice. The plan called for gradual industrial growth with extensive state regulation. The bulk of the plan concerned agricultural development, irrigation, and overcoming power shortages. Taking socioeconomic inequalities as the source of India's poverty, planners hoped to initiate more equitable land reform and redistribution and sought to enact ceilings on landholdings.

Foreign Exchange Regulation Act (FERA), 1973: This piece of legislation regulates foreign investment and currency exchange. It determines the amount of foreign equity allowed in an industry, the types of activities permitted, the use of real estate, the remittance of dividends and profits abroad, foreign exchange trading, and access to banking facilities. In 1993, an amendment loosened some of the regulations on borrowing, acquisitions, and sales by foreign firms as well as opening branch offices in order to put them at the same level as Indian firms.

Fourth Five-Year Plan (1969–74): After a three-year hiatus from state planning, this plan sought to redress problems caused by the first three plans, such as inflation, slow industrial expansion, low increases in agricultural production, and dependence on foreign aid. Planners strove to increase exports to improve balance of payments, limit inflationary financing, and stay within budgetary constraints. Public investment in industry was reduced, leaving a larger role for private industry. Private industry was encouraged through investment incentives, the lifting of price controls, and greater access to credit. In agriculture, development projects in rural electrification and irrigation were undertaken, and farmers were helped in moving from subsistence to commercial cultivation.

Green Revolution: An economic and agricultural strategy developed by Western aid organizations in conjunction with Indian planners to improve agricultural production through the introduction of new technologies and high yielding varieties of cereal grains. Initiated in the late 1960s, this "revolution" was meant to make India self-sufficient in food production, utilizing greater quantities of fertilizers, new methods of cultivation, and mechanization. Certain regions were targeted for growth, and although the Green Revolution significantly increased food production, it did heighten economic disparities between regions and rural peoples.

Indian National Congress (INC): Founded in 1885 to represent the economic and political interests of Indian elite and professional groups to the British government, the INC became the premier political organization of the nationalist movement. Under Gandhi's leadership in the 1920s it moved from an elite party to a mass-based party. It emerged as the major ruling party after independence in 1947 and has governed 44 out of 50 years. The Congress is largely a centrist party, and it has attempted to incorporate low-caste and minority groups in its ranks. Under the leadership of Jawaharlal Nehru, it has favored active state involvement in economic planning and decision making. The party split twice in factional struggles, in 1969 and 1977, but Indira Gandhi gained the loyalty of the bulk of the party in both cases. The Congress now officially carries the name Congress (I) for Indira. After the assassination of Rajiv Gandhi in 1991, the Congress returned to the center as a minority government under P. V. Narasimha Rao but lost this position in the May 1996 elections.

Industrial Policy, 1991: The new industrial policy of the Rao government sought to attract foreign investment in high priority industries (such as power, chemicals, electronics, transportation, telecommunications, and fuel and oil refineries). This policy allowed direct foreign investment of up to 51% in these industries and permitted 100% ownership if the entire output was exported. The government also encouraged the use of high technology, particularly in the computer and information technology sectors.

Industrial Policy Resolution, 1948: A document that sought to clarify the new Indian government's economic policies and objectives. The government recognized the need for a mixed economy and reserved national monopolies for only the munitions, atomic energy, and railroad industries. While extant industries held by private firms could continue, the government had the exclusive rights to initiate projects in six other industries–coal, iron and steel, aircraft manufacturing, shipbuilding, telephone and telegraph, and minerals–yet could seek the aid of the private sector if necessary. Moreover, the government could regulate and license 18 other industries of national importance. This resolution reassured foreign investors and the domestic business community that their holdings would not be confiscated or nationalized. In 1956, the government expanded the list of public sector holdings to capital and intermediate goods. Some services, such as the insurance business, were gradually nationalized by 1973.

Industries Development and Regulative Act (IDRA), 1951: This act implemented the Industrial Policy Resolution, which guides private sector participation in industrial development. It required that a new industrial unit or any expansion in an extant industry be licensed by the central government. It further provided that the central government could determine or set guidelines for the price or quantity of production.

Integrated Rural Development Programme (IRDP): Introduced during the 1978–80 Janata period, the IRDP absorbed several of the existing poverty alleviation programs, such as the Small Farmer Development Agency (SFDA) and the Marginal Farmer and Agricultural Labour (MFAL) program, among others. The focus of these programs was on permanently raising the standard of living of the poorest above the poverty line by training them in sustainable skills or by providing assets to increase their productivity. Most analysts agree that implementation of IRDP programs has been poor.

Intensive Agricultural District Programme, 1960s: An intensive agricultural development program initiated by the Indian government in conjunction with the Ford Foundation. In contrast to earlier agricultural development plans that sought to raise production in all areas generally, this program departed from previous attempts to induce agricultural development by concentrating resources, such as improved technology, higher yield seed varieties, and irrigation, in the potentially most productive districts.

International Monetary Fund (IMF): Established in 1945 as a financial agency affiliated with the United Nations (UN), the International Monetary Fund (IMF) is responsible for regulating balance of payments and stabilizing exchange rates among UN members. It grants loans to nations experiencing negative payment balances or unstable currencies. It has had a strong impact on developing nations in which the IMF has often required economic restructuring as a condition of receiving loans.

Janata Dal: A recent coalition of north Indian non-Communist left and lower-caste-based parties leftover from the 1977 Janata Party, the Janata Dal joined forces with several regional parties in the south and the northeast to contest the 1989 national elections as the National Front. Though it did not gain an electoral majority, the Janata Dal leader, former Congressman V. P. Singh, was asked to form a government in parliament with the outside support of the Bharatiya Janata Party (BJP) and the Communist Party of India–Marxist (CPM). During his tenure, V. P. Singh pressed for reforms favoring reservations for lower castes in government employment, which sparked riots and controversy throughout the country. The BJP withdrew its support in 1990, and V. P. Singh resigned. Another Janata Dal leader, Chandra Shekhar, headed another minority government, this time supported by the Congress, which lasted only four months, until new elections were called in March 1991. The Janata Dal is a major partner in the current National Front government.

Janata Party: A coalition of leftist, regional, caste-based, and right-wing parties that united to win the national elections in 1977. The Janata Party ("People's Party") capitalized on widespread disenchantment with the policies of Indira Gandhi (particularly the measures taken during the emergency) and remained in office until factional struggles caused the coalition to disintegrate in 1980. Morarji Desai (an estranged Congressman) served as prime minister until 1979, when he was removed by Charan Singh of the Socialist Party, who could not maintain the coalition. In office, the Janata Party followed a centrist course similar to the Congress, though it gave more focus to agricultural development.

License-Permit Raj: A phrase used to describe the system of allotting industrial and commercial permits to expand or initiate production ventures under government regulations, a process that has been highly politicized, bureaucratized, and much derided. *Raj* is the Hindi word for "rule" or "sovereignty" and has historically been used to refer to British rule, hence the perjorative usage.

Mandal Commission: Chaired by Parliamentarian B. P. Mandal, the Mandal Commission was a government-appointed study of affirmative action policies for backward and disadvantaged castes and tribal people. Released in 1980, it proposed that the central government reserve 27% of its positions, and 27% of university admissions, for backward and disadvantaged castes. In August 1990, Prime Minister V. P. Singh announced that he would implement these measures, but this policy was met with tremendous protest and resistance, resulting in his eventual resignation.

Monopolies and Restrictive Trade Practices (MRTP) Act, 1969: This act built on the regulations in the 1951 Industrial Policy. It sought to limit monopolistic tendencies among larger firms, creating additional licensing and permit requirements.

Multinational Corporation (MNC): A term used to denote a corporation that operates in nations other than its home country, such as General Motors, Sony, or Philips Electronics. These operations can take the form of direct investment, joint ventures with local partners, or equity in domestic corporations. In recent years, direct investment and joint ventures in India by foreign corporations has been made easier under the New Economic Policy, which has loosened foreign exchange, ownership, and licensing regulations in order to attract foreign investment.

National Development Council (NDC): Founded in 1952 by Nehru in order to give a voice to state chief ministers in the national planning process. This council gave them an opportunity to review, oppose, or make recommendations for the economic and social policies proposed by the Planning Commission. The NDC grew to greatly influence Planning Commission policies.

National Front: A national coalition of 13 parties containing left, regional, and lower-caste-based parties that successfully contested the 1989 and 1996 national elections. The Front positions itself as an alternative to the upper-caste bias of the Congress and the communalist platform of the Bharatiya Janata Party (BJP). Major partners in the Front are the Janata Dal, the Telugu Desam Party (TDP) of Andhra Pradesh, the Dravida Munnetra Kazhagam (DMK) and the All-India Anna Dravida Munnetra Kazhagam (AIADMK) of Tamil Nadu, the Asom Gana Parishad (AGP) of Assam, and other smaller regional parties. The current National Front government under Prime Minister H. D. Deve Gowda of the Karnataka Congress Party also includes the National Conference of Kashmir and the Communist Party of India (CPI) of West Bengal, the first time a Communist party has participated in a central government. Neither Front government has diverted significantly from the economic policies of the previous Congress government.

Naxalites: Followers of a radical and violent agrarian protest movement that broke out in 1967 in the Naxalbari subdivision in the state of West Bengal, where land reforms had not been implemented. Groups of tribal cultivators, led by Communist Party activists, confronted area landlords who held land illegally or refused to open fallow fields for cultivation. The death of a policeman in one confrontation precipitated violent retaliation by the police with the backing of the Communist-led state government. By the end of the year, most of the Naxalite leaders had been killed or arrested. The Naxalite movement is seen as the progenitor of several other radical agrarian movements that now operate in Andhra Pradesh and Bihar and other areas of great social and economic inequality between tribal and low-caste groups and landowners.

New Economic Policy, 1991: A package of new economic policies put forth by the Rao government in 1991 in response to the fiscal crises faced at that time. This strategy promoted liberalization, privatization, and increased competitiveness with the ultimate goal of globalizing the Indian economy. This policy opened doors for investment by MNCs as many industries were delicensed and certain foreign exchange requirements were lifted.

Nonresident Indian (NRI): This is a common term used to refer to nonresident Indians settled abroad. In recent years, the government of India has sought to attract direct investment by NRIs through special investment schemes and interest rates.

Panchayat: Literally, the "council of five," "panchayati raj" is the constitutionally mandated system of local self-governance at the village, district, and development-block level. These elected councils received emphasis from the central government in the late 1950s and early 1960s in an effort to devolve some state powers to the local level in order to encourage development and agricultural programs. The panchayats were an essential part of early economic and social planning to bring in agriculturists and the rural poor into participatory government, but in recent years, they have proven ineffective.

Planning Commission: Created initially to serve as an advisory board to the government on issues of planning and development after independence, the Planning Commission in the 1950s took a central role in balancing the allocation of development resources to state and central government and designing the five-year plans. The Planning Commission consists of prominent economists, planning and development experts, and members of the prime minister's cabinet (normally the finance minister). The Planning Commission was most influential as a centralized planning body with significant political support during Nehru's tenure, but it has declined in significance since the 1970s due to conflicts between state governments and the center over development priorities. The Planning Commission was instrumental in introducing such projects as the Community Development Program and for advocating balanced development leading to the gradual demise of socioeconomic inequalities.

Poverty Alleviation Programs (PAPs): Common phrase to describe a variety of government-sponsored programs in the 1970s designed to ameliorate or eliminate the causes and effects of poverty. Some programs targeted immediate situations, such as supplying temporary wage employment during droughts or other natural disasters. Other programs have sought long-term solutions to poverty by seeking to permanently raise the standard of living of the rural poor to above the poverty line through improvements in agricultural production, skills training, and loans and investments. Development planners hoped to enable the rural poor to remain self-employed by introducing "productive assets." Further, many of the PAPs were linked with social development programs, such as efforts in improving family and community health, education, and the position of women.

Reserve Bank of India (RBI): The central bank of India, responsible for ensuring saving, approving foreign investment projects, regulating foreign exchange, and managing monetary policy.

Scheduled Castes and Tribes: Terms for economically or socially disadvantaged castes and tribes, so-called because of their presence on an official "list" or "schedule" identifying them as such. Under the Constitution, they receive preferential treatment or special protection. Scheduled Castes and Tribes make up about 25% of India's population. Though Scheduled Castes were once known as "Untouchables," the term *Dalit* (the downtrodden) is preferred and in common use.

Second Five-Year Plan (1956–61): Under the guidance of P. C. Mahalanobis and Nehru, this plan shifted gears from the agricultural emphasis of the First Five-Year Plan and focused on rapid and intensive industrial development through central planning and public ownership. This plan set the pattern for future industrial policy, emphasizing import-substitution strategies and state ownership of heavy and capital goods industries. This plan called for a smaller role for the private sector, which the state increasingly regulated. Its ultimate goal was to turn India into a self-sufficient economy, increase national income, and reduce unemployment.

Securities and Exchange Board of India (SEBI): Government agency responsible for protecting investors and the development and regulation of Indian capital markets, including the Mumbai (Bombay) Stock Exchange.

Shiv Sena: A regional political party in Maharashtra, the Shiv Sena was founded in the 1960s by a former cartoonist, Bal Thackery, as an urban anti-immigrant party (particularly against migrants to Bombay from south India). It reinvented itself as an extreme Hindu nationalist party in the 1970s and 1980s and has fought state elections in partnership with the Bharatiya Janata Party (BJP) in Maharashtra; it has a very limited following outside of its home state. In the 1995 state elections, the BJP-Shiv Sena combination defeated a beleaguered Congress party, and Murli Manohar Joshi of the Sena is the current chief minister.

South Asian Association for Regional Cooperation (SAARC): Consortium of South Asian nations (India, Pakistan, Bangladesh, Sri Lanka, Maldives, and Bhutan) formed in 1985 through initiatives made by Rajiv Gandhi and Pakistan's President Zia. Focuses on issues such as terrorism, narcotics, telecommunications, tourism, regional development and regional concerns. In April 1993, member nations agreed to the South Asian Preferential Trade Agreement (SAPTA). Inaugurated November 1995, SAPTA established favorable regulations on trade between member nations for certain items.

Swadeshi: A major plank of the nationalist movement popularized by Gandhi, *swadeshi* (of one's own country) enjoined the boycott of British goods in favor of indigenously made products to encourage economic self-sufficiency. In particular, Gandhi campaigned for wearing only homespun cloth, or *khadi*, which soon became the identifiable characteristic of Congress workers.

Third Five-Year Plan (1961–66): The third plan continued on similar lines as the second, concentrating on industrial development and expansion led by the state. Political conflicts made the introduction of large-scale agrarian land reforms unviable, and general investment in agricultural development declined.

Zamindars: Local landholding potentates who acquired the right to collect revenues or taxes from an area of land. Under the British, they resembled landed gentry and sometimes styled themselves as little kings, or rajas. Zamindari privileges were abolished after independence.

Zila Parishad: In the panchayati system, *zila parishads* are the district councils.

Appendix 3

Personalities

Prime Ministers and Presidents of India

Prime Minister	*Party*	*Dates*
Jawaharlal Nehru	Congress	1947–64
Lal Bahadur Shastri	Congress	1964–66
Indira Gandhi	Congress	1966–77
Morarji Desai	Janata	1977–79
Charan Singh	Janata (S)	1979–80
Indira Gandhi	Congress	1980–84
Rajiv Gandhi	Congress	1984–89
Vishwanath Pratap Singh	National Front	1989–90
Chandra Shekhar	Janata Dal (S)	1990–91
P. V. Narasimha Rao	Congress	1991–96
Atal Behari Vajpayee	BJP	15 May 1996–28 May 1996
H. D. Deve Gowda	National Front	1 June 1996–11 April 1997
Inder Kumar Gujral	National Front	21 April 1997–

Presidents of India	*Date*
Rajendra Prasad	1950
Rajendra Prasad	1952
Rajendra Prasad	1957
S. Radhakrishnan	1962
Zakir Hussain (died 1969)	1967
V. V. Giri	1969
Fakhruddin Ali Ahmed (died 1977)	1974
Neelam Sanjiva Reddy	1977
Zail Singh	1982
R. Venkataraman	1987
Shankar Dayal Sharma	1992

Birla Group: A merchant family from Rajasthan who migrated to Calcutta in the early twentieth century and built an industrial dynasty; the union of the four Birla brothers was incorporated in 1918. Though they first came as traders and speculators in the opium trade, quick profits during World War I allowed them to move into the textile (jute and cotton) industry and challenge the British monopoly. The most prominent brother, Ghanshyamdas (1894–1983), or GD, was a close associate of Gandhi during the freedom movement. Growth during the 1950s and 1960s transformed Birla Brothers into a diversified international corporation with holdings in tea, chemicals, minerals, transportation, cement, steel, and paper. Today, the Birla Group is a much more decentralized operation, though each holding is financially linked to others; companies are spread across the descendants of the four brothers, and there is little central control.

Chidambaram, Palaniappa (1945–): Commerce minister in P. V. Narasimha Rao's cabinet (1991–96) and then finance minister under H. D. Deve Gowda. Originally a Congress member of parliament, he resigned to join a splinter Congress group from his home state of Tamil Nadu, which aligned with the National Front for the 1996 elections. As finance minister, he increased the pace of liberalization in the 1997 budget, cutting tax rates and lowering import duties, hoping to spark demand and investment in an inflationary and lagging economy. This budget also gave tax breaks to Telecom and industrial ventures and loosened price controls on agricultural products. In the prime minister reshuffling of April 1997, Chidambaram resigned from the cabinet, but he is now finance minister once again.

Gandhi, Indira (1917–84): Daughter of Jawaharlal Nehru and prime minister (1966–77, 1980–84). Indira Gandhi learned her political skills at the foot of her father during the nationalist movement and followed him into government service after independence. She was minister of information in Lal Bahadur Shastri's cabinet and succeeded him as prime minister on his death. Originally thought to be easily malleable by other Congress members, Mrs. Gandhi quickly came into her own by consolidating her authority and creating a reliable and loyal power base. Declaring a war on poverty (*Garibi Hatao*), she pursued land reform policies, placed ceilings on personal income and corporate profits, and nationalized the banks. In foreign policy, she grew closer to the Soviet Union and championed the nonalignment movement. She proved a strong leader in the 1971 war against Pakistan, but a series of economic setbacks put her administration in trouble. Increasing agitation against her rule led her to declare a state of emergency in 1975, restricting civil rights, censoring the press, and arresting much of the opposition. This move lost her a lot of credibility and cost her the 1977 elections; but, immensely popular with the masses, she returned to power in 1980. Forced to deal with a series of ethnic and regional crises in her second tenure as prime minister, she was assassinated by her Sikh bodyguards in retaliation for her military assault against Sikh extremists in the Golden Temple.

Gandhi, Mohandas K. (1869–1948): India's most prominent national and spiritual leader. Trained as a lawyer in London, Gandhi was drawn to politics by his experience with racial discrimination in South Africa, where he first practiced. After achieving small but important victories for the Indian community there, he returned to India to wide public recognition in 1915. Increasingly involved in religious and social issues, Gandhi's thought linked political protest against British colonialism with internal moral and spiritual improvement. He focused on the elimination of caste discrimination, the uplift of the poor, and the use of nonviolence and civil disobedience in the independence movement. His focus on the rural poor endeared him to the masses as no leader had done before. His economic thought stressed small-scale development through the use of indigenous technology and methods, forsaking industrial ventures based on Western models. Though he never held a formal political office, he was the de facto leader of the nationalist movement and remains the celebrated "Father of the Nation" to this day. He was assassinated on the 30 January 1948 by Hindu nationalists angered over his conciliatory policies toward the new Muslim state of Pakistan.

Gandhi, Rajiv (1944–91): Son of Indira Gandhi and prime minister (1984–89). After the tragic death of his brother Sanjay in 1982, Rajiv retired from his job as a commercial pilot and reluctantly joined politics as his mother's closest adviser. Upon her death, he was swept into power in a wave of sympathy in the 1984 elections. During his tenure, he introduced moderate economic liberalization, which did spark growth. In foreign policy, he became a prominent worldwide advocate for India and South Asia. He made significant efforts to settle regional conflicts by meeting with South Asian leaders and establishing the South Asian Association for Regional Cooperation (SAARC). He committed the Indian Peace-Keeping Force (IPKF) to Sri Lanka as a measure to resolve the civil war between Tamil separatist groups and the Sri Lankan government. Popular disaffection with corruption among his ministers led to his defeat at the polls in 1989. His popularity was resurging when he was assassinated on the campaign trail by a member of a Tamil separatist group in 1991.

Gowda, H. D. Deve (1933–): Prime minister from 1 June 1996 to 11 April 1997. The son of farmers in the southern state of Karnataka, Gowda was a fixture of state politicals until his escalation to prime minister as a consensus candidate of the National Front. His tenure was plagued by party infighting and maneuvering by the Congress Party, which supported him from the outside. He was able to improve relations with India's neighbors and ensure continuing economic growth. Squabbles with the Congress made him increasingly ineffective, and Gowda was forced to resign in April 1997.

Gujral, Inder Kumar (1919–): National Front prime minister from 21 April 1997. Born in Pakistan before independence, Gujral migrated with his family to India in the mass exodus in 1947. Though originally a member of the Communist Party during the nationalist movement, he served as a minister in Indira Gandhi's cabinet for over a decade but then defected to the Janata Party during the 1980s. He was Minister for External Affairs twice, most recently in the Gowda cabinet. As foreign minister, he stressed a "good neighbor" policy with India's smaller border countries and impressed observers by resuming bilateral talks with Pakistan. As prime minister, he has pledged to eliminate corruption, continue the economic reforms started in the early 1990s, and normalize relations with Pakistan. It remains to be seen if this consensus candidate will be able to maintain unity in the fractious National Front.

Mahalanobis, P. C. (1893–1972): A classmate and longtime friend of Nehru, Mahalanobis was a physicist by profession. He rose from the position of director of the Indian Statistical Institute to become the architect of the Second Five-Year Plan. He favored intensive public investment in industry at the expense of agriculture to build a sustainable economic base. He served as statistical adviser to Nehru's cabinet (1955–58) and as a member of the Planning Commission from 1959.

Nehru, Jawaharlal (1889–1964): Political leader of the nationalist movement and prime minister from 1947 until his death in 1964. Born into a prominent family in north India, Nehru studied law and politics at Oxford. He was influenced by Fabian Socialism and the methods of state planning practiced by the Soviet Union in the 1930s. He was also a committed secularist and proponent of India's prominent place in the world after gaining independence. These ideals were evident in his state-led economic plans, his secularist policies, and his refusal to be drawn into Cold War battles, which he believed inhibited the development of postcolonial nations. In his time, he was a global figure, known as a spokesman for postcolonial nations and as a proponent of the nonalignment movement. He established early close relations with Communist China (which were betrayed in China's 1962 brief invasion of northern India). Under his policies, India developed a strong industrial base at the expense of agricultural growth; Nehru set the precedent for strong state involvement in economic growth and regulation.

Rao, P. V. Narasimha (1921–): Prime minister from June 1991 to May 1996. Brought into national politics after a stint as the chief minister of Andhra Pradesh (1972–73) by Indira Gandhi, he rose through the Congress ranks, serving as Minister of External Affairs, Home, Defense, and Human Resource Development. Though he had unofficially retired from politics, he became the Congress' consensus candidate for prime minister after Rajiv Gandhi's assassination. His economic legacy is the implementation of an economic liberalization program, which encouraged foreign investment and reduced regulations on industrial growth and expansion. His tenure also saw an drastic increase in communal conflicts, especially the Ayodhya incident and the Bombay bomb blasts of 1992–93. He left office in 1996 after the defeat of the Congress Party by the National

Front coalition. He resigned from the leadership of the Congress Party in September 1996 after corruption investigations.

Singh, Manmohan (1933–): Finance minister in the Rao government (1991–96). Considered by many to be an apolitical technocrat, the Oxford-educated economist served as Reserve Bank of India (RBI) governor, as chief of the Planning Commission, and as an economic consultant in the World Bank. He is the main architect of India's liberalization policy, designed to initiate restructuring that would make India eligible for IMF loans and integrate the country into the world market. His legacy was to loosen foreign exchange regulations, restructure capital markets, liberalize industrial licensing, reduce barriers to foreign investment, and raise taxes on some expensive consumer goods and services.

Tata and Sons: The first and foremost of India's industrial families. Established by Sir Jamsetji Tata (1839–1904), Tata holdings moved from speculation and trading in nineteenth-century Bombay to an international corporation specializing in heavy industry by World War II. Jamsetji bought his first cotton mill in 1869, acquiring other mills and becoming the largest landowner in Bombay by the 1890s. His biggest venture was the Tata Iron and Steel Company (TISCO), which was established in 1907 in Jamshedpur, Bihar, after his death. Tata Iron and Steel Company is the largest steel producer in India and is associated with other heavy industries such as hydroelectric power, trucks, steam and diesel locomotives and automobiles (Tata Engineering and Locomotive Company–TELCO), cement, airlines (which were later nationalized as Air India and Indian Airlines), and chemicals. Tata and Sons have resisted nationalization for most of their holdings; TELCO is one of the largest private companies in India.

Vajpayee, Atal Bihari (1925–): Prime minister of a Bharatiya Janata Party (BJP) government for 13 days in May 1996. Originally a member of the Congress during the nationalist movement and through independence, Vajpayee parted ways with the Congress during Indira Gandhi's tenure. He served as Minister for External Affairs during the 1977–79 Janata Party government. He was one of the original founders of the Hindu nationalist BJP, but he is considered to be one of the more moderate elements. Political jockeying during his short stint as prime minister prevented the implementation of any but the most token policies.

Appendix 4

Bibliography

Ahluwalia, Isher Judge, *Industrial Growth in India: Stagnation since the Mid-Sixties,* New Delhi: Oxford University Press, 1985

An evaluation of recent discussions of the industrial decline in India in the 1960s and 1970s that looks critically at state policies in licensing and infrastructural investment. This study echoes other calls for liberalization in the 1980s.

Ahluwalia, Isher Judge, *Productivity and Growth in Indian Manufacturing,* Delhi and New York: Oxford University Press, 1991

A useful quantitative study of India's industrial performance since independence that shows the decline in productivity in manufacturing in the 1960s and 1970s and the rise in productivity in the first half of the 1980s.

Alagh, Yoginder K., *Indian Development Planning and Policy,* New Delhi: Vikas, 1995

This study reviews the planning process in India from the perspective of an academic economist involved in the implementation of policy.

Alagh, Yoginder K., et al., editors, *Sectoral Growth and Change,* New Delhi: Her-Anand Publications, 1993

Explores the direction in which economic liberalism in the 1980s, compared to the earlier period of planned economic development and protectionism, affects India's technological capability. The author argues in favor of a relatively free environment of the market for technology import accompanied by positive state intervention to encourage domestic R and D by Indian firms.

Alvares, Claude Alphonso, *Science, Development, and Violence: The Revolt against Modernity,* Delhi and Oxford: Oxford University Press, 1992

A polemical look at the hidden social and environmental costs of development.

Andersen, Walter K., and Shridhar D. Damle, *The Brotherhood in Saffron: The Rashtriya Swayamsevak Sangh and Hindu Revivalism,* New Delhi: Vistar, and Boulder, Colo.: Westview Press, 1987

A respected study of the rise and influence of the Rashtriya Swayamsevak Sangh (RSS), the force behind Hindu nationalism.

Arndt, H. W., *Economic Development: The History of an Idea,* Chicago: University of Chicago Press, 1987

A sympathetic and knowledgeable review of the foundational ideas and chief contributors to the field of economic development.

Arora, B., and D. V. Verney, editors, *Multiple Identities in a Single State: Indian Federalism in Comparative Perspective,* New Delhi: Centre for Policy Research, Centre for the Advanced Study of India (Philadelphia), and Konark Publishers, 1995

Recent essays on India's political economy, democracy, and federalism.

Ashe, Geoffrey, *Gandhi,* London: Heinemann, and New York: Stein and Day, 1968

This biography of the political and spiritual leader of the independence movement remains a classic, in large part for revealing the complexity and humanity of a man who became a myth and a symbol of modern India.

——, *The Economy of India,* London: Weidenfeld and Nicolson, and Boulder, Colo.: Westview Press, 1984

A nontechnical introduction to India's economic performance and policies from independence until the early 1980s. This book also provides a succinct review of critical policy questions, such as the then pervasive nature of import and industrial controls and regulation and their effect on efficiency, growth and poverty, and the debate between proponents and opponents of greater outward orientation of the economy.

Balasubramanyam, V. N., *International Transfer of Technology to India,* New York: Praeger, 1973

An empirical study of 20 Indian firms, this study evaluates India's foreign technical collaboration agreements, charting successes and pitfalls.

Bardhan, Pranab K., *The Political Economy of Development in India,* Oxford and New York: Basil Blackwell, 1984

A major analysis of the political economy of India's economic policy making from a critical left perspective.

Basu, Alaka Malwade, *Culture, the Status of Women, and Demographic Behaviour: Illustrated with the Case of India,* Oxford and New York: Oxford University Press, 1992

Considers aspects of women's status, maternal characteristics, child health, and fertility. In terms of fertility, focusing on low-income slum dwellers in Delhi. It emphasizes cultural differences between women of north and south Indian origins in the initiation, pace, and termination of childbearing.

Bhaduri, Amit, and Deepak Nayyar, *The Intelligent Person's Guide to Liberalization,* New Delhi: Penguin Books, 1996

The recent reforms in India are discussed in accessible language without sacrificing detail and scope.

Bhagwati, Jagdish N., *India in Transition: Freeing the Economy,* Oxford and New York: Oxford University Press, 1993

Three lectures critical of India's planning legacy, providing a highly intelligible survey of developments leading up to the liberalization, 1991.

Bhagwati, Jagdish N., and Padma Desai, *India, Planning for Industrialization: Industrialization and Trade Policies since 1951,* London and New York: Oxford University Press, 1970

One of the first studies to take a critical look at India's licensing policies for industry. The authors maintain a balanced treatment of the system's shortcomings and its missed opportunities.

Bhagwati, Jagdish N., and T. N. Srinivasan, *India's Economic Reform,* New Delhi: Associated Chambers of Commerce and Industry of India, 1993

A short work from two leading economists that provides a brief appraisal of the reform process in India. This study was undertaken on behalf of the Finance Ministry of the Government of India.

Brass, Paul R., *Caste, Faction, and Party in Indian Politics,* volume 1, *Faction and Party,* Delhi: Chanakya Publications, 1984

A collection of the author's previously published essays, including important works on the socialist movement, factionalism, and agrarian populism in Uttar Pradesh.

Brass, Paul R., *The Politics of India since Independence,* 2nd edition, Cambridge and New York: Cambridge University Press, 1994

An indispensable analysis of Indian politics over nearly 50 years. Brass provides an examination of agrarian policy and change, national and regional political institutions, foreign policy, and bureaucratic decision making. He further critically investigates India's political and economic development.

Brass, Paul R., and Marcus Franda, editors, *Radical Politics in South Asia,* Cambridge, Mass.: MIT Press, 1973

A collection of essays about radical political traditions in South Asia, with excellent pieces on socialist movements.

Byres, Terence J., editor, *The State and Development Planning in India,* Delhi and New York: Oxford University Press, 1994

An important collection with different viewpoints, primarily of the progressive variety.

Chakravarty, Sukhamoy, *Development Planning: The Indian Experience,* Oxford and New York: Oxford University Press, 1987

A good, readable overview of the basics of the Indian planning process from the point of view of a professional economist and onetime member of the Planning Commission.

——, *Selected Economic Writings, Part III: Development*

Strategies in India and Abroad, Delhi, Oxford, and New York: Oxford University Press, 1993

Essays by a noted economist who played a major role in making economic policy under Indira and Rajiv Gandhi.

Chanda, Asok, *Federalism in India: A Study of Union State Relations,* London: George Allen and Unwin, 1965

An institutional history of Indian federalism and a discussion of the constitutional provisions relating to center-state relations.

Chandhok, H. L., and the Policy Group, *India Database: The Economy, Annual Time Series Data,* volumes 1 and 2, New Delhi: Living Media, 1990

An invaluable and comprehensive systematization of India's principal economic series.

Chandra, Bipan, *The Rise and Growth of Economic Nationalism in India: Economic Policies of Indian National Leadership, 1880–1905,* New Delhi: People's Publishing House, 1966

A famous study of the formation of the Indian nationalist elite from one of India's premier historians

Chatterjee, Partha, *Nationalist Thought and the Colonial World: A Derivative Discourse?* London: Zed Books for the United Nations University, and Minneapolis, Minn.: University of Minnesota Press, 1986

A challenging postcolonial analysis of the relationship between modernity, colonialism, and the nationalist movement in India.

Collins, Larry, and Dominique Lapierre, *Freedom at Midnight,* London: Book Club Associates, and New York: Simon and Schuster, 1975

A novelistic treatment of the transfer of power from Britain to India and Pakistan at partition in 1947. A good, readable introduction to modern India and many of the figures of the nationalist movement.

Dandekar, V. M., *The Indian Economy 1947–92,* New Delhi and London: Sage, 1994

A study of agriculture's place in India's economy since independence.

Dantwala, M. L., editor, *Indian Agricultural Development since Independence,* New Delhi: Oxford and IBH, 1986

Articles by leading Indian scholars examining various aspects of agricultural development in postindependence India.

Dasgupta, Jyotirindra, *Language Conflict and National Development: Group Politics and National Language Policy in India,* Berkeley: University of California Press, 1970

An authoritative study of the evolution of language policy and regional political development.

Das Gupta, Monica, Lincoln C. Chen, and T. N. Krishnan, editors, *Women's Health in India: Risk and Vulnerability,* Bombay: Oxford University Press, 1995

Essays on women's health in India. Topics include child mortality, reproductive health, maternal mortality, risks of STD and HIV transmission to women, widowhood, and health status of older women.

Desai, Meghnad, Susanne Hoeber Rudolph, and Ashok Rudra, editors, *Agrarian Power and Agricultural Productivity in South Asia,* Delhi and New York: Oxford University Press, and Berkeley: University of California Press, 1984

Articles by eight leading scholars on the relationship between political economic power and productivity in Indian agriculture, focusing critically on the role of local power in inhibiting growth and development.

Drèze, Jean, and Amartya Sen, *India, Economic Development, and Social Opportunities,* Delhi: Oxford University Press, 1995

A major review of the broader social consequences of India's economic development. It links poverty, population, health, and education with issues of social and gender inequality.

Drèze, Jean, and Amartya Sen, editors, *Indian Development: Selected Regional Perspectives,* Delhi and New York: Oxford University Press, 1997

Discusses key issues facing India as privatization proceeds, focusing on regional case studies.

Echeverri-Gent, John, *The State and the Poor: Public Policy and Political Development in India and the United States,* Berkeley: University of California Press, 1993

A study of rural employment generation programs in Maharashtra and West Bengal.

Economic and Political Weekly

One of the most useful publications on the Indian Economy, covering a variety of subjects. While having an editorial point of view to the left, a wide variety of views are expressed in the articles included.

Encarnation, Dennis J., *Dislodging Multinationals: India's Strategy in Comparative Perspective,* Ithaca, N.Y.: Cornell University Press, 1989

Analyzes the changing relations among multinationals, the state, and local enterprises in India in the first four decades of its independence. In a comparative study, the author demonstrates how Indian enterprises and states greatly overcame the bargaining imbalance with MNCs.

Erdman, Howard, *The Swatantra Party and Indian Conservatism,* London: Cambridge University Press, 1967

The major study of the only recognizably conservative national party in Indian political history.

Financial Express and *Financial Times*

Two of the leading daily financials in India. The in-depth articles and contributed op-ed pieces represent many points of view.

Fox, Richard G., *Gandhian Utopia: Experiments with Culture,* Boston: Beacon Press, 1989

Fox's anthropological analysis of Gandhi and Gandhian philosophy examines the cultural roots and context of his thought.

Franda, Marcus, *Small Is Politics: Organizational Alternatives in India's Rural Development,* New Delhi: Wiley Eastern Limited, 1979

An excellent overview of the Janata party period (1977–79) with emphasis on development proposals.

Frankel, Francine R., *India's Political Economy, 1947–1977: The Gradual Revolution,* Princeton, N.J.: Princeton University Press, 1978

A meticulous examination of government policy and the planning of the Indian economy after independence that examines the debates and politics of the planning process. A standard.

Frankel, Francine, and M. S. A. Rao, editors, *Dominance and State Power in Modern India: Decline of a Social Order,* 2 volumes, Delhi and Oxford: Oxford University Press, 1989, 1990

A comprehensive survey of state politics in all the major states, focusing on the rise of backward castes and middle peasants.

Galanter, Marc, *Competing Equalities: Law and the Backward Classes in Modern India,* Delhi: Oxford University Press, 1984

An important study of the history and evolution of "backward caste" reservations from the perspective of a legal scholar.

Ganguly Thukral, Enakshi, editor, *Big Dams, Displaced People: Rivers of Sorrow, Rivers of Change,* New Delhi and Newbury Park, Calif.: Sage, 1992

A critical look at water-resources development policy since independence.

Gopal, Sarvepalli, *Jawaharlal Nehru: A Biography,* 2 volumes, Delhi: Oxford University Press, 1975

The definitive biography of Nehru, which uses many sources, including personal letters, for the first time.

Graham, Bruce D., *Hindu Nationalism and Indian Politics: The Origins and Development of the Bharatiya Jana Sangh,* Cambridge and New York: Cambridge University Press, 1990

An excellent study of Hindu nationalism up to the 1960s.

Guha, Ramachandra, *The Unquiet Woods: Ecological Change and Peasant Resistance in the Himalaya,* Delhi and New York: Oxford University Press, 1989

Discusses the connections between colonial and postindependence forestry development policy.

Hanson, A. H., *The Process of Planning: A Study of India's Five-Year Plans, 1950–1964,* London: Oxford University Press, 1966

A carefully reasoned and judicious account of the making of India's first three five-year plans that centers on administrative processes rather than political infighting or economic results.

Hanson, James A., and Samuel S. Lieberman, *India: Poverty, Employment, and Social Services,* Washington, D.C.: World Bank, 1990

An in-depth study of social policies in India in the spheres of education, health, and poverty alleviation.

Hardgrave, Robert L., and Stanley Kochanek, *India: Government and Politics in a Developing Nation,* 5th edition, Fort Worth, Tex.: Harcourt Brace College Publishers, 1993

A comprehensive introduction to Indian politics and contemporary political events. An excellent textbook survey.

Harrison, Selig S., *The Widening Gulf: Asian Nationalism and American Policy,* New York: Free Press, 1978

A veteran observer takes a critical look at the American encounter with nationalism in Asia.

Higgins, Benjamin, *Economic Development: Problems, Principles, and Policies,* New York: W. W. Norton, 1968

An important textbook that evaluates development policies in the 1960s and 1970s.

Huntington, Samuel, *Political Order in Changing Societies,* New Haven, Conn.: Yale University Press, 1968

A seminal work in the field of comparative politics that looks at social and political upheavals in developing societies.

India and Modernity: Decentering Western Perspectives, special issue of *Thesis Eleven* 39 (1994)

A collection of valuable essays, particularly Ashis Nandy's "Culture, Voice, and Development: A Primer for the Unsuspecting" and Sudhir Chandra's "'The Language of Modern Ideas': Reflections on an Ethnological Parable." The second half of Chandra's article examines an important and telling 1945 exchange of letters between Gandhi and Nehru on the future course of development in India.

India Briefing, edited by Philip Oldenburg, Boulder, Colo.: Westview Press

The *India Briefings* are excellent overviews of contemporary politics and the major events of the year.

Jaffrelot, Christophe, *The Hindu Nationalist Movement and Indian Politics: 1925 to the 1990s,* London: Hurst, and New York: Columbia University Press, 1996

The most recent major work on Hindu nationalism, with excellent material on the strategies of the Bharatiya Janata Party and special reference to the state of Madhya Pradesh.

Jain, L. C., *Grass without Roots: Rural Development under Government Auspices,* New Delhi and Beverly Hills, Calif.: Sage, 1985

A perspective that maintains that bureaucratic development planning interfered with Gandhian village-level policies.

Jalan, Bimal, *India's Economic Crisis: The Way Ahead,* Delhi: Oxford University Press, 1991

An investigation of recent crises in Indian economic planning.

Jalan, Bimal, editor, *The Indian Economy: Problems and Prospects,* New Delhi and New York: Viking, 1992

A very good collection of essays on current economic issues by an adviser on economic policy.

Jeffery, Roger, and Alaka M. Basu, editors, *Girls' Schooling, Women's Autonomy, and Fertility Change in South Asia,* New Delhi: Sage, 1996

Essays advising women's education in South Asia. Theo authors investigate the relations between female education and fertility.

Jha, Prem Shankar, *India, A Political Economy of Stagnation,* Delhi: Oxford University Press, 1980

An eminent journalist argues that India's economic stagnation after the mid-1960s was related to the rise to dominance of the intermediate class, consisting of the self-employed such as small-scale industrialists, traders, and rich peasants.

Joshi, Vijay, and I. M. D. Little, *India: Macroeconomics and Political Economy 1964–1991,* Washington, D.C.: World Bank, 1994

Traces the roots of the 1991 reforms through the macroeconomic policies of Indira and Rajiv Gandhi.

——, *India's Economic Reforms, 1991–2001,* Oxford: Oxford University Press, 1996

An excellent overall review of the issues of economic reform in India in the 1990s.

Khan, Rasheeduddin, *Federal India: A Design for Change,* New Delhi: Vikas, 1992

This book places the problem of federalism in the global context and focuses attention on the problems of federal polity in India. It is a policy-oriented study covering themes such as the sociocultural dimensions of India's federal polity, political federalism, and the pattern of federal nation building in India.

Kidron, Michael, *Foreign Investments in India,* London and New York: Oxford University Press, 1965

A study of foreign investments in India prior to and immediately after independence.

Kohli, Atul, *Democracy and Discontent: India's Growing Crisis of Governability,* Cambridge and New York: Cambridge University Press, 1990

An influential work that maintains that the Congress Party has suffered institutional decline since the 1960s.

Kohli, Atul, *The State and Poverty in India: The Politics of Reform,* Cambridge: Cambridge University Press, 1987

A landmark comparative study of the relationship between party organization and effectiveness at implementing redistributive policies in Karnataka, West Bengal, and Uttar Pradesh. Kohli presents a critical analysis of the Indian government's failure to significantly reduce poverty after 40 years of independence.

Kohli, Atul, editor, *India's Democracy: An Analysis of Changing State-Society Relations,* Princeton, N.J.: Princeton University Press, 1988

A major collection of essays on the nature of Indian democracy in the 1980s.

Kolenda, Pauline, *Caste in Contemporary India: Beyond Organic Solidarity,* Menlo Park, Calif.: Benjamin/Cummings, 1978

An excellent short survey of the various dimensions of caste.

Kothari, Rajni, *Growing Amnesia: An Essay on Poverty and the Human Consciousness,* New Delhi and New York: Viking, 1993

A work that links state socialist Nehruvian development and the liberalization policies of recent years. The author offers practical alternatives to recent trends in economic thought.

—, *Rethinking Development: In Search of Humane Alternatives,* Delhi: Ajanta Publications, 1988

Offers an extended description of other possibilities for a less authoritarian, less violent development politics in India.

Kumar, Dharma, editor, *Cambridge Economic History of India,* Cambridge and New York: Cambridge University Press, 1983

An encyclopedia review of Indian economic history.

Kumar, Girish, and Buddhadeb Ghosh, *West Bengal Panchayat Elections 1993: A Study in Participation,* New Delhi: Institute of Social Sciences and Concept Publishing Company, 1996

This book, based on field surveys in West Bengal during and after the May 1993 panchayat elections, offers a critical analysis of the extent of popular participation in the panchayat election process and the working of the panchayat institutions. The authors base their analysis primarily on the state's successful and long-standing experience with local self-governance since 1978.

Kurien, C. T., *Global Capitalism and the Indian Economy,* London: Sangam, 1994

A very good review of economic and social issues arising form the reform process.

Lal, D., *The Hindu Equilibrium,* volume 1, *Cultural Stability and Economic Stagnation,* Oxford: Oxford University Press, 1988

One of the best economic histories of the Indian economy and its current problems.

Lewis, John Prior, *India's Political Economy: Governance and Reform,* Delhi: Oxford University Press, 1995

A study of Indian policy making, and especially foreign aid and development in India, by one of the most knowledgeable American political economists.

Lewis, W. Arthur, *The Theory of Economic Growth,* London: Ruskin House, and Homewood, Ill.: R. D. Irwin, 1955

Magisterial synthesis of the incentive, institutional, and input factors in economic growth. Coupled with Lewis's "Economic Development with Unlimited Supplies of Labor," in *The Manchester School* 22 (May 1954): 139–91, this volume set the terms of reference for the analysis and planning of economic growth in the less-developed world.

Lipton, Michael, and John Toye, *Does Aid Work in India? A Country Study of the Impact of Official Development Assistance,* London: Routledge, 1989

This is a revision and updating of a previous work called *Does Aid Work?* Lipton and Toye are skeptical of what development assistance does for the Indian economy.

Lucas, Robert E. B., and Gustav F. Papanek, editors, *The Indian Economy: Recent Development and Future Prospects,* Boulder, Colo.: Westview Press, 1988

A good compilation on liberalization under Rajiv Gandhi.

Maddison, Angus, *Class Structure and Economic Growth: India and Pakistan since the Moghuls,* London: Allen and Unwin, 1971, and New York: Norton, 1972

Maddison's study, now dated, still provides a clear comparison of economic development in India and Pakistan.

Malik, Yogendra K., and V. B. Singh, *Hindu Nationalists in India: The Rise of the Bharatiya Janata Party,* Boulder, Colo.: Westview Press, 1994

A political history of the BJP and an analysis of the circumstances giving rise to the party as a major political force in the 1980s and the 1990s.

Mansingh, Surjit, *India's Search for Power: Indira Gandhi's Foreign Policy 1966–1982,* New Delhi and Beverly Hills: Sage, 1984

A thorough study of India's foreign policy during its most important period.

Marglin, Frederique Apffel, and Stephen Apffel Marglin, editors, *Dominating Knowledge: Development, Culture, and Resistance,* Oxford and New York: Oxford University Press, 1990

A look at development policy from the perspective of discourse based on analysis.

Masani, Minocheher Rustom, *Picture of a Plan,* Bombay and New York: Oxford University Press, 1945

A primer for adults on the so-called Bombay Plan of 1943. Masani later grew disillusioned with state-dominated development, embracing more free-market approaches.

Mathur, S. N., *Nyaya Panchayats as Instruments of Justice,* New Delhi: Institute of Social Sciences and Concept Publishing Company, 1997

This book looks at the village courts in the emerging context as a system of decentralized justice in relation to the broad objective of local governance.

Miller, Barbara D., *The Endangered Sex: Neglect of Female Children in North India,* 2nd edition, Delhi: Oxford University Press, 1997

Examines regional and class differences behind rural India's unbalanced juvenile sex ratios. Considers history of direct female infanticide in India, contemporary patterns of intrahousehold allocations of food and medical care, marriage practices and dowry, and women's labor force participation.

Morris, Morris David, *Measuring the Changing Condition of the World's Poor: The Physical Quality of Life Index, 1960–1990,* Providence, R.I.: Thomas J. Watson Institute for International Studies, Brown University, 1996

A pathbreaking statistical review of world social development since 1960 and India's achievements in that area.

Mukarji, Nirmal, and Balveer Arora, editors, *Federalism in India: Origins and Development,* New Delhi: Vikas, 1992

Essays analyzing the dynamics of federal processes in different spheres, especially in the post-Nehruvian Phase. It provides an interdisciplinary perspective on the mainsprings of India's federal polity. The book contains ten papers and an epilogue.

Nandy, Ashis, *At the Edge of Psychology: Essays in Politics and Culture,* Delhi: Oxford University Press, 1980

An important book about political economy from a perspective outside the disciplines traditionally devoted to it.

Nandy, Ashis, *Traditions, Tyranny, and Utopias: Essays in the Politics of Awareness,* Delhi and New York: Oxford University Press, 1987

This book is very useful for exploring the relationship between colonialism and the dominant political culture (and hence the development culture) of postindependence India.

Nayar, Baldev Raj, *India's Mixed Economy: The Role of Ideology and Interest in Its Development,* Bombay: Popular Prakashan, and London: Sangam, 1989

A comprehensive analysis of the state and public sector that establishes the ideological origins of the public sector.

——, *The Modernization Imperative and Indian Planning,* New Delhi: Vikas, 1972

Demonstrates the importance of national power considerations in the evolution of the economic strategy of the Second Five-Year Plan.

——, *The Political Economy of India's Public Sector: Policy and Performance,* Bombay: Popular Prakashan, and London: Sangam, 1990

Provides paired comparisons between private and public sector enterprises in the steel and aluminum industries. Also discusses the politics of liberalization under Rajiv Gandhi.

Nayyar, Deepak, editor, *Industrial Growth and Stagnation: The Debate in India,* Bombay and New York: Oxford University Press, 1994

A collection of important articles of varied viewpoints from the influential left-wing *Economic and Political Weekly* on the industrial stagnation during the 1960s and 1970s.

Rosen, George, *Contrasting Styles of Industrial Reform: China and India in the 1980s,* Chicago: University of Chicago Press, and Bombay: Vora, 1992

A review of India's industrial reform goals and achievements in the 1980s and a comparison with those of China.

——, *Democracy and Economic Change in India,* Bombay: Vora, and Berkeley: University of California Press, 1966

Excellent review of India's economic development through the early 1960s. Continuing an evaluation of rural and urban winners and losers, and the political implications.

——, *Western Economists and Eastern Societies: Agents of Change in South Asia, 1950–1970,* Baltimore, Md.: Johns Hopkins University Press, 1985

Useful for the history of ideas and the people who formulated them. Rosen is less skeptical than others about the overall project of development.

Rudolph, Lloyd I., and Susanne Hoeber Rudolph, *In Pursuit of Lakshmi: The Political Economy of the Indian State,* Chicago: University of Chicago Press, 1987

One of the best analyses of the political economy of India's development since independence, marked by empathy and critical reflection. The authors focus on the bureaucratic institutions and interest groups that dominate political life.

Rudolph, Lloyd I., and Susanne Hoeber Rudolph, *The Modernity of Tradition: Political Development in India,* Chicago: University of Chicago Press, 1967

An important earlier contribution to the social science critique of Western modernity in and for India.

Sandesara, J. C., *Industrial Policy and Planning, 1947–91: Tendencies, Interpretations, and Issues,* New Delhi: Sage, 1992

Based on quantitative data, the study provides a balanced assessment of the debate on industrial growth and stagnation. The author considers Indian industrial growth to be satisfactory and highlights the role of growth alone in meeting social objectives without special targeting.

Shiva, Vandana, *The Violence of the Green Revolution: Ecological Degradation and Political Conflict in Punjab,* Dehra Dun: Natraj, 1989

An important but controversial attack on Green Revolution technology.

Siebeck, Wolfgang E., editor *Strengthening Protection of Intellectual Property in Developing Countries: A Survey of the Literature,* Washington, D.C.: World Bank, 1990

Provides a review of the vast literature–theoretical and empirical–on patent economics and other instruments of intellectual property. While focusing on industrial countries, the authors stress the need for further research on developing countries.

Sisson, Richard, and Ramashray Roy, editors, *Diversity and Dominance in Indian Politics,* 2 volumes, New Delhi and Newbury Park, Calif.: Sage, 1990

A collection of recent essays on regional party systems and the place of various social groups in the Congress.

Smith, Donald Eugene, *India as a Secular State,* Princeton, N.J.: Princeton University Press, 1963

An authoritative and thorough work on state and political party policy toward religion.

Srinivasan, T. N., editor, *Agriculture and Trade in China and India: Policies and Performance since 1950,* San Francisco: International Center for Economic Growth, ICS Press, 1994

An excellent comparative study of agriculture and foreign trade in China and India that addresses the agriculture and foreign trade sectors within the context of Indian planning and recent reforms.

Srinivasan, T. N., and Pranab K. Bardhan, editors, *Rural Poverty in South Asia,* New York: Columbia University Press, and Delhi: Oxford University Press, 1988

This volume contains contributions by many of the leading authorities on rural poverty in South Asia that investigate the impact of international trade policies on rural areas.

Stiles, Kendall W., *Negotiating Debt: The IMF Lending Process,* Boulder, Colo.: Westview Press, 1991

This book has excellent institutional details of the IMF lending process. The loan to India in 1981 is included as a case study.

Subha, K., *Karnataka Panchayat Elections 1995: Process, Issues, and Membership Profile,* New Delhi: published for the Institute of Social Sciences by Concept, 1997

This book critically examines the important issues that influenced the voting pattern in the 1993 and 1995 panchayat elections in the state of Karnataka and the level of popular participation in the *gram* (village), *taluk* (block), and *zilla* (district) panchayats. It is based on extensive field investigations in all the districts of the state.

Sundrum, R. M., *Growth and Income Distribution in India: Policy and Performance since Independence,* New Delhi and Beverly Hills, Calif.: Sage, 1986

A sound and authoritative account of growth and income distribution policies.

Tomlinson, B. R., *The Economy of Modern India: 1860–1970,* Cambridge and New York: Cambridge University Press, 1993

An excellent survey of the evolution of the Indian economy from colonial times through the first decades of independence. Rich in historical detail, this work lies between positions established by British colonialist historians and by Indian nationalists.

Vaidyanathan, A., *The Indian Economy: Crisis, Response, and Prospects,* New Delhi: Orient Longman, and London: Sangam, 1995

An excellent short review of India's current economic problems and policies, both overall and in the agricultural area.

Weiner, Myron E., *Party-Building in a New Nation: The Indian National Congress,* Chicago: University of Chicago Press, 1967

The landmark study of the functioning of the Congress organization at the local level.

———, *Party Politics in India: The Development of a Multi-Party System,* Princeton, N.J.: Princeton University Press, 1957

An early look at the development of the multiparty system in India, before and after independence.

Wood, John R., editor, *State Politics in Contemporary India: Crisis or Continuity?* Boulder, Colo., and London: Westview Press, 1984

An important collection of studies of state politics during Indira Gandhi's last term in office.

World Bank, *Economic Developments in India: Achievements and Challenges,* Washington, D.C.: World Bank, 1995

This report is particularly strong on legal and administrative changes at the center and state level as they affect labor, the power, transport, and communications sectors, and education. The report also provides a review of recent macro developments in the Indian economy.

World Bank, *India: Five Years of Stabilization and Reform and the Challenges Ahead,* A World Bank Country Study, Washington, D.C.: World Bank, 1996

This study applauds the reforms carried out so far in India and suggests directions for further reform in trade and the public and financial sectors.

Wyon, John B., and John E. Gordon, *The Khanna Study: Population Problems in the Rural Punjab,* Cambridge: Center for Population Studies and Harvard University Press, 1967

Classic report on the design, implementation, and outcome of Harvard's multiyear, multivillage attempt to promote the use of cosmopolitan family planning methods in India's Punjab state.

Michael Newbill received his M.A. in South Asian history and politics from the University of Wisconsin, Madison. He is currently the research associate for South Asia and China at the Henry L. Stimson Center in Washington, D.C.

Index

Index